KAPLAN

800

Run to 800

Get the 800

Come on !!!

Sharon

Other Kaplan Books for College-Bound Students

College Admissions and Financial Aid

Conquer the Cost of College
Parent's Guide to College Admissions
The Unofficial, ~~Un~~biased Insider's Guide to the 320 Most Interesting Colleges
The *Yale Daily News* Guide to Succeeding in College

Test Preparation

ACT
AP Biology
AP Calculus AB
AP Chemistry
AP English Literature & Composition
AP Chemistry
AP Macroeconomics/Microeconomics
AP Physics
AP Statistics
AP U.S. Government & Politics
AP U.S. History
PSAT
SAT & PSAT
SAT Vocabulary Words Flip-O-Matic
SAT Math Mania
SAT Math Workbook
SAT Verbal Velocity
SAT Verbal Workbook
SAT II: Biology E/M
SAT II: Chemistry
SAT II: Literature
SAT II: Mathematics
SAT II: Spanish
SAT II: World History
SAT II: Writing

KAPLAN®

SAT* II PHYSICS

2003–2004 Edition

By Hugh Henderson

Simon & Schuster

New York · London · Singapore · Sydney · Toronto

Kaplan Publishing
Published by Simon & Schuster
1230 Avenue of the Americas
New York, New York 10020

For bulk sales to schools, colleges, and universities, please contact: Order Department, Simon and Schuster, 100 Front Street, Riverside, NJ 08075. Phone: (800) 223-2336. Fax: (800) 943-9831.

Contributing Editor: Seppy Basili
Project Editors: Eileen Mager
Cover Design: Cheung Tai
Production Editor: Maude Spekes
Production Manager: Michael Shevlin
Editorial Coordinator: Déa Alessandro
Executive Editor: Del Franz

Manufactured in the United States of America.
Published simultaneously in Canada.

April 2003
10 9 8 7 6 5 4 3 2 1

ISBN: 0-7432-4124-X

TABLE OF CONTENTS

Section Three: Ready, Set, Go!

Section Four: Full-Length Practice Tests

ABOUT THE AUTHOR

Hugh Henderson received his bachelor of science degree in physics from Stephen F. Austin State University, and earned a master's degree in physics and teaching from the University of Texas—Dallas in 1991. He is a former member of the Physics Test Development Committee for the Advanced Placement Program®, and has led numerous workshops in the United States and abroad for teachers and students in science curriculum and test preparation. He has authored several articles in science education, and has taught physics at Plano Senior High School in Plano, Texas since 1985.

The author would like to give special thanks to Connie Wells for her invaluable suggestions in the preparation of this book, and to his wife Rachel, "the evidence of God's grace in my life."

kaptest.com/publishing

The material in this book is up-to-date at the time of publication. However, the College Entrance Examination Board and Educational Testing Service (ETS) may have instituted changes in the test after this book was published. Be sure to carefully read the materials you receive when you register for the test.

If there are any important late-breaking developments—or any changes or corrections to the Kaplan test preparation materials in this book—we will post that information online at **kaptest.com/publishing.** Check to see if there is any information posted there regarding this book.

For more than 50 years, Kaplan has prepared students to take SATs. Our team of researchers and editors knows more about SAT preparation than anyone else, and you'll find their accumulated experience and knowledge in this book. As you work your way through the chapters, we'll show you precisely what knowledge and skills you'll need in order to do your very best on the SAT II: Physics Subject Test. You'll discover the most effective way to tackle each type of question, and you'll reinforce your studies with lots of practice questions. At the beginning of the physics review section, you'll find a lengthy diagnostic practice test, and each chapter in this section ends with a short quiz. At the back of the book you'll find three full-length, formatted tests with answer keys, scoring instructions, and detailed explanations. In addition, the Ready, Set, Go! section contains helpful tips on beating stress while you prepare for the test and on pulling off a victory on Test Day.

The first practice test is a diagnostic test, so that you can assess areas that need improvement. Use the other practice tests to gauge your improvement. Save the last practice test for a "dress rehearsal."

THE CLASSIC PLAN

If possible, work your way through this book bit by bit over the course of a few weeks. Cramming the week before the test is not a good idea; you probably won't absorb much information, and it's sure to make you more anxious.

If you find that your anxiety about the test is interfering with your ability to study, start off by reading the Stress Management chapter in this book. It provides many practical tips to help you stay calm and centered. Use these tips before, during, and after the test.

Learn the Basics

The first thing you need to do is find out what's on the SAT II: Physics Test. In the first section of this book, The Basics, we'll provide you with background information about the SAT II: Physics Subject Test and what it's used for. We'll also give you the lowdown on the kinds of questions that are typically seen on the test and how best to tackle them.

Physics Review

Once you have the big picture, it's time to focus on physics. The Physics Review section gives you a succinct review of the physics you'll need to know on Test Day. Each chapter within this section deals with a major subdivision of physics and focuses on concepts crucial to a full understanding of the subject.

The review section begins with a diagnostic test. If your time is limited, you can use the diagnostic test to bypass the material you already know well enough and to zero in on what you need to work on. Each chapter in this section also ends with a follow-up quiz, complete with answers and explanations. When you feel you have mastered the material in a chapter, take the follow-up quiz to make sure. Don't forget to use the glossary at the end of the book to brush up on vital definitions.

Take the Practice Tests

At the back of the book are three full-length physics practice tests. The best way to use these tests is to take them under testlike conditions. Don't just drop in and do a random question here and there. Use these tests to gain experience with the complete testing experience, including pacing and endurance. You can do these tests at any time. You don't have to save them all until after you've read this whole book. Just be sure to save at least one test for your dress rehearsal some time in the last week before Test Day.

Take a Break Before Test Day

A day or two before the test, be sure to review section three of this book. It includes essential "how-to" tips that will help you get the upper hand on test day. Then, relax! Read a book or watch a movie. And most important, get a good night's sleep. How you approach the days leading up to the test really does matter!

On the morning of the test, eat a light breakfast (nothing too heavy to make you sleepy!) and quickly review a few questions if you feel like it (just enough to get you focused). Walk into the test center with confidence—you're ready for the challenge!

THE EMERGENCY PLAN

What if you have only two or three weeks until the test—or even less time than that—and you haven't started prepping yet? Don't worry! Kaplan's *SAT II: Physics* has been designed to work for students in your situation, too. If you go through two chapters in the Physics Review section every day, you can finish this book in a couple of weeks. If you run out of time, at least read the sidebars in each chapter: The suggestions in these sidebars can really streamline your study strategy as well as ensure that you cover the vital concepts before Test Day.

Be sure to read chapter two, Getting Ready for the SAT II: Physics Test, to find out what to expect on Test Day. Then take as many of the practice tests as you can—under timed conditions. Review your results, with special attention to the questions you missed.

SAT II Emergency FAQs

Q: *It's two days before the test and I'm clueless. What should I do?*

A: First of all, don't panic. If you only have a day or two to prepare for the test, then you don't have time to prepare thoroughly. But that doesn't mean you should just give up. There's still a lot you can do to improve your potential score. First and foremost, you should become familiar with the test by reading chapter two. And if you don't do anything else, take one of the full-length practice tests in this book under reasonably testlike conditions. When you finish the practice test, check your answers and look at the explanations to the questions you didn't get right.

Q: *I don't get to use a calculator on the SAT II: Physics Exam! Does that mean I have to worry about learning math?*

A: Not exactly. The exam is designed in such a way that you really shouldn't need a calculator. But you will be expected to know basic algebra, geometry, and trigonometry. If you're worried about math, you can find a thorough review of the math skills you'll need for the exam in chapter three of this book.

Q: *My parents are upset with me for waiting till the last minute to study, and now I can't concentrate. Help!*

A: Take a deep breath! Anxiety and stress are the enemies of all test takers—no matter how much time they have to prepare. Turn to section three of this book and read through Kaplan's suggestions for managing stress. Do the suggested exercises. And don't forget to think positively!

Q: *The exam is tomorrow. Should I stay up all night studying physics formulas?*

A: The best thing to do right now is to try to stay calm. Read chapter 25, The Final Countdown, to find out the best way to survive, and thrive, on Test Day. If you haven't yet taken a practice test, do it now, just to familiarize yourself with the exam. And get a good night's sleep.

Q: *I don't feel confident. Should I just guess?*

A: You should always guess if you can do so intelligently. The SAT II: Physics Exam does have a wrong answer penalty, but that doesn't mean you should never guess. If you can rule out even ONE answer choice, you should guess, because you have significantly increased your chances of guessing correctly. Also, on questions that appear early in a section, more obvious answers will tend to be correct, so you can guess more confidently on those questions.

Q: *So it's a good idea to panic, right? RIGHT?*

A: No! No matter how prepared you are for the exam, stress will hurt your performance, and it's really no fun. Stay confident, and don't cram. So . . . breathe. Stay calm, and remember, it's just a test.

A SPECIAL NOTE TO INTERNATIONAL STUDENTS

If you are an international student considering attending an American university, you are not alone. 547,867 international students pursued academic degrees at the undergraduate, graduate, or professional school level at U.S. universities during the 2000-2001 academic year, according to the Institute of International Education's *Open Doors* report. Almost 50 percent of these students were studying for a bachelor's or first university degree. This number of international students pursuing higher education in the United States is expected to continue to grow. Business, management, engineering, and the physical and life sciences are particularly popular majors for students coming to the United States from other countries.

If you are not a U.S. citizen and you are interested in attending college or university in the United States, here is what you'll need to get started.

- If English is not your first language, you'll probably need to take the TOEFL (Test of English as a Foreign Language) or provide some other evidence that you are proficient in English. Colleges and universities in the United States will differ on what they consider to be an acceptable TOEFL score. A minimum TOEFL score of 213 (550 on the paper-based TOEFL) or better is often required by more prestigious and competitive institutions. Because American undergraduate programs require all students to take a certain number of general education courses, all students—even math and computer science students—need to be able to communicate well in spoken and written English.
- You may also need to take the SAT or the ACT. Many undergraduate institutions in the United States require both the SAT and TOEFL of international students.
- There are over 2,700 accredited colleges and universities in the United States, so selecting the correct undergraduate school can be a confusing task for anyone. You will need to get help from a good advisor or at least a good college guide that gives you detailed information on the different schools available. Since admission to many undergraduate programs is quite competitive, you may want to select three or four colleges and complete applications for each school.
- You should begin the application process at least a year in advance. An increasing number of schools accept applications year round. In any case, find out the application deadlines and plan accordingly. Although September (the fall semester) is the traditional time to begin university study in the United States, you can begin your studies at many schools in January (the spring semester).
- In addition, you will need to obtain an I-20 Certificate of Eligibility from the school you plan to attend if you intend to apply for an F-1 Student Visa to study in the United States.

Kaplan International Programs

If you need more help with the complex process of university admissions, assistance preparing for the SAT, ACT, or TOEFL, or help building your English language skills in general, you may be interested in Kaplan's programs for international students. Kaplan International Programs were designed to help students and professionals from outside the United States meet their educational and career goals. At locations throughout the United

States, international students take advantage of Kaplan's programs to help them improve their academic and conversational English skills, raise their scores on the TOEFL, SAT, ACT, and other standardized exams, and gain admission to the schools of their choice. Our staff and instructors give international students the individualized attention they need to succeed. Here is a brief description of some of Kaplan's programs for international students:

General Intensive English

Kaplan's General Intensive English classes are designed to help you improve your skills in all areas of English and to increase your fluency in spoken and written English. Classes are available for beginning to advanced students, and the average class size is 12 students.

TOEFL and Academic English

This course provides you with the skills you need to improve your TOEFL score and succeed in an American university or graduate program. It includes advanced reading, writing, listening, grammar, and conversational English. You will also receive training for the TOEFL using Kaplan's exclusive computer-based practice materials.

SAT Test Preparation Course

The SAT is an important admission criterion for American colleges and universities. A high score can help you stand out from other applicants. This course includes the skills you need to succeed on each section of the SAT, as well as access to Kaplan's exclusive practice materials.

Other Kaplan Programs

Since 1938, more than 3 million students have come to Kaplan to advance their studies, prepare for entry to American universities, and further their careers. In addition to the above programs, Kaplan offers courses to prepare for the ACT, GMAT, GRE, MCAT, DAT, USMLE, NCLEX, and other standardized exams at locations throughout the United States.

Applying to Kaplan International Programs

To get more information, or to apply for admission to any of Kaplan's programs for international students and professionals, contact us at:

Kaplan International Programs
700 South Flower, Suite 2900
Los Angeles, CA 90017, USA
Phone (if calling from within the United States): 1-800-818-9128
Phone (if calling from outside the United States): 213-452-5800
Fax: 213-892-1364
Website: www.kaplaninternational.com
Email: world@kaplan.com

Kaplan is authorized under federal law to enroll nonimmigrant alien students. Kaplan is accredited by ACCET (Accrediting Council for Continuing Education and Training). Test names are registered trademarks of their respective owners.

section one

THE BASICS

CHAPTER ONE

About the SAT II: Subject Tests

You're serious about going to the college of your choice. You wouldn't have opened this book otherwise. You've made a wise choice, because this book can help you to achieve your goal. It'll show you how to score your best on the SAT II: Physics Subject Test. But before turning to the physics review section, let's look at the SAT II as a whole.

FREQUENTLY ASKED QUESTIONS

The following background information about the SAT II is important to keep in mind as you get ready to prep for the SAT II: Physics Subject Test. Remember, though, that sometimes the test makers change the test policies after a book has gone to press. The information here is accurate at the time of publication but it's a good idea to check the test information on the College Board website at www.collegeboard.com.

Originally, SAT stood for *Scholastic Aptitude Test*. Then, when the test changed in the mid-1990s, the official name was changed to *Scholastic Assessment Test*. Finally, in 1997, the test makers announced that SAT no longer stands for anything, officially.

What Is the SAT II?

The SAT II is actually a set of more than 20 different Subject Tests. These tests are designed to measure what you have learned in such subjects as Literature, American History and Social Studies, Biology, and Spanish. Each test lasts one hour and consists entirely of multiple-choice questions, except for the Writing Test, which has a 20-minute essay section and a 40-minute multiple-choice section. On any one test date, you can take up to three Subject Tests.

How Does the SAT II Differ from the SAT I?

SAT I is largely a test of verbal and math skills. True, you need to know some vocabulary and some formulas for the SAT I; but it's designed to measure how well you read and think rather than how much you remember. The SAT II tests are very different. They're designed to measure what you know about specific disciplines. Sure, critical reading and thinking skills play a part on these tests, but their main purpose is to determine exactly what you know about writing, math, history, physics, and so on.

Colleges use your SAT II scores in both admissions and placement decisions.

How Do Colleges Use the SAT II?

Many people will tell you that the SATs (I and II alike) measure only your ability to perform on standardized exams—that they measure neither your reading and thinking skills nor your level of knowledge. Maybe they're right. But these people don't work for colleges. Those schools that require SATs feel that they are an important indicator of your ability to succeed in college. Specifically, they use your scores in one or both of two ways: to help them make admissions and/or placement decisions.

Like the SAT I, the SAT II tests provide schools with a standard measure of academic performance, which they use to compare you with applicants from different high schools and different educational backgrounds. This information helps them to decide whether you're ready to handle their curriculum.

SAT II scores may also be used to decide what course of study is appropriate for you once you've been admitted. A low score on the Writing Test, for example, might mean that you have to take a remedial English course. Conversely, a high score on an SAT II: Mathematics Test might mean that you'll be exempted from an introductory math course.

Many colleges require you to take certain SAT II tests. Check with all of the schools you're interested in applying to before deciding which tests to take.

Which SAT II Tests Should I Take?

The simple answer is: those that you'll do well on. High scores, after all, can only help your chances for admission. Unfortunately, many colleges demand that you take particular tests, usually the Writing Test and/or one of the Mathematics Tests. Some schools will give you a degree of choice in the matter, especially if they want you to take a total of three tests. Before you register to take any tests, therefore, check with the colleges you're interested in to find out exactly which tests they require. Don't rely on high school guidance counselors or admissions handbooks for this information. They might not give you accurate or current information.

You can take up to three SAT II tests in one day. The Writing Test must be taken first.

When Are the SAT II Tests Administered?

Most of the SAT II tests, including Physics, are administered six times a year: in October, November, December, January, May, and June. A few of the tests are offered less frequently. Due to admissions deadlines, many colleges insist that you take the SAT II no later than December or January of your senior year in high school. You may even have to take it sooner if you're interested in applying for "early admission" to a school. Those schools that use scores for placement decisions only may allow you to take the SAT II as late as May or June of your senior year. You should check with colleges to find out which test dates are most appropriate for you.

How Do I Register for the SAT II?

The College Board® administers the SAT II tests, so you must sign up for the tests with them. The easiest way to register is to obtain copies of the *SAT Registration Bulletin* and *Taking the SAT II: Subject Tests*. These publications contain all of the necessary information, including current test dates and fees. They can be obtained at any high school guidance office or directly from the College Board.

You can also reregister by telephone if you have previously registered for an SAT I or SAT II test. If you choose this option, you should still read the College Board publications carefully before you make any decisions.

Want to register for the SAT II or get more info? Ask your school counselor's office for the SAT Registration Bulletin, which contains a Registration Form, test dates, fees, and instructions.

By mail: Mail in the Registration Form to the College Board.

Online: With a credit card, you can register online at www.collegeboard.com.

By phone: You can register by phone *only if* you have registered for an SAT test in the past.

College Board SAT Program
P.O. Box 6200
Princeton, NJ 08541-6200
(609) 771-7600
www.collegeboard.com

How Are the SAT II Tests Scored?

Like the SAT I, the SAT II tests are scored on a 200–800 scale.

What's a "Good" Score?

That's tricky. The obvious answer is: the score that the colleges of your choice demand. Keep in mind, though, that SAT II scores are just one piece of information that colleges will use to evaluate you. The decision to accept or reject you will be based on many criteria, including your high school transcript, your SAT I scores, your recommendations, your personal statement, your interview (where applicable), your extracurricular activities, and the like. So, failure to achieve the necessary score doesn't automatically mean that your chances of getting in have been damaged.

The mean score of the students who took the SAT II: Physics Test in 2000 was 640.

Gather your test materials the day before the test. You'll need:

- Your admission ticket
- A proper form of I.D.
- Some sharpened No. 2 pencils
- A good eraser

What Should I Bring to the SAT II?

It's a good idea to get your test materials together the day before the tests. You'll need an admission ticket; a form of identification (check the *Registration Bulletin* to find out what is permissible); a few sharpened No. 2 pencils; and a good eraser. (Note that calculators are not allowed on any of the SAT II tests except for Math Level IC and Math Level IIC.) If you'll be registering as a standby, collect the appropriate forms beforehand. Also, make sure that you know how to get to the test center.

SAT II MASTERY

Now that you know a little about the SAT II tests, it's time to let you in on a few basic test taking skills and strategies that can improve your performance on them. You should practice these skills and strategies as you prepare for the SAT II.

Use the Structure of the Test to Your Advantage

The SAT II tests are different from the tests that you're used to taking. On your high school tests, you probably go through the questions in order. You probably spend more time on hard questions than on easy ones, since hard questions are generally worth more points. And you often show your work, since your teachers tell you that how you approach questions is as important as getting the right answers.

None of this applies to the SAT II tests. You can benefit from moving around within the tests, hard questions are worth the same as easy ones, and it doesn't matter how you answer the questions—only what your answers are.

The SAT II tests are highly predictable. Because the format and directions of the SAT II tests remain unchanged from test to test, you can learn the setup of each test in advance. On Test Day, the various question types on each test shouldn't be new to you.

One of the easiest things you can do to help your performance on the SAT II tests is to understand the directions before taking the test. Since the instructions are always the same, there's no reason to waste a lot of time on Test Day reading them. Learn them beforehand as you work through this book and the College Board publications.

Not all of the questions on the SAT II tests are equally difficult. The questions often get harder as you work through different parts of a test. This pattern can work to your benefit. Try to be aware of where you are in a test.

When working on more basic problems, you can generally trust your first impulse—the obvious answer is likely to be correct. As you get to the end of a test section, you need to be a bit more suspicious. Now the answers probably won't come as quickly and easily; if they do, look again because the obvious answers may be wrong. Watch out for answers that just "look right." They may be *distractors*: wrong answer choices deliberately meant to entice you.

Learn SAT II directions as you prepare for the tests. That way, you'll have more time to spend answering the questions on Test Day.

There's no mandatory order to the questions on the SAT II. You're allowed to skip around on the SAT II tests. High scorers know this fact. They move through the tests efficiently. They don't dwell on any one question, even a hard one, until they've tried every question at least once.

When you run into questions that look tough, circle them in your test booklet and skip them for the time being. Go back and try again after you've answered the easier ones if you've got time. After a second look, troublesome questions can turn out to be remarkably simple.

If you've started to answer a question but get confused, quit and go on to the next question. Persistence might pay off in high school classes, but it usually hurts your SAT II scores. Don't spend so much time answering one hard question that you use up three or four questions' worth of time. That'll cost you points, especially if you don't even get the hard question right.

Do the questions in the order that's best for you. Skip hard questions until you've gone through every question once. Don't pass up the opportunity to score easy points by wasting time on hard questions. Come back to them later.

You can use the so-called guessing penalty to your advantage. You might have heard it said that the SAT II has a "guessing penalty." That's a misnomer. It's really a *wrong-answer penalty*. If you guess wrong, you get a small penalty. If you guess right, you get full credit.

The fact is, if you can eliminate one or more answer choices as definitely wrong, you'll turn the odds in your favor and actually come out ahead by guessing. The fractional points that you lose are meant to offset the points you might get "accidentally" by guessing the correct answer. With practice, however, you'll see that it's often easy to eliminate *several* answer choices on some of the questions.

Don't guess, unless you can eliminate at least one answer choice. Don't leave a question blank unless you have absolutely no idea how to answer it.

The answer grid has no heart. It sounds simple, but it's extremely important: Don't make mistakes filling out your answer grid. When time is short, it's easy to get confused going back and forth between your test booklet and your grid. If you know the answers, but misgrid, you won't get the points. Here's how to avoid mistakes.

Always circle the questions you skip. Put a big circle in your test booklet around any question numbers that you skip. When you go back, these questions will be easy to relocate. Also, if you accidentally skip a box on the grid, you'll be able to check your grid against your booklet to see where you went wrong.

A common mistake is filling in all of the questions with the right answers–in the wrong spots. Whenever you skip a question, circle it in your test booklet and make doubly sure that you skip it on the answer grid as well.

Always circle the answers you choose. Circling your answers in the test booklet makes it easier to check your grid against your booklet.

Grid five or more answers at once. Don't transfer your answers to the grid after every question. Transfer them after every five questions. That way, you won't keep breaking your concentration to mark the grid. You'll save time and gain accuracy.

Approaching SAT II Questions

Apart from knowing the setup of the SAT II tests that you'll be taking, you've got to have a system for attacking the questions. You wouldn't travel around an unfamiliar city without a map, and you shouldn't approach the SAT II without a plan. What follows is the best method for approaching SAT II questions systematically.

Try to think of the answer to a question before you shop among the answer choices. If you've got some idea of what you're looking for, you'll be less likely to be fooled by "trap" choices.

Think about the questions before you look at the answers. The test makers love to put distractors among the answer choices. Distractors are answers that look like they're correct, but aren't. If you jump right into the answer choices without thinking first about what you're looking for, you're much more likely to fall for one of these traps.

Guess—when you can eliminate at least one answer choice. You already know that the "guessing penalty" can work in your favor. Don't simply skip questions that you can't answer. Spend some time with them in order to see whether you can eliminate any of the answer choices. If you can, it pays for you to guess.

Pace yourself. The SAT II tests give you a lot of questions in a short period of time. To get through the tests, you can't spend too much time on any single question. Keep moving through the tests at a good speed. If you run into a hard question, circle it in your test booklet, skip it, and come back to it later if you have time.

Work quickly on easier questions to leave more time for harder questions. But not so quickly that you make careless errors. And it's okay to leave a few questions blank if you have to–you can still get a high score.

You don't have to spend the same amount of time on every question. Ideally, you should be able to work through the easier questions at a brisk, steady clip, and use a little more time on the harder questions. One caution: Don't rush through basic questions just to save time for the harder ones. The basic questions are points in your pocket, and you're better off not getting to some harder questions if it means losing easy points because of careless mistakes. Remember, you don't earn any extra credit for answering hard questions.

Locate quick points if you're running out of time. Some questions can be done more quickly than others because they require less work or because choices can be eliminated more easily. If you start to run out of time, look for these quicker questions.

When you take the SAT II tests, you have one clear objective in mind: to score as many points as you can. It's that simple. The rest of this book is dedicated to helping you to do that on the SAT II: Physics Subject Test.

CHAPTER TWO

Getting Ready for the SAT II: Physics Test

Now that you know the basics about the SAT II: Subject Tests, it's time to focus on the Physics Test. What's on it? How is it scored? After reading this chapter, you'll know what to expect on Test Day.

CONTENT

The SAT II: Physics Subject Test expects you to have a mastery of the concepts and principles covered in a one-year, college-prep physics class. This one-hour exam consists of 75 multiple-choice questions covering the topics in the outline below.

The multiple-choice questions will require you to recall and understand the terms, principles, and concepts listed below and apply these to solve specific physics problems. You will also need to show your understanding of simple algebraic, trigonometric, and graphical relationships, such as equations, sine, cosine, tangent, slope, and area under a curve, as well as ratios and proportions. You will not be allowed to use a calculator on this exam, but don't worry; the math should be nothing more than simple calculations—nothing more difficult than multiplication or division. You will be provided with any physical constants you may need, along with the values of relevant trigonometric relationships such as sine, cosine, and tangent. Occasionally, you will be given some laboratory data and asked to draw conclusions about the data or the analysis of the data, such as interpreting a graph. Of course, the SAT II: Physics Test is limited in the amount of lab skills it can test, since the format of the test includes only multiple-choice questions.

Topics	Percent of Test
Mechanics	**34–38%**
Kinematics velocity, acceleration, one-dimensional motion, projectile motion	
Dynamics force, Newton's laws, statics	
Energy and Momentum potential and kinetic energy, work, power, impulse, conservation laws	
Circular Motion and Rotation uniform circular motion, centripetal force, torque, angular momentum	
Vibrations simple harmonic motion, mass on a spring, simple pendulum	
Gravity law of gravitation and orbits	
Electricity and Magnetism	**22–26%**
Electric Fields, Forces, and Potentials Coulomb's law, induced charge, field and potential of groups of point charges, charged particles in electric fields	
Magnetic Fields and Forces permanent magnets, fields caused by currents, particles in magnetic fields	
Electromagnetic Induction induced currents and fields	
Circuits and Circuit Elements capacitance, resistance, Ohm's law, Joule's law, DC circuits with resistance and capacitance	
Waves	**15–19%**
General Wave Properties speed, frequency, wavelength, Doppler effect	
Reflection and Refraction Snell's law, changes in wavelength and speed	
Interference, Diffraction, and Polarization single-slit diffraction, double-slit interference, standing wave patterns	
Ray Optics image formation in lenses and mirrors	

Heat, Kinetic Theory, and Thermodynamics	**8–12%**
Thermal Properties mechanical equivalent of heat, temperature, specific and latent heats, thermal expansion, heat transfer	
Gases and Kinetic Theory ideal gas law from molecular properties	
Laws of Thermodynamics first and second laws, internal energy, heat engine efficiency	
Modern Physics	**8–12%**
Quantum Phenomena photons, photoelectric effect, the uncertainty principle	
Atomic Physics Rutherford and Bohr models, atomic energy levels, atomic spectra	
Nuclear and Particle Physics radioactivity, nuclear or particle reactions	
Special Relativity mass-energy equivalence, limiting velocity	
Contemporary Physics astrophysics, biophysics, superconductivity	
Miscellaneous	**2–4%**
Measurement	
Math Skills	
Laboratory Skills	
History of Physics	
Questions of a general nature that overlap several topics	

The review chapters in this book follow these topics closely. As you can see, the SAT II: Physics Subject Test covers a broad range of topics. It requires you to think about these topics in ways you may not have done before. As a result, it's likely that some test questions will involve topics that you did not cover in your physics class. Don't be alarmed; there is so much to physics that you cannot possibly cover everything in a year. If, while taking the diagnostic test in this book, you discover that there are areas you haven't covered in school, plan to spend a little extra time on those chapters that cover your areas of weakness.

There are three levels of concept application used throughout the SAT II: Physics Test. *Recall questions*, comprising 20 to 33 percent of the test (15–25 questions), require you to simply remember a definition or concept. *Single-concept questions* require you to apply a single concept to a question in order to arrive at the answer. These questions typically comprise 40 to 53 percent of the test (30–40 questions). *Multiple-concept questions* require you to think through and relate (synthesize) several concepts together to arrive at an answer to a question. These are often the most difficult of the three levels of concept application, and comprise 20 to 33 percent of the test (15–25 questions).

SCORING INFORMATION

This exam is scored in a range from 200–800 (in multiples of ten), just like a section of the SAT I. Your raw score is calculated by subtracting 1/4 of the number of questions you got wrong from the number of questions you got right. If you answered 55 questions correctly and 17 questions incorrectly, having skipped 3 questions, then your raw score would be:

Number correct:	55
1/4 × Number incorrect:	− 4.25
Raw score:	50.75 (rounded to 51)

Three basic skills are tested on the SAT II: Physics Test:

Recalling information

Applying knowledge

Synthesizing information

This raw score is then compared to all the other test takers' scores to come up with a scaled score. This scaling takes into account any slight variations between test administrations. On a recent administration, it was possible to miss 12 questions and still receive a scaled score of 800. A raw score of 51 out of 75 on a 1995 SAT II: Physics Test translated into a 730. The average raw score on the May 2000 exam was a 35 out of 75, which translated to a scaled score of 635. So basically, you can miss a few questions and still receive a competitive score.

Our test strategies won't make up for a weakness in a given area, but they will help you to manage your time effectively and maximize points.

QUESTION TYPES

The multiple-choice questions on the SAT II: Physics Test are divided up into two types: *classification questions* and *five-choice completion questions*. Make sure you feel comfortable with all question types and their directions before Test Day. Don't waste time rereading familiar directions when you are being timed!

On Test Day, do classification questions first: They require less reading and will give you the most points for your time invested.

Classification Questions

Classification questions consist of five lettered choices, typically ideas, physical laws, graphs, or some other type of data presentation. Following the five choices will be three to five statements that can be functions of the choices, definitions, descriptive characteristics, or conditions that would favor that data set. Any of the five choices may be used more than once, so do not eliminate an answer just because you have used it.

Read through the directions and attempt to answer questions 1–3 below. Check your answers against the explanations that follow the question set.

Directions: Each set of lettered choices below refers to the numbered questions or statements immediately following it. Select the one lettered choice that best answers each question or best fits each statement, and then fill in the corresponding oval on the answer sheet. A choice may be used once, more than once, or not at all in each set.

Sample Questions 1–3 relate to the following:

(A) Mass
(B) Amplitude
(C) Length
(D) Equilibrium position
(E) Restoring force

1. For a pendulum swinging at a small angle, which of the above affects the period of the pendulum?
2. Which of the above is the point at which the potential energy of the pendulum is maximum?
3. Which of the above is the point at which the kinetic energy of the pendulum is maximum?

Don't eliminate an answer choice just because you've used it. Answer choices on classification questions can be used more than once.

As a pendulum swings at a small angle, its period depends only upon the length of the pendulum. The mass of the pendulum bob and the amplitude of swing (assuming small angles) do not change the time for one complete swing. The potential energy of the pendulum is maximum when the pendulum is the highest distance above its equilibrium position, which corresponds to its amplitude, or maximum displacement from its equilibrium position. The kinetic energy of the pendulum is maximum as it swings through the equilibrium position, since the pendulum speeds up as it moves toward the equilibrium position and slows down after it passes the equilibrium position. Thus, the correct answers for this question set are: 1 (C); 2 (B); 3 (D).

Five-Choice Completion Questions

There are three types of these questions, all of which use the same set of directions.

Directions: Each of the incomplete statements or questions below is followed by five suggested completions or answers. In each case, select the one that is best and fill in the corresponding oval on the answer sheet.

Type 1 Questions

These have a unique solution, often the only correct answer or the best answer. Sometimes, however, the most inappropriate answer will be correct; these question types will have the words NOT, EXCEPT, or LEAST somewhere in the stimulus.

Sample Questions 4–5:

4. An object released from rest at time $t = 0$ falls freely under the influence of gravity.

   What is the speed of the object at $t = 2$ seconds?

   (A) 2 m/s
   (B) 5 m/s
   (C) 10 m/s
   (D) 20 m/s
   (E) 40 m/s

5. Which of the following is NOT an example of a vector?

   (A) force
   (B) velocity
   (C) displacement
   (D) mass
   (E) momentum

Question 4 tests your understanding of the velocity and acceleration of a falling object. As an object falls freely, it gains 10 m/s of speed for each second it falls. Thus after two seconds, its speed is 20 m/s. You could also use the equation $v = gt$, where $g = 10\ \text{m/s}^2$ and $t = 2$ s. Thus, answer (D) is correct.

Question 5 distinguishes between a vector and a scalar. A vector is a quantity having both magnitude and direction, such as force, velocity, displacement, and momentum. Mass is a scalar, having no direction. Therefore, answer (D) is correct.

Type 2 Questions

These typically have three to five Roman numeral statements following each question. One or more of these statements may be the correct answer. Following each statement, there are five lettered choices with various combinations of the Roman numerals. You must select the combination that includes all of the correct answers and excludes all of the incorrect answers.

Sample Question 6:

6. A candle may be placed on the principal axis at the following distances from the center of a convex lens:

   I. at the focal length

   II. at 3/2 the focal length

   III. less than the focal length

   IV. greater than twice the focal length

   Which of the above will produce an image that is larger than the actual candle?

   (A) I only
   (B) I, II, and III only
   (C) II and III only
   (D) II, III, and IV only
   (E) I, II, III, and IV

A convex lens produces a real image when the candle is placed at a distance greater than the focal length, and an enlarged virtual image when it is placed at a distance less than the focal length. No image is produced when the candle is placed at the focal length; thus statement I is not correct. If the candle is placed between the focal length and twice the focal length, the image is also enlarged, but the image is smaller if the candle is placed at a distance greater than twice the focal length. Therefore, statements II and III will produce an image larger than the candle, and the correct answer is (C).

Type 3 Questions

These questions are also organized in sets, but center on an experiment, chart, graph, or other experimental data presentation. They assess how well you apply science to unfamiliar situations. Each question is independent of the others; in addition, these questions are typically found in the latter part of the test. Most students find them to be the most difficult questions on the test.

Type 3 questions test your ability to identify a problem; evaluate experimental situations; suggest hypothesis; interpret data such as graphs or mathematical expressions; make inferences and draw conclusions; check the logical consistency of hypotheses based on your observations; convert information to graphical form; apply mathematical relationships; and select the appropriate procedure for further study.

Sample Questions 7–9 are based on the following radioactive decay equation:

$$^{226}_{88}Ra \rightarrow {}^{A}_{Z}X + {}^{4}_{2}He$$

7. The type of radiation emitted in the above decay is

   (A) alpha
   (B) beta
   (C) gamma
   (D) neutrino
   (E) electron

8. The atomic number of the daughter atom which correctly completes the equation is

   (A) 226
   (B) 224
   (C) 222
   (D) 86
   (E) 84

9. The mass number of the daughter atom which correctly completes the equation is

   (A) 226
   (B) 224
   (C) 222
   (D) 86
   (E) 84

In this radioactive decay, the element radium emits a helium nucleus, which is also known as an alpha particle. Because an alpha particle consists of 2 protons and 2 neutrons, it has an atomic number of 2 (number of protons) and a mass number of 4 (number of protons and neutrons). Thus, a radium atom undergoing alpha decay loses two atomic numbers, leaving 86, and four mass numbers, leaving 222. The correct answers are: 7 (A); 8 (D); 9 (C).

KAPLAN STRATEGIES

There are ways to approach this test that will allow you to maximize your score. Read through these strategies before you begin your diagnostic practice test. Try to internalize each of them so that on Test Day they will be second nature to you. If you accomplish this, you will be rewarded with a higher score on your SAT II: Physics Test.

1. Do classification questions first; they require less reading and you'll get the most points for your time invested.

2. Next, do the Type 1 and Type 2 five-choice completion questions. Again, you will get more points with less time invested.

3. This test emphasizes general trends and basic physics concepts, so you probably won't see a question/graph that would take a rocket scientist 30 minutes to figure out. Look for trends and wide variations in graphs. If a value or a plot is vastly different from the others, it is likely that there will be a question about it. Also be prepared to use the slope of a graph or area under a curve to determine quantities.

4. Look for opposing answers in the answer selections. If two answers are close in wording or if they contain opposite ideas, there is a strong possibility that one of them is the right answer.

5. By the same token, if two answers mean basically the same thing, then they both cannot be correct and you can eliminate both answer choices.

6. Use the structure of a Type 3 Roman numeral question to your advantage. Eliminate choices as soon as you find them to be inconsistent with the truth or falsehood of a statement in the stimulus. Similarly, consider only those choices that include a statement that you've already determined to be true.

7. Predict your answer before you go to the answer choices so you don't get persuaded by the wrong answers. This helps protect you from persuasive or tricky incorrect choices. Most wrong answer choices are logical twists on the correct choice.

8. Eliminate answers and guess.

9. Think, don't compute!

Now you're ready to tackle our diagnostic practice test. This test will probe your mastery of the various physics topics on the SAT II: Physics Test. Use it to identify areas in which you need to refresh your knowledge, and plan to review the chapters that deal with these topics carefully.

section two

PHYSICS REVIEW

PRACTICE TEST 1

Diagnostic Test

HOW TO TAKE THIS DIAGNOSTIC TEST

Before taking this diagnostic test, find a quiet room where you can work uninterrupted for one hour. Make sure you have several No. 2 pencils with erasers.

Use the answer sheet provided to record your answers.

Time yourself. Spend no more than one hour on the 75 questions. Once you start this practice test, don't stop until you've reached the one-hour time limit.

Do not use a calculator. Calculators are not allowed on the SAT II: Physics Test.

You'll find an answer key, score conversion table, and complete explanations following the test.

Good luck!

Answer Sheet for Practice Test 1

1. (A) (B) (C) (D) (E)
2. (A) (B) (C) (D) (E)
3. (A) (B) (C) (D) (E)
4. (A) (B) (C) (D) (E)
A 5. (A) (B) (C) (D) (E)
6. (A) (B) (C) (D) (E)
7. (A) (B) (C) (D) (E)
8. (A) (B) (C) (D) (E)
9. (A) (B) (C) (D) (E)
10. (A) (B) (C) (D) (E)
11. (A) (B) (C) (D) (E)
12. (A) (B) (C) (D) (E)
13. (A) (B) (C) (D) (E)
14. (A) (B) (C) (D) (E)
15. (A) (B) (C) (D) (E)
16. (A) (B) (C) (D) (E)
17. (A) (B) (C) (D) (E)
18. (A) (B) (C) (D) (E)
19. (A) (B) (C) (D) (E)
20. (A) (B) (C) (D) (E)
21. (A) (B) (C) (D) (E)
22. (A) (B) (C) (D) (E)
23. (A) (B) (C) (D) (E)
24. (A) (B) (C) (D) (E)
A 25. (A) (B) (C) (D) (E)
26. (A) (B) (C) (D) (E)
27. (A) (B) (C) (D) (E)
28. (A) (B) (C) (D) (E)
29. (A) (B) (C) (D) (E)
30. (A) (B) (C) (D) (E)
B 31. (A) (B) (C) (D) (E)
D 32. (A) (B) (C) (D) (E)
33. (A) (B) (C) (D) (E)
34. (A) (B) (C) (D) (E)
35. (A) (B) (C) (D) (E)
36. (A) (B) (C) (D) (E)
37. (A) (B) (C) (D) (E)
38. (A) (B) (C) (D) (E)
39. (A) (B) (C) (D) (E)
40. (A) (B) (C) (D) (E)
41. (A) (B) (C) (D) (E)
B 42. (A) (B) (C) (D) (E)
43. (A) (B) (C) (D) (E)
B 44. (A) (B) (C) (D) (E)
45. (A) (B) (C) (D) (E)
46. (A) (B) (C) (D) (E)
47. (A) (B) (C) (D) (E)
48. (A) (B) (C) (D) (E)
49. (A) (B) (C) (D) (E)
50. (A) (B) (C) (D) (E)
D 51. (A) (B) (C) (D) (E)
52. (A) (B) (C) (D) (E)
53. (A) (B) (C) (D) (E)
54. (A) (B) (C) (D) (E)
55. (A) (B) (C) (D) (E)
A 56. (A) (B) (C) (D) (E)
57. (A) (B) (C) (D) (E)
58. (A) (B) (C) (D) (E)
59. (A) (B) (C) (D) (E)
60. (A) (B) (C) (D) (E)
61. (A) (B) (C) (D) (E)
62. (A) (B) (C) (D) (E)
63. (A) (B) (C) (D) (E)
64. (A) (B) (C) (D) (E)
65. (A) (B) (C) (D) (E)
66. (A) (B) (C) (D) (E)
B 67. (A) (B) (C) (D) (E)
68. (A) (B) (C) (D) (E)
69. (A) (B) (C) (D) (E)
70. (A) (B) (C) (D) (E)
71. (A) (B) (C) (D) (E)
72. (A) (B) (C) (D) (E)
73. (A) (B) (C) (D) (E)
74. (A) (B) (C) (D) (E)
75. (A) (B) (C) (D) (E)

Raw: $67 - \frac{8}{4} = 67 - 2 = 65$

Scaled: 800

Practice Test 1: Diagnostic Test

Part A

Directions: Each set of lettered choices below relates to the numbered questions immediately following it. Select the one lettered choice that best answers each question. A choice may be used once, more than once, or not at all in each set.

Questions 1–3 relate to the diagram below and the choices that follow.

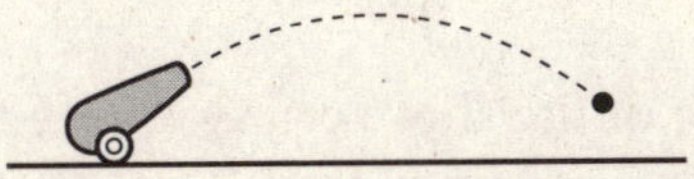

A projectile is launched at an angle from level ground. Assume air resistance is negligible.

(A) horizontal velocity
(B) vertical velocity
(C) horizontal acceleration
(D) vertical acceleration
(E) angle at which the projectile is launched

1. Which of the above choices is NOT constant throughout the flight of the projectile?

2. Which of the above choices is zero throughout the flight of the projectile?

3. Which of the above choices changes direction during the flight of the projectile?

Questions 4–7 relate to the following equations or physical principles that might be used to solve certain problems.

(A) Newton's first law (law of inertia)
(B) Newton's second law ($F_{net} = ma$)
(C) Newton's third law (For every action there is an equal and opposite reaction)
(D) Newton's law of universal gravitation
(E) conservation of linear momentum

Select the choice that should be used to provide the best and most direct solution to each of the following problems.

4. Two lead spheres each of mass m are 3 cm apart and are not in contact. What force do they apply to each other?

5. A jar remains at rest on a table when a piece of paper is pulled out from under it.

6. A ball of clay is shot at a block of wood on a table and the two stick together. What was the initial speed of the clay?

7. Two teams pull against each other on a rope that is stretched across a stream. One team pulls with 1,000 N of force. What is the tension in the rope?

CONTINUE ON THE NEXT PAGE

Questions 8–9 relate to a small particle entering a magnetic field B which is directed out of the page as shown.

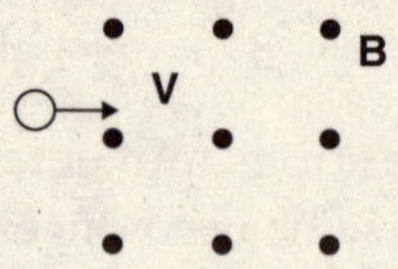

(A) proton
(B) electron
(C) neutron
(D) x-ray
(E) photon of red light

8. Which of the following particles would follow the path shown below while moving through the magnetic field?

9. Which of the following particles would follow the path shown below while moving through the magnetic field?

Questions 10–13 relate to the following.

(A) reflection
(B) refraction
(C) diffraction
(D) constructive interference
(E) destructive interference

10. Which of the above occurs when two waves that are in phase with each other meet at the same time in the same medium?

11. Which of the above creates two angles that are equal to each other?

12. Which of the above causes a wave to change its speed and wavelength?

13. Which of the above occurs when a wave passes through a small opening?

CONTINUE ON THE NEXT PAGE

Part B

Directions: Each of the questions or incomplete statements below is followed by five answer choices. Select the one that is best in each case.

14. Each of the figures below shows the forces acting on a particle, where each force is the same magnitude. In which figure can the particle's velocity be constant?

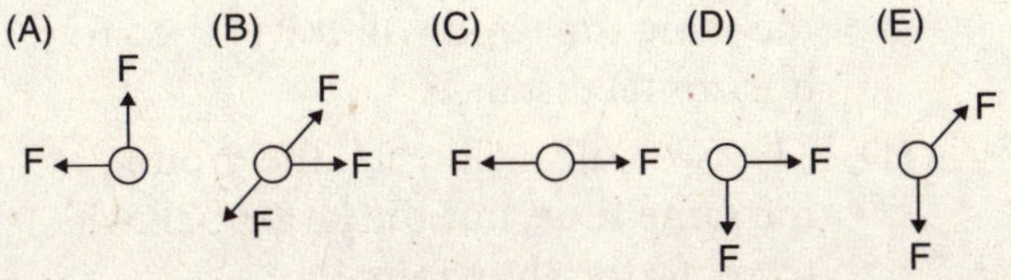

15. The water in a river is running due west. A boy in a boat tries to cross the river by rowing due south. The motion of the boat relative to the shore is

    (A) due south.
    (B) due north.
    (C) due west.
    (D) southwest.
    (E) northeast.

Questions 16–17 refer to the circuit shown below.

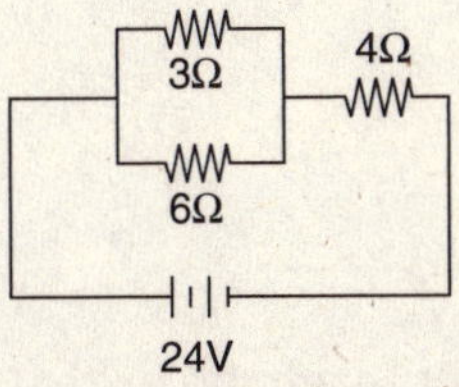

16. The current through the 4 Ω resistor is

    (A) 2 A
    (B) 4 A
    (C) 6 A
    (D) 8 A
    (E) 12 A

17. The voltage across the 4 Ω resistor is

    (A) 4 V
    (B) 8 V
    (C) 12 V
    (D) 16 V
    (E) 24 V

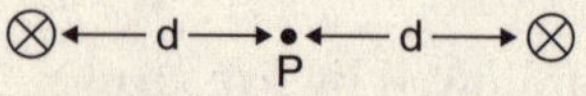

18. Two parallel wires are located a distance $2d$ apart. Each wire has a current of 1 A flowing into the page as shown above. The direction of the net magnetic field at point P halfway between the wires is:

 (A) →
 (B) ←
 (C) ↑
 (D) ↓
 (E) The net magnetic field at point P is zero.

19. A 500 kg car is initially at rest and is pushed by a constant horizontal force for a distance of 30 meters at which point the car has a kinetic energy of 18,000 J. The magnitude of the force is most nearly

 (A) 500 N
 (B) 600 N
 (C) 1,500 N
 (D) 15,000 N
 (E) 540,000 N

20. Two objects collide on a surface with negligible friction. In the absence of external forces, which of the following statements is true?

 (A) The total momentum before the collision is always equal to the total momentum after the collision.
 (B) Both momentum and kinetic energy are conserved in the collision.
 (C) The two objects must stick together.
 (D) The speed of each object must remain the same before and after the collision.
 (E) The momentum of each object must remain the same before and after the collision.

21. A ball with a large mass and a ball with a small mass roll off a horizontal table at exactly the same time with exactly the same velocity. Neglecting air resistance, which of the following is true?

 (A) The larger ball will reach the ground first, since it has more weight.
 (B) The two balls will reach the ground at the same time, but the smaller ball will travel farther horizontally.
 (C) The two balls will reach the ground at the same time and will travel the same horizontal distance.
 (D) The two balls will reach the ground at the same time, but the larger ball will travel farther horizontally.
 (E) The smaller ball will land first, since it has less weight.

22. An electric generator produces current by the process of

 (A) capacitance.
 (B) electromagnetic induction.
 (C) an electric motor.
 (D) a battery.
 (E) resistance.

CONTINUE ON THE NEXT PAGE

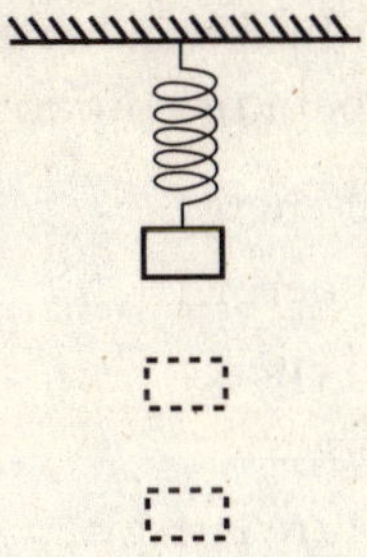

23. A mass is attached to the end of a fixed spring as shown above. The spring is compressed upward and then released. Immediately after the spring is released, which of the following increase?

(A) kinetic energy, gravitational potential energy, elastic potential energy
(B) kinetic energy and gravitational potential energy
(C) kinetic energy only
(D) gravitational potential energy and elastic potential energy
(E) elastic potential energy only

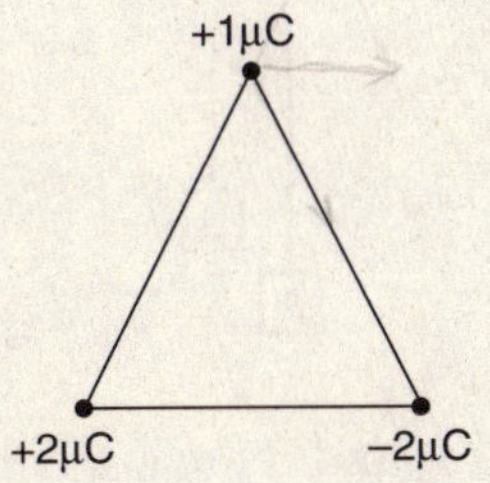

24. Three charges of +1 microcoulomb, +2 microcoulombs, and −2 microcoulombs are fixed at the vertices of an equilateral triangle as shown above. The net force acting on the +1 microcoulomb charge is directed in which direction?

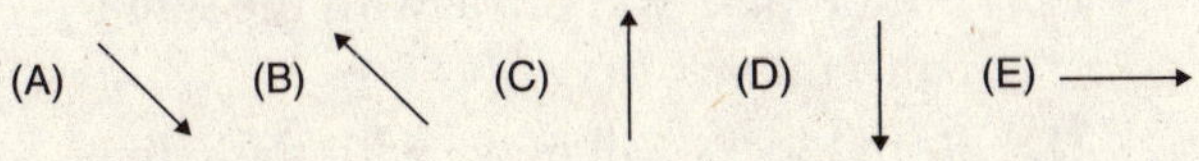

Questions 25–26:

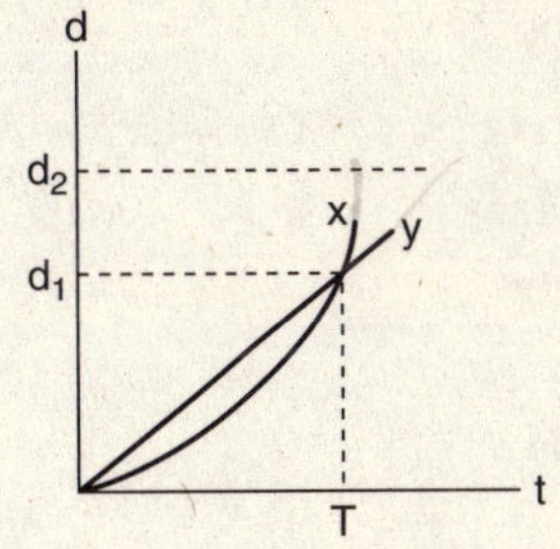

Two trains, X and Y, begin moving on parallel tracks from the same starting position at a train station. The graph above shows their respective distances from the starting point as functions of time.

25. At time T,

(A) train X is moving faster than train Y.
(B) train Y is moving faster than train X.
(C) trains X and Y are moving at the same speed.
(D) train X is ahead of train Y.
(E) train Y is ahead of train X.

26. Which of the following statements is true?

(A) Train Y moves faster than train X at all points on the graph.
(B) Train X moves faster than train Y at all points on the graph.
(C) Train X will reach the distance d_2 before train Y.
(D) Train Y will reach the distance d_2 before train X.
(E) The trains will reach the distance d_2 at the same time.

CONTINUE ON THE NEXT PAGE

27. If a rotating amusement park ride completes 6 revolutions in one minute, its frequency of rotation is

(A) 360 Hz
(B) 60 Hz
(C) 10 Hz
(D) 1 Hz
(E) 0.1 Hz

28. Pluto orbits the Sun in an elliptical orbit. Which of the following statements is true of Pluto's orbit?

(A) The gravitational force acting between Pluto and the Sun must point in the direction of Pluto's velocity.
(B) The gravitational force between Pluto and the Sun is greater when Pluto is farthest from the Sun.
(C) The speed of Pluto is greatest when Pluto is closest to the Sun.
(D) The speed of Pluto is greatest when Pluto is farthest from the Sun.
(E) The gravitational force between the Sun and Pluto is constant at all points in Pluto's orbit.

29. The pitch of a middle-C sound wave has a frequency of 256 Hz. If the speed of sound in air is 350 m/s, what is the wavelength of middle-C in air?

(A) 89,600 m
(B) 350 m
(C) 94 m
(D) 1.4 m
(E) 0.73 m

30. The energy transferred between two bodies due to a temperature difference is called

(A) potential energy.
(B) nuclear energy.
(C) chemical energy.
(D) heat energy.
(E) temperature energy.

31. An electron of mass 9.1×10^{-31} kg is orbiting in a circular path of radius 100 meters with a speed of 3×10^{7} m/s in a magnetic field. Which of the following is the best estimate of the order of magnitude of the magnetic force needed to maintain this orbit?

(A) 10^{-22} N
(B) 10^{-17} N
(C) 10^{-10} N
(D) 10^{-6} N
(E) 10^{-2} N

32. Two blocks A and B are attached to a string that passes over a pulley of negligible mass and friction as shown above. The two blocks do not have equal mass. Which of the following statements is true?

(A) Block A is more massive than block B.
(B) Block B is more massive than block A.
(C) Both blocks will move with the same constant velocity.
(D) Both blocks will move with the same constant acceleration.
(E) Block A must have a greater acceleration than block B.

CONTINUE ON THE NEXT PAGE

Questions 33–34 relate to the circuit shown below.

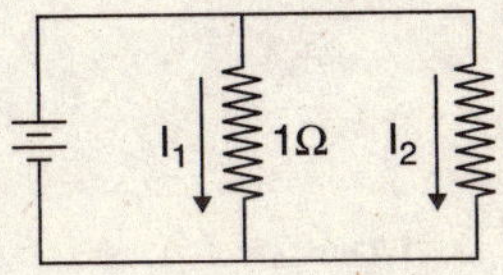

33. The current I_1 in the 1 Ω resistor is related to the current I_2 in the 2 Ω resistor by which of the following equations?

(A) $I_1 = \frac{1}{2}I_2$

(B) $I_1 = \frac{2}{3}I_2$

(C) $I_1 = I_2$

(D) $I_1 = \frac{3}{2}I_2$

(E) $I_1 = 2I_2$

34. The voltage V_1 across the 1 Ω resistor is related to the voltage V_2 across the 2 Ω resistor by the equation

(A) $V_1 = \frac{1}{2}V_2$

(B) $V_1 = \frac{2}{3}V_2$

(C) $V_1 = V_2$

(D) $V_1 = \frac{3}{2}V_2$

(E) $V_1 = 2V_2$

35. Sound has the highest speed in which of the following?

(A) steel
(B) wood
(C) water
(D) air
(E) helium

36. A converging lens has a focal length of 20 cm. A candle is placed at 30 cm from the lens, and an image is formed 60 cm from the lens. The magnification is

(A) 0.5
(B) 0.67
(C) 1.5
(D) 2.0
(E) 3.0

37. The image formed by a plane mirror alone is always

(A) real.
(B) virtual.
(C) larger than the object.
(D) smaller than the object.
(E) inverted.

38. Which of the following statements is true of two waves that interfere with each other?

(A) They always create a larger wave than either of the individual waves.
(B) The larger wave always cancels the smaller wave.
(C) The displacements of the individual waves are added at each point.
(D) An antinode is always created.
(E) A node is always created.

39. How much energy does a 2,400-Watt microwave oven use in 3 minutes?

(A) 432,000 J
(B) 7,200 J
(C) 800 J
(D) 24 J
(E) 13.3 J

CONTINUE ON THE NEXT PAGE

Questions 40–41 refer to the electroscope shown below.

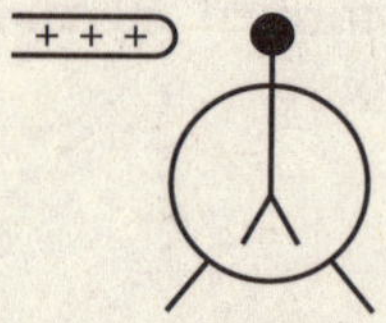

40. A positively charged rod is brought near the knob of an uncharged electroscope and the leaves of the electroscope diverge, as shown. The leaves diverge because

(A) they are both negatively charged.
(B) they are both positively charged.
(C) one is positive and one is negative.
(D) one leaf is neutral and one is positive.
(E) both leaves are neutral.

41. As the positively charged rod is held near the knob, the electroscope is grounded with a wire. Then the grounding wire and the positively charged rod are removed. The leaves of the electroscope

(A) diverge and are negatively charged.
(B) diverge and are positively charged.
(C) converge and are negatively charged.
(D) converge and are positively charged.
(E) converge and are neutral.

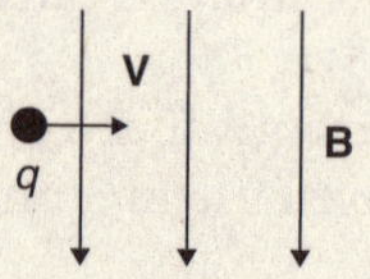

42. A positive charge moves through a magnetic field of magnitude 10^{-2} T with a speed of 10^{6} m/s when it experiences a force of 10^{-14} N. The magnitude of the charge is

(A) 10^{-22} C
(B) 10^{-18} C
(C) 10^{-10} C
(D) 10^{-6} C
(E) 10^{-2} C

$F = quB$

$q = \frac{F}{vB} = 10^{-18}$ C

Questions 43–44 relate to the standing wave in a string shown below.

A standing wave is produced in a string when a periodic wave is passed through the string and is reflected off the fixed end. The wavelength of the wave in the string shown above is 0.5 m, and the frequency of vibration is 120 Hz.

43. The speed of the wave is

(A) 240 m/s
(B) 120 m/s
(C) 60 m/s
(D) 30 m/s
(E) 20 m/s

$V = \frac{0.5}{\frac{1}{120}} = 60$

44. The length of the string is

(A) 2.0 m
(B) 1.0 m
(C) 0.5 m
(D) 0.25 m
(E) 0.1 m

CONTINUE ON THE NEXT PAGE

Questions 45–46:

The heat of fusion of ice is 3.3×10^5 J/kg and the specific heat of water is 4.2×10^3 J/kg°C. Heat is added to a 3 kg block of ice, and continues to be added to raise the temperature of the resulting water to 30° C.

45. How much heat is needed to melt the block of ice?

(A) 1.26×10^4 J
(B) 4.2×10^5 J
(C) 3.3×10^5 J
(D) 6.6×10^5 J
(E) 9.9×10^5 J

46. How much heat is required to raise the liquid water from 0° C to 30° C?

(A) 1.26×10^4 J
(B) 4.2×10^3 J
(C) 3.8×10^5 J
(D) 6.6×10^5 J
(E) 9.9×10^5 J

47. A beam of light is incident on a piece of glass, and most of the beam is refracted through the glass. Which of the following properties of the light does NOT change as it is refracted?

(A) speed
(B) frequency
(C) wavelength
(D) direction
(E) intensity

Questions 48–49:

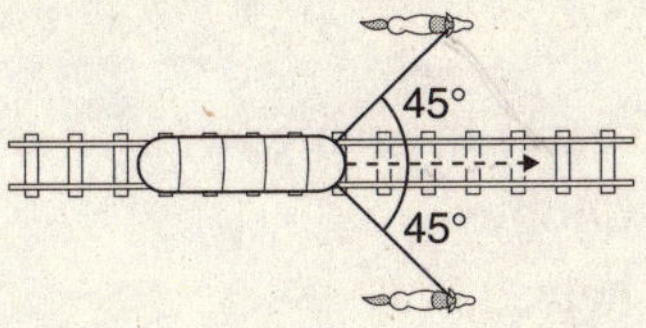

A railroad car of mass 1,000 kg is being pulled on tracks of negligible friction by ropes tied to two horses, causing the car to move along the dashed horizontal line as shown above. The net force acting on the car is 500 N. (sin 45° = cos 45° = 0.7)

48. What is the tension in each rope?

(A) 1,000 N
(B) 500 N
(C) 350 N
(D) 250 N
(E) 100 N

49. What is the acceleration of the car?

(A) 5 m/s^2
(B) 3.5 m/s^2
(C) 2.5 m/s^2
(D) 2.0 m/s^2
(E) 0.5 m/s^2

50. The photoelectric effect is best explained by the

(A) wave model of light.
(B) photon model of light.
(C) interference of light waves.
(D) diffraction of light waves.
(E) Heisenberg uncertainty principle.

CONTINUE ON THE NEXT PAGE

51. If the average speed of the molecules of an ideal gas is doubled, the temperature of the gas

(A) remains the same.
(B) is doubled.
(C) is halved.
(D) is quadrupled. ✓
(E) is quartered.

Questions 52–53:

A heat engine uses heat to do 50 J of work, and then exhausts 100 J of energy into a cold reservoir.

52. The heat added to the heat engine is

(A) 150 J
(B) 100 J
(C) 50 J
(D) 2 J
(E) zero

53. The efficiency of the heat engine is

(A) 25%
(B) 33%
(C) 50%
(D) 75%
(E) 100%

$\frac{50}{150}$

Questions 54–55:

A ball of mass m is thrown horizontally at a vertical wall with a speed v and bounces off elastically and horizontally.

54. What is the magnitude of the change in momentum of the ball?

(A) mv
(B) $2mv$
(C) $\frac{mv}{2}$
(D) $\frac{mv}{4}$
(E) zero

55. Which of the following force vs. time graphs below best represents the force acting on the ball while it is in contact with the wall?

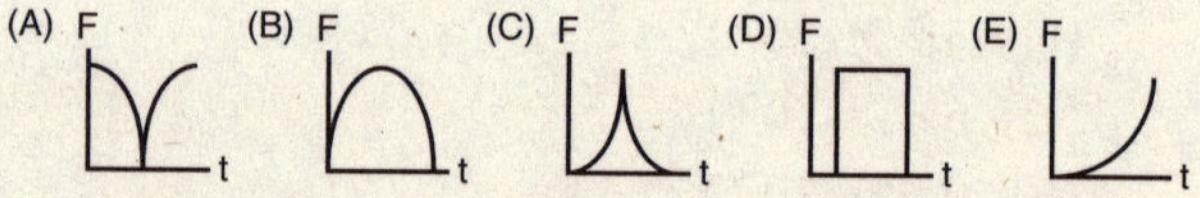

56. A periodic wave is sent through a tight spring. If the tension in the spring is reduced, the wavelength of the wave will

(A) decrease.
(B) increase.
(C) remain the same.
(D) become zero.
(E) equal the length of the spring.

CONTINUE ON THE NEXT PAGE

Questions 57–58:

A child on a skateboard crosses a line on the sidewalk traveling with a speed of 2 m/s when he begins accelerating at a constant rate of 2 m/s^2.

57. What will be the child's speed after 3 s?

    (A) 8 m/s
    (B) 9 m/s
    (C) 12 m/s
    (D) 15 m/s
    (E) 18 m/s

2+6=8

58. How far past the line on the sidewalk will the child be after 3 s?

    (A) 8 m
    (B) 9 m
    (C) 12 m
    (D) 15 m
    (E) 18 m

$S = vt + \frac{1}{2}at^2$
$= 6 + \frac{1}{2} \times 2 \times 9$
$= 15$

59. A series of alternate bright and dark bands of light appear on a screen. What most likely caused this pattern on the screen?

    (A) Light passing through a converging lens
    (B) Light passing through a diverging lens
    (C) Light passing through two very narrow slits
    (D) Photons emitted from a metal
    (E) Photons in an emission spectrum

60. A constant force acts on a mass. If the mass were tripled but the force remained constant, the acceleration of the mass would be

    (A) nine times as much.
    (B) three times as much.
    (C) unchanged.
    (D) one-third as much.
    (E) one-ninth as much.

Questions 61–62 refer to the diagram below, which represents a mass on a spring fixed to a wall sliding on a surface of negligible friction. The mass oscillates between points A and C, and B is halfway between A and C.

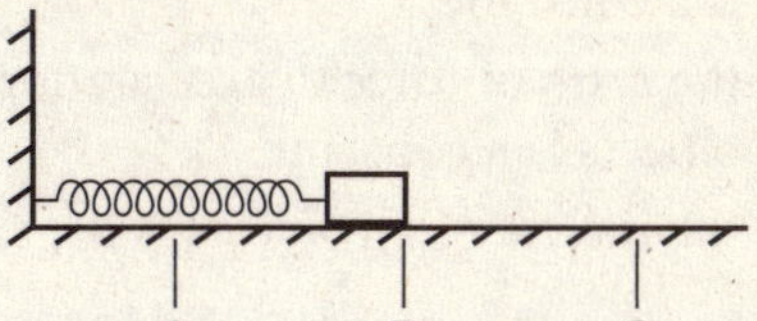

61. The restoring force acting on the mass is greatest

    (A) at B.
    (B) at A and C.
    (C) between A and B.
    (D) between B and C.
    (E) at A, B, and C.

62. When the mass passes through B, it has

    (A) kinetic energy but no potential energy.
    (B) potential energy but no kinetic energy.
    (C) kinetic energy and potential energy.
    (D) no kinetic or potential energy.
    (E) total energy equal to zero.

63. As the moon orbits the earth in a circular orbit,

    (A) its velocity is parallel to the gravitational force between the moon and Earth.
    (B) its speed increases.
    (C) its acceleration increases.
    (D) its acceleration is constant in both magnitude and direction.
    (E) its acceleration is constant in magnitude but variable in direction.

64. Neglecting air resistance, all objects fall with the same acceleration because

(A) the gravitational force is the same on each.
(B) their weights are equal.
(C) the ratio of weight to mass for all objects is a constant.
(D) the ratio of force to acceleration for all objects is a constant.
(E) the inertia of all objects is the same.

65. A boy facing forward in a tall bus throws a ball straight up at the same instant the bus begins to accelerate forward. Which of the following best describes the subsequent motion of the ball?

(A) The ball goes up and then straight back down into the boy's hand.
(B) The ball goes up and lands in front of the boy.
(C) The ball goes up and lands behind the boy.
(D) The ball stops in the air as the bus is accelerating and "floats" until the bus stops accelerating.
(E) The ball goes up and to the right of the boy.

66. On a cold morning, a metal floor feels colder to your feet than a wood floor because

(A) metal conducts heat more readily than wood.
(B) wood conducts heat more readily than metal.
(C) metal has a higher specific heat than wood.
(D) the temperature of the wood is higher than the temperature of the metal.
(E) the temperature of the metal is higher than the temperature of the wood.

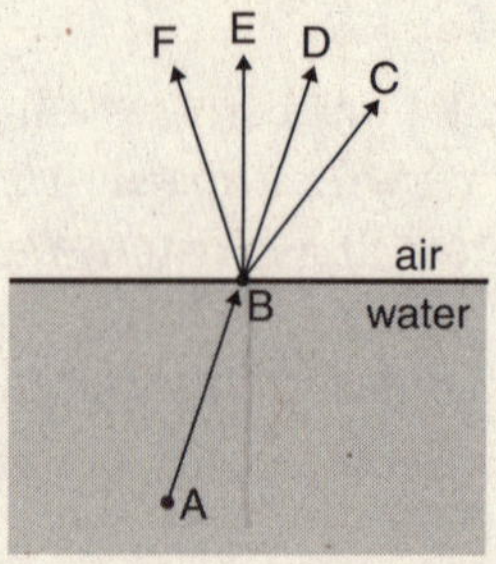

67. A ray of light is shined from beneath the water to the air above the water as shown above. Which of the following rays is the refracted ray?

(A) AB
(B) BC
(C) BD
(D) BE
(E) BF

Questions 68–69 refer to the choices below.

I. Sound waves
II. Visible light waves
III. X-rays

68. Which of the above waves can be refracted?

(A) I only
(B) II only
(C) I and II only
(D) I and III only
(E) I, II, and III

69. Which of the above waves CANNOT be polarized?

(A) I only
(B) II only
(C) I and II only
(D) I and III only
(E) I, II, and III

CONTINUE ON THE NEXT PAGE

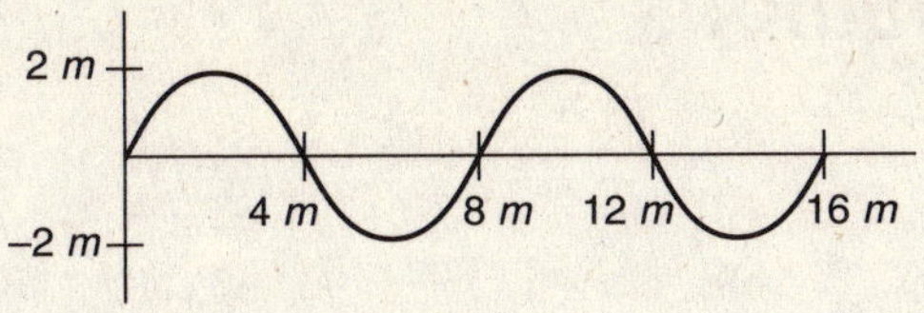

70. The amplitude of the wave shown above is

(A) 0.5 m
(B) 1.0 m
(C) 2.0 m
(D) 4.0 m
(E) 8.0 m

71. The half-life of a particular element is 6 days. What percent of the original sample of the element remains after 24 days?

(A) 100%
(B) 50%
(C) 25%
(D) 12.5%
(E) 6.25%

72. Which of the following is NOT an isotope of uranium?

(A) $^{235}_{92}U$
(B) $^{236}_{92}U$
(C) $^{238}_{92}U$
(D) $^{239}_{92}U$
(E) $^{235}_{91}U$

Questions 73–74: A ball on the end of a string is being swung in a *vertical* circle as shown below. Points I, II, and III are labeled on the circle.

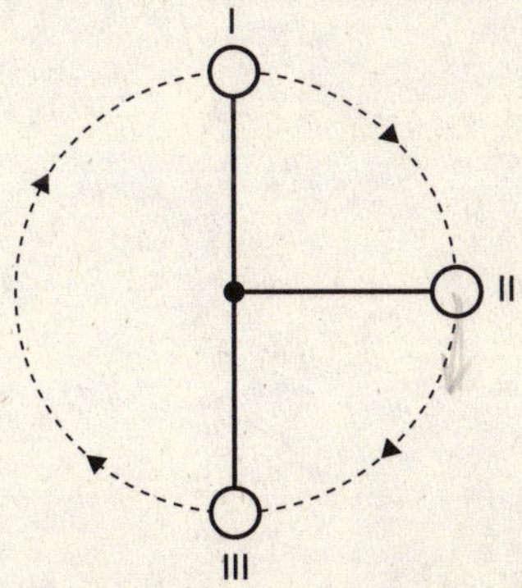

73. Which of the following statements is true?

(A) The tension in the string is greater at point III than at point I.
(B) The tension in the string is greater at point I than at point III.
(C) The tension at point I is equal to the tension at point III.
(D) The tension in the string is greatest at point II.
(E) The tension in the string is the same at points I, II, and III.

74. If the string were to break at point II, what would be the path of the ball?

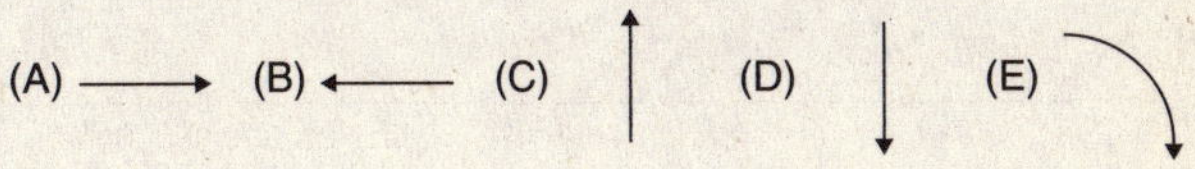

75. A brick is dropped from a high scaffold and strikes the ground 4 seconds later. How high is the scaffold?

(A) 20 m
(B) 40 m
(C) 60 m
(D) 80 m
(E) 100 m

END OF TEST

YOU MAY GO BACK AND CHECK YOUR WORK ON THIS TEST

PRACTICE TEST 1: ANSWERS AND EXPLANATIONS

1. B
2. C
3. B
4. D
5. A
6. E
7. C
8. B
9. A
10. D
11. A
12. B
13. C
14. C
15. D
16. B
17. D
18. E
19. B
20. A
21. C
22. B
23. C
24. E
25. A
26. C
27. E
28. C
29. D
30. D
31. B
32. D
33. E
34. C
35. A
36. D
37. B
38. C
39. A
40. B
41. A
42. B
43. C
44. B
45. E
46. C
47. B
48. C
49. E
50. B
51. D
52. A
53. B
54. B
55. B
56. A
57. A
58. D
59. C
60. D
61. B
62. A
63. E
64. C
65. C
66. A
67. B
68. E
69. A
70. C
71. E
72. E
73. A
74. D
75. D

Computing Your Practice Test Scores

Step 1: Calculate your raw score.

Raw score = # of correct answers – (# of wrong answers × .25)

Step 2: Find your scaled score.

In the Score Conversion Table below, find your raw score (rounded to the nearest whole number). The score directly to the right of it will be your scaled score.

Note: Don't take these scores too literally. Practice test conditions cannot precisely mirror real test conditions. Your scores on the practice test will give you a rough idea of your range on the actual exam.

Raw	Scaled	Raw	Scaled	Raw	Scaled	Raw	Scaled	Raw	Scaled	Raw	Scaled
75	800	59	780	43	690	27	590	11	480	–5	380
74	800	58	770	42	680	26	580	10	480	–6	370
73	800	57	770	41	670	25	580	9	470	–7	370
72	800	56	760	40	670	24	570	8	470	–8	360
71	800	55	760	39	660	23	570	7	460	–9	350
70	800	54	750	38	650	22	560	6	450	–10	350
69	800	53	750	37	650	21	550	5	450	–11	340
68	800	52	740	36	640	20	540	4	440	–12	330
67	800	51	730	35	640	19	540	3	430	–13	330
66	800	50	730	34	630	18	530	2	430	–14	320
65	800	49	720	33	630	17	530	1	420	–15	310
64	800	48	720	32	620	16	520	0	410	–16	310
63	800	47	710	31	610	15	510	–1	410	–17	300
62	790	46	700	30	610	14	510	–2	400	–18	290
61	790	45	700	29	600	13	500	–3	390	–19	290
60	780	44	690	28	600	12	490	–4	390		

1. B

The vertical velocity decreases on the way up and increases on the way down.

2. C

Since the horizontal velocity is constant throughout the flight, the horizontal acceleration is zero.

3. B

The vertical velocity is directed up during the first half of the flight, and down during the second half of the flight.

4. D

The force between the masses can be calculated by Newton's law of universal gravitation.

5. A

The jar's inertia causes it to stay at rest as the paper is pulled out from under it.

6. E

The momentum of the clay before the collision is equal to the momentum of the clay and wood block after the collision.

7. C

The action force of a team pulling on the rope is equal and opposite to the force the rope pulls on the team, which is equal to the tension in the rope.

8. B

By the second right-hand rule, if you put your thumb in the direction of the velocity of the charge and your fingers in the direction of the magnetic field (outward), your palm will face the direction of the force acting on the charge. In this case, only your left hand will match these vectors, and thus the charge must be negative, an electron.

9. A

By the right-hand rule, the charge must be opposite the charge in question 8. Thus the charge is positive, a proton.

10. D

When two waves are in phase with each other, the crest of one wave meets the crest of the other wave, and constructive interference occurs.

11. A

The law of reflection states that the angle of incidence is equal to the angle of reflection as measured from the normal line.

12. B

When a wave passes from one medium into another, it refracts, changing both its wavelength and speed.

13. C

When a wave squeezes through a small opening, part of the wave lags behind, and the wave bends or diffracts.

14. C

The net force on the particle is zero, and thus it can be moving with a constant velocity according to the law of inertia.

15. D

As the boy rows south, the river carries his boat to the west, causing his resultant velocity to be southwest.

16. B

The total current in the circuit passes through the 4Ω resistor. The total resistance in the circuit is the sum of the parallel combination of the 3Ω and 6Ω resistors (which is 2Ω), and the 4Ω resistor. Since the total resistance is 6Ω , the total current in the circuit is:

$$I = \frac{V}{R_{total}} = \frac{24V}{6\Omega} = 4\text{A}$$

which is also the current through the 4Ω resistor.

17. D

By Ohm's law, V = IR = (4A)(4Ω) = 16 V.

18. E

By the right hand rule, the wire on the left creates a magnetic field which is directed to the top of the page at point P, and the wire on the right creates a magnetic field which is directed down to the bottom of the page, and the two cancel each other out.

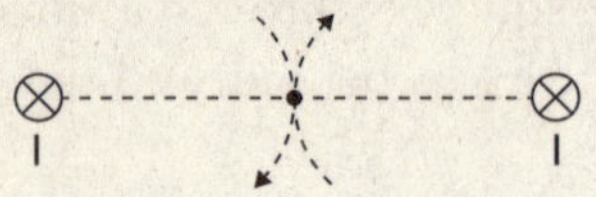

19. B

The work done on the car is equal to the change in its kinetic energy, and force, work, and displacement are related to each other by the equation

$$F = \frac{W}{d} = \frac{KE}{d} = \frac{18{,}000J}{30m} = 600N$$

20. A

Since there are no external forces acting on the system, the total momentum of the system must remain constant.

21. C

All masses will fall at the same acceleration and land at the same time, and since the mass also does not affect the horizontal distance traveled, they will land at the same place at the same time.

22. B

In an electric generator, a coil of wire is rotated in a magnetic field, producing a current by electromagnetic induction.

23. C

As the mass falls, its kinetic energy increases initially, but the potential energy decreases until the mass reaches the equilibrium position.

24. E

The +1μC charge is pulled downward and to the right toward the negative charge, and at the same time is pushed upward and to the right by the positive charge, producing a net force on the +1μC charge which is directed to the right.

25. A

The slope of both graphs represents speed, and the slope of train X is greater than the slope of train Y at time T.

26. C

If we extend the lines for trains X and Y, we see that train X reaches d_2 at an earlier time than train Y.

27. E

Frequency in Hertz is equal to the number of revolutions per second. In this case, we have 6 revolutions/60 seconds = 0.1 rev/s or 0.1 Hz.

28. C

The law of conservation of angular momentum states that the closer an orbiting planet is to the Sun, the smaller its orbital radius, and the greater its velocity.

29. D

$$\lambda = \frac{v}{f} = \frac{350 \text{ m/s}}{256 \text{ Hz}} = 1.4 \text{ m}$$

30. D

Heat energy flows from a body of higher temperature to a body of lower temperature.

31. B

The magnetic force needed to keep the electron orbiting in a circle is a centripetal force. The equation for centripetal force is:

$$F = \frac{mv^2}{r}$$

When the given values are substituted into this equation, the force, to the nearest order of magnitude, is 10^{-17} N.

32. D

We don't know which way the two blocks will accelerate since we don't know which block is more massive, but since they are connected to each other by a string, they must have the same acceleration.

33. E

Since the 1 Ω resistor has half the resistance of the 2 Ω resistor in this parallel circuit, twice the current passes through the 1 Ω resistor as the 2 Ω resistor, $I_1 = 2I_2$.

34. C

Since the two resistors are in parallel with each other, they have the same voltage across them.

35. A

Sound travels at the highest speed through more dense material, and steel is the most dense of the choices.

36. D

The magnification is found by the ratio of the image distance to object distance, 60 cm/30 cm = 2.

37. B

A plane mirror alone always creates a virtual image that is the same size as the object, but left-right reversed.

38. C

The principle of superposition states that the displacements of the waves which coincide at each point can be added to find the height of the wave created by constructive or destructive interference.

39. A

Energy = (power)(time in seconds) = (2,400 W)(180 s) = 432,000 J.

40. B

Although the electroscope is neutral, electrons in the metal are attracted to the rod and go to the top of the electroscope, leaving the leaves with a deficiency of electrons and positively charged.

41. A

Electrons are attracted from the ground to the rod, and therefore more electrons deposit themselves on the electroscope giving it a net negative charge. We could also think of the positive charges being repelled from the rod and escaping to the ground through the grounding wire, leaving the electroscope negatively charged.

42. B

$F = qvB$ implies that

$$q = \frac{F}{vB} = \frac{10^{-14}\text{ N}}{(10^6\text{ m/s})(10^{-2}\ T)} = 10^{-18}\ C$$

43. C

$$v = f\lambda = (120\text{ Hz})(0.5\text{ m}) = 60\text{ m/s}$$

44. B

Two "loops" make up one wavelength of 0.5 m, so there are two wavelengths in the string which is 1.0 meter long.

45. E

To melt the ice, it takes heat energy equal to $Q = mH_f$, where H_f is the heat of fusion for ice, giving $Q = (3 \text{ kg})(3.3 \times 10^5 \text{ J/kg}) = 9.9 \times 10^5$ J.

46. C

The heat required to raise the temperature by 30° C is:

$$Q = mc\Delta T$$
$$= (3 \text{ kg})(4.2 \times 10^3 \text{ J/kg °C})(30° \text{ C} - 0° \text{ C})$$
$$= 3.8 \times 10^5 \text{ J}$$

47. B

Frequency depends of the source of the light, not on the medium through which it passes.

48. C

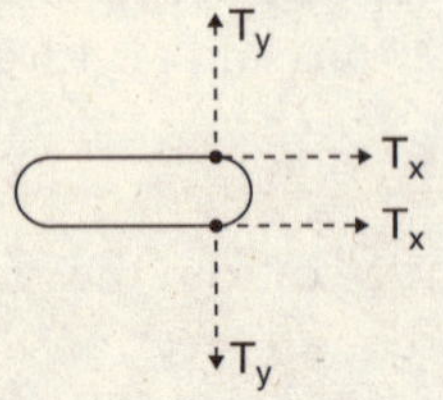

The net force is made up of the horizontal component of each tension, each component being equal to $T_x = T\cos 45 = 0.7T$. Since there are two horizontal components making up the net horizontal force of 500 N, we can write $2(0.7)T = 500$ N, and $T = 350$ N (approximately).

49. E

The acceleration of the car can be found by Newton's second law:

$$a = \frac{F_{net}}{m} = \frac{500 \text{ N}}{1{,}000 \text{ kg}} = 0.5 \text{ m/s}^2$$

50. B

Each photon can give its energy to an electron and give the electron enough energy to escape the metal.

51. D

Temperature is proportional to the average kinetic energy of the molecules, and kinetic energy is proportional to the square of speed. Thus, doubling the speed of the molecules will give us four times the temperature.

52. A

The heat added is equal to the work done by the engine plus any heat exhausted. Thus, the heat added = 50 J + 100 J = 150 J.

53. B

The efficiency is the ratio of the work done to the heat added (× 100%).

$$\text{efficiency} = \frac{\text{Work}}{\text{heat added}} = \frac{50 \text{ J}}{150 \text{ J}} \times 100\% = 33\%$$

54. B

The change in momentum of the ball is $p_f - p_i$, and since the ball reverses direction when it strikes the wall, the change in momentum would be $mv - (-mv) = 2mv$. The change in momentum cannot be zero, since the wall exerts an external force on the ball.

55. B

The force starts out small as the ball first makes contact with the wall, then gets larger as the ball is squashed against the wall, then smaller again as it rebounds from the wall.

56. A

The wave travels at a higher speed in a tight spring, and at a lower speed in a loose spring. With reduced tension in the spring, the wavelength will be smaller since the speed is smaller, and wavelength is proportional to speed by $v = f\lambda$.

57. A

$$v_f = v_i + at = (2 \text{ m/s}) + (2 \text{ m/s}^2)(3 \text{ s}) = 8 \text{ m/s}$$

58. D

$s = v_i t + \frac{1}{2}at^2 = (2\ \text{m/s})(3\ \text{s}) + \frac{1}{2}(2\ \text{m/s}^2)(3\ \text{s})^2$
$= 15\ \text{m}$

59. C

The series of bright and dark lines are antinodes and nodes created by the interference of light waves after passing through two narrow slits.

60. D

Newton's second law states that $F_{net} = ma$, so to keep F_{net} constant, if we triple the mass, the acceleration will be reduced to one-third as much. In other words, if force is constant, acceleration is inversely proportional to the mass.

61. B

The restoring force exerted by the spring on the mass is greatest when the stretch of the spring is greatest, at the amplitudes A and C.

62. A

Since point B is the equilibrium position for the mass on the spring, the spring force at B is zero, and the mass has reached its maximum speed, and the all of the energy in the system is kinetic energy.

63. E

The magnitude of the acceleration remains constant, but since it must continually adjust its direction to point toward the center of the orbit, it is always changing direction.

64. C

By Newton's second law, the ratio of the weight of an object to its mass is the acceleration due to gravity, and is constant, $W/m = g$.

65. C

As the ball rises, it continues forward at a constant velocity according the law of inertia, but the bus accelerates out from under it, making it appear to the boy that the ball goes over his head and lands behind him.

66. A

The heat from your feet is more readily conducted to the metal because the metal is a good conductor, leaving your feet with a lack of heat and feeling cold.

67. B

As the ray of light exits the water into the air, it bends away from the normal, since the ray is passing from a more dense medium (water) to a less dense medium (air).

68. E

When passing from one medium to another, all three of the choices would refract, changing their speed and wavelength.

69. A

Sound is a longitudinal wave, and only transverse waves can be polarized, that is, made to vibrate in only one plane.

70. C

The amplitude of a wave is the maximum displacement from the baseline to a crest, or from the baseline to a trough, in this case, 2.0 m.

71. E

A time of 24 days represents 4 half-lives, since each half-life is 6 days. If we cut 100% in half four times, we get 6.25% of the sample remaining.

72. E

Isotopes of an element can have various mass number (number of protons and neutrons), but must have the same atomic number (number of protons). Thus, there is no uranium with an atomic number of 91.

73. A

At point III, the tension in the string must be enough to support the weight of the ball *and* provide the centripetal force.

74. D

At point II, the ball is instantaneously moving straight downward, and if the string breaks at this point, the ball will move downward.

75. D

$$s = \tfrac{1}{2}gt^2 = \tfrac{1}{2}(10\ \text{m/s}^2)(4\ \text{s})^2 = 80\ \text{m}$$

CHAPTER THREE

Measurement, Scalars, and Vectors

The purpose of this chapter is to briefly review some of the math skills you will need in order to be successful on the SAT II: Physics Test, including exponents, powers of ten, the metric system and SI units, scientific notation, graphs, trigonometry, scalars, and vectors. If you are confident in your math skills in these areas, you may want to skip ahead to the next chapter. Otherwise, let's get warmed up!

指数

EXPONENTS

For any nonzero number a and any integer n:

exponent of zero: $a^0 = 1$

exponent of one: $a^1 = a$

negative exponent: $a^{-n} = \frac{1}{a^n}$

If a, b, x, y, and z are all integers:

product of powers: $(a^x)(a^y) = a^{x+y}$

power of powers: $(a^x)^y = a^{xy}$

quotient of powers: $\frac{a^x}{a^y} = a^{x-y}$

power of a product: $(ab)^x = a^x b^x$

power of a monomial: $(a^x b^y)^z = a^{xz} b^{yz}$

the nth root of powers: $\sqrt[y]{a^x} = a^{x/y}$

Powers of Ten

The rules for exponents listed above can also be applied when $a = 10$. For example:

$10^0 = 1$

$10^1 = 10$

$10^2 = 100$

$10^3 = 1{,}000$, and so on. Also:

$10^{-1} = \frac{1}{10}$

$10^{-2} = \frac{1}{100}$

$10^{-3} = \frac{1}{1{,}000}$, and so on.

Scientific Notation

In physics, very large and very small numbers are often used. It is convenient for us to write these numbers as a number between 1 and 10 (the *mantissa*) times a power of ten. For example, the mass of the Earth is 6,000,000,000,000,000,000,000,000 kg, or a 6 followed by 24 zeroes. We can write this as 6×10^{24} kg in scientific notation, and we refer to 6 as the *mantissa* and 24 as the *power of ten*. By the same token, the mass of an electron in kg is a decimal point followed by 30 zeroes and then 911. In scientific notation, we would write the mass of an electron as 9.11×10^{-31} kg.

The Metric System and Systeme Internationale (SI) Units

The metric system is a system of units of different sizes which are related by powers of ten. The base units of the metric system which are commonly used on the SAT II: Physics Test are listed in Table 3.1.

Table 3.1: SI Base Units of the Metric System

Base quantity	Base unit	Symbol
length	meter	m
mass	kilogram	kg
time	second	s
temperature	Kelvin	K
amount of substance	mole	mol
electric current	ampere	A

All other units used on the SAT II: Physics Test are derived from these base units. Several of the powers of ten have been given names in the metric system. These metric prefixes are placed before a unit to indicate the power of ten which is the fraction or multiplier of that unit. Common examples of these metric prefixes are listed in Table 3.2 below.

Table 3.2: Common Metric Prefixes

Prefix	Symbol	Power of ten	Example
nano	n	10^{-9}	nanosecond (ns)
micro	μ	10^{-6}	microfarad (μF)
milli	m	10^{-3}	milligram (mg)
centi	c	10^{-2}	centimeter (cm)
kilo	k	10^{3}	kilogram (kg)
mega	M	10^{6}	megawatt (MW)

GRAPHING

As you prepare for the SAT II: Physics Test, you might need to review some of your graphing skills. You will need to be able to interpret linear and parabolic graphs, and find the slope of a line and the area under a curve.

Linear Graphs

A linear graph results when we plot two quantities which are proportional to each other. The change in the quantity on the y-axis is called the *rise*, and the change in the quantity on the x-axis

> A linear graph has a constant slope, meaning the two quantities are proportional to each other.

is called the *run*. Thus, the rise divided by the run is a constant for a linear graph, and we call this constant the *slope*, denoted by the letter m. We write:

$$\text{slope } m = \frac{rise}{run} = \frac{\Delta y}{\Delta x}$$

The slope of a linear graph has units equal to the rate of change of y with respect to x.

Example: A simple circuit consisting of a power supply, wires of negligible resistance, and a resistor is connected and several voltages and currents are measured. A graph of voltage (in volts) versus current (in amperes) for a simple circuit is plotted below.

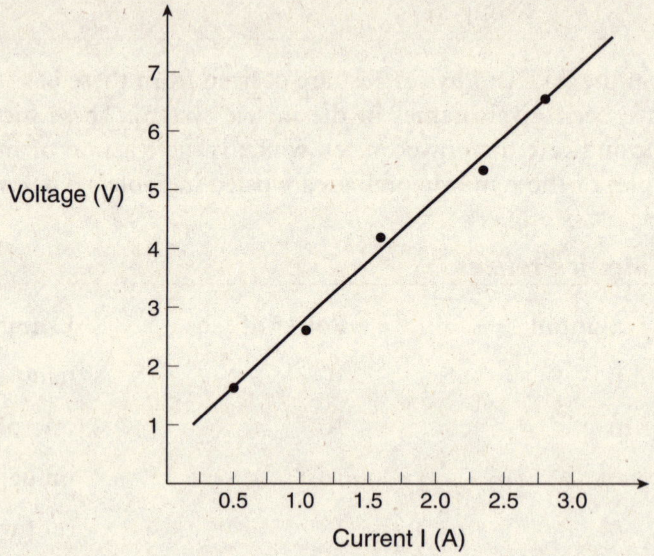

We see that we can draw the best-fit straight line through the data points. Remember, a best-fit line represents the average of the data, so points should be distributed evenly both above and below the line. The slope of the line can be found by choosing two points that are on the line. In this case, we have chosen the points (0.5 A, 2 V) and (2.5 A, 6 V):

$$\text{slope } m = \frac{rise}{run} = \frac{\Delta V}{\Delta I} = \frac{V_2 - V_1}{I_2 - I_1} = \frac{6V - 2V}{2.5A - 0.5A} = 2\frac{V}{A} \text{ or } 2\Omega$$

The ratio of the voltage to the current is the *resistance* of the circuit. We say that voltage and current are proportional to each other, and the resistance is the slope of a voltage vs. current graph. We will return to circuits in chapter 12.

Area Under a Curve

Another common way to analyze a graph is to find the area under the curve. Remember, a curve is a generic word for any line resulting from graphing two quantities, and can mean a curved line or a straight line. The area under a curve is equal to the product of the two quantities on the axes for a particular interval. For example, consider the graph of velocity (in meters per second) vs. time (in seconds) below:

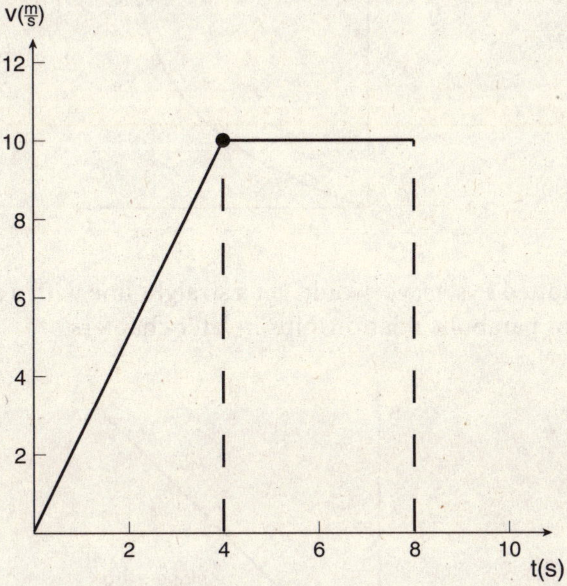

If we consider the time interval from 0 to 8 s, we can find the area under the curve (graph). The area from 0 to 4 s is the area of a triangle with a base of 4 s and a height of 10 m/s. The area from 4 s to 8 s is a rectangle with a base of 4 s and a height of 10 m/s. The total area under the graph is the sum of the area of the triangle and the area of the rectangle:

Total area = ½(base)(height) + (base)(height) = ½(4 s)(10 m/s) + (4 s)(10 m/s) = 60 m

Note that the units for the area is meters, which indicates that the area under a velocity vs. time graph is *displacement*. Motion is discussed further in chapter 4.

The slope of a graph is the *ratio* of the change in the quantities, whereas the area under the curve is the *product* of the change in the quantities.

Parabolic Graphs

Often, two quantities in physics are not proportional to each other, but one quantity depends on the *square* of the other. For example, for an object falling freely from rest, the total distance fallen is proportional to the square of the time it has been falling. The equation that relates distance fallen to the time of fall is

$$s = \tfrac{1}{2}gt^2$$

where g is the acceleration of the falling object due to gravity, about 10 m/s^2. Since s is proportional to t^2 (and not to t), the graph of s vs. t will be a parabola.

A *parabola* results when one quantity is proportional to the square of the other.

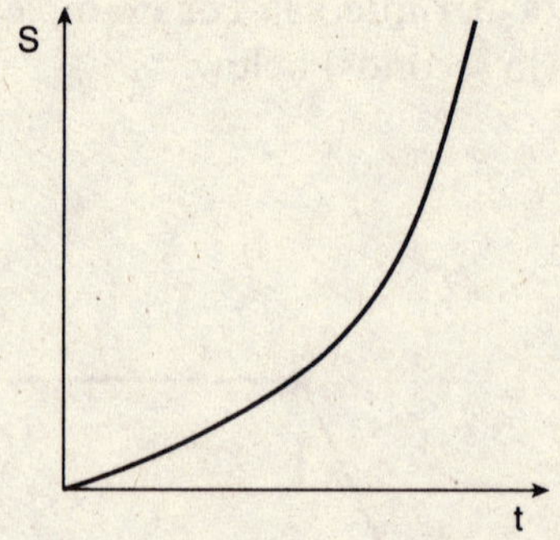

Of course, if we plotted *s* vs. t^2, we would get a straight line with a slope of $\frac{1}{2}g$. We will review several examples of parabolic relationships in later chapters.

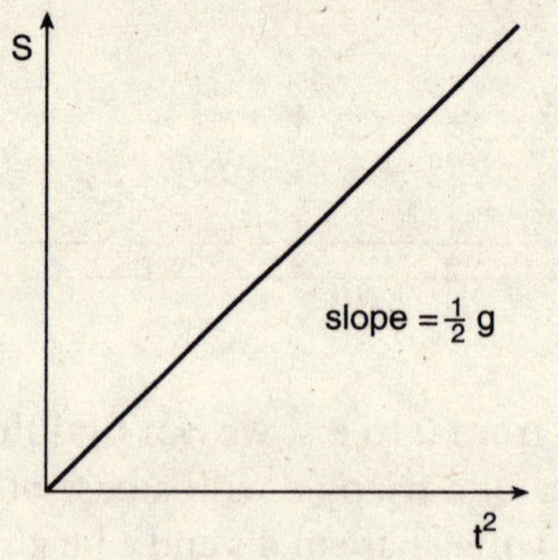

TRIGONOMETRY

Trigonometry is the study of triangles, and often right triangles. The lengths of the sides of a right triangle can be used to define some useful relationships, called the *sine*, *cosine*, and *tangent*, abbreviated *sin*, *cos*, and *tan*, respectively. Take a look at the right triangle below:

The functions *sin*, *cos*, and *tan* are defined as:

$$\sin\theta = \frac{\textit{opposite side}}{\textit{hypotenuse}} = \frac{a}{c}$$

$$\cos\theta = \frac{\textit{adjacent side}}{\textit{hypotenuse}} = \frac{b}{c}$$

$$\tan\theta = \frac{\textit{opposite side}}{\textit{adjacent side}} = \frac{a}{b}$$

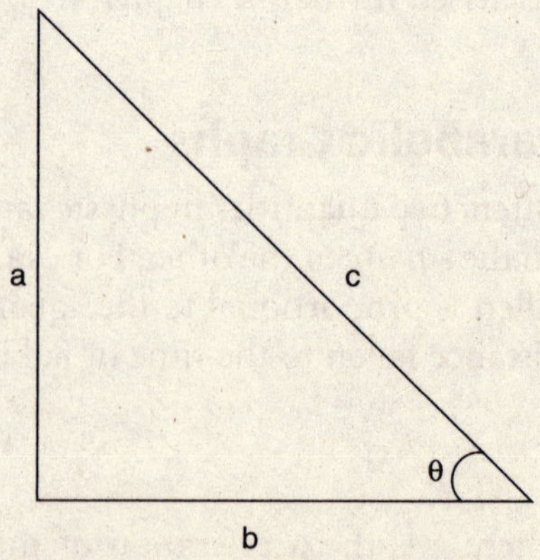

Also, the three sides are related to each other by the Pythagorean Theorem:

$$c^2 = a^2 + b^2$$

Since you will not be allowed to use a calculator on the SAT II: Physics Test, trigonometric values will often be given to you with the questions that might use them. The relationships above will be particularly helpful when dealing with vectors.

SCALARS AND VECTORS

A *scalar* is a quantity which has no direction associated with it, such as mass, volume, time, and temperature. We say that scalars have only magnitude, or size. A mass may have a magnitude of 2 kilograms, a volume may have a magnitude of 5 liters, and so on. But a *vector* is a quantity that has both magnitude (size) and direction (angle). For example, if someone tells you they are going to apply a 10 newton force on you, you would want to know the direction of the force, that is, whether it will be a push or a pull. So, force is a vector, since direction is important in specifying a force. See chapter 5 for more on forces. Other examples of vectors are velocity, acceleration, and momentum.

A *scalar*, like mass, has only size. A *vector*, like force, has both size and direction.

Another example of a vector is *displacement*. Displacement **s** is the straight-line distance between two points, and is a vector which points from an object's initial position toward its final position. Let's say a hiker walks 6 kilometers due west, then 8 km due north, then 10 km due east. What is the hiker's displacement from the origin?

A vector can be represented by an arrow whose length gives an indication of its magnitude (size), with the arrow tip pointing in the direction of the vector. We represent a vector by a letter written in bold type. For this example, we list the displacement vectors like this:

$\mathbf{s}_1$ = 6 km west
$\mathbf{s}_2$ = 8 km north
$\mathbf{s}_3$ = 10 km east

We can graphically add vectors to each other by placing the tail of one vector onto the tip of the previous vector. For our hiker displacement example, we can graphically add the second displacement vector to the first, and the third displacement vector to the second:

Scalars : mass, volume, time, temperature ...
Vectors : force, velocity, acceleration, momentum, displacement ...

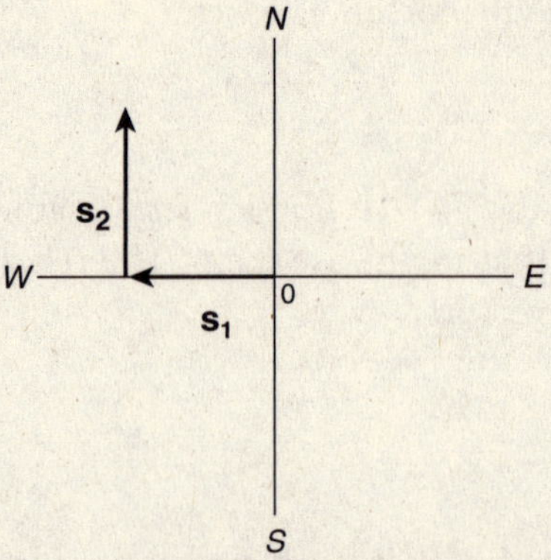

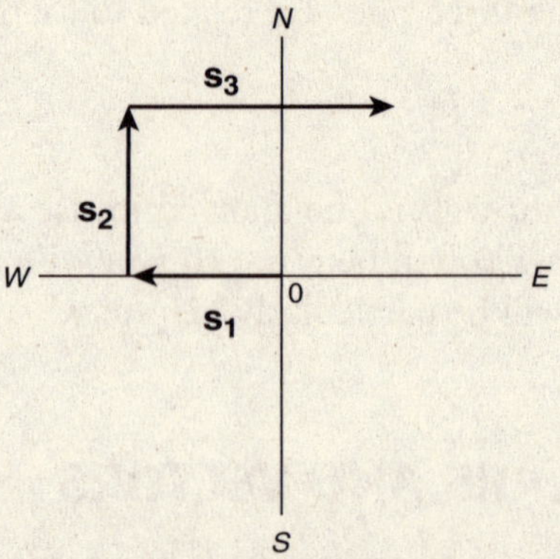

The *resultant vector* is the displacement from the origin to the tip of the last vector. In other words, the *resultant* is the *vector sum* of the individual vectors, and can replace the individual vectors and end up with the same result. Of course, just adding the lengths of the vectors together will not achieve the same result. Adding 6 km, 8 km, and 10 km gives 24 km, which is the total distance traveled, but not the straight-line displacement from the origin. We see in the diagram below that the resultant displacement is 8.9 km from the origin at an angle of 27° east of the north axis.

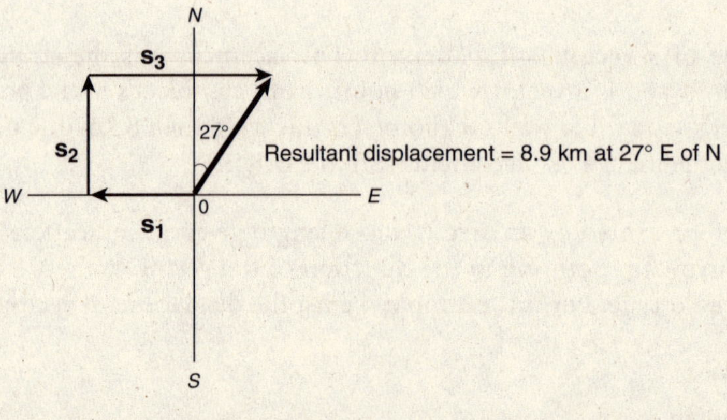

We could have added the displacement vectors in any order and achieved the same resultant. We say that the addition of vectors is *commutative*.

A *vector* is any quantity that can be added like displacements (that is, tip-to-tail) to get the *resultant*.

The *equilibrant* is the vector which can cancel or balance the resultant vector. In this case, the equilibrant displacement is the vector which can bring the hiker back to the origin. Thus, the equilibrant is always equal and opposite to the resultant vector.

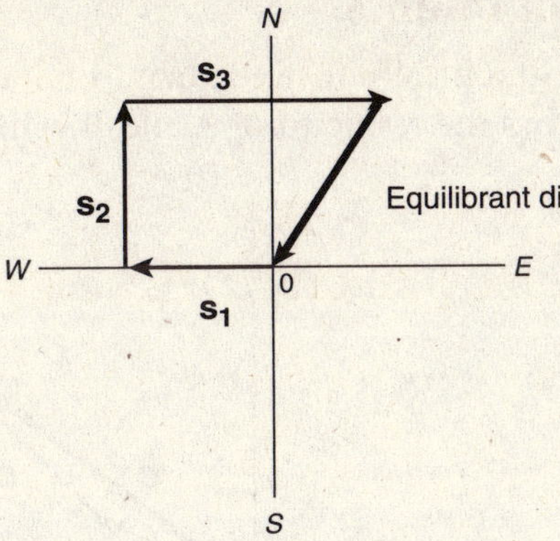

Vector Components

We may also work with vectors mathematically by breaking them into their components. A vector component is the projection or shadow of a vector onto the *x*- or *y*-axis. For example, let's say we have two vectors **A** and **B** shown below:

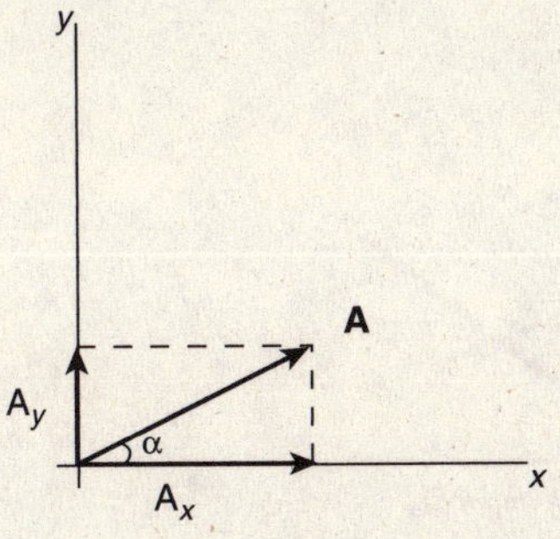

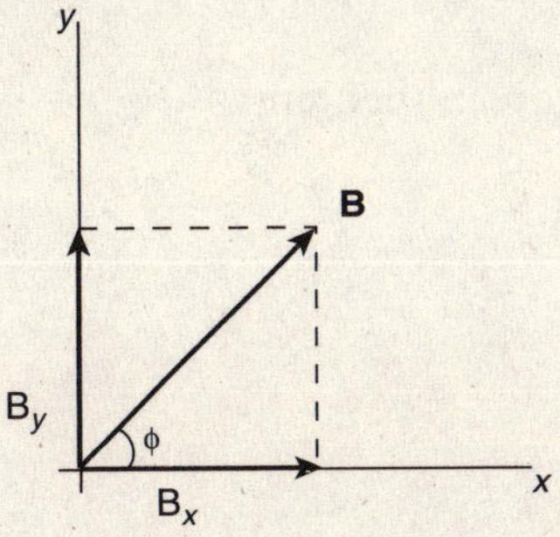

We will call the projection of vector **A** onto the *x*-axis its *x*-component, $\mathbf{A_x}$. Similarly, the projection of **A** onto the *y*-axis is $\mathbf{A_y}$. The vector sum of $\mathbf{A_x}$ and $\mathbf{A_y}$ is **A**, and, since the magnitude of **A** is the hypotenuse of the triangle formed by legs A_x and A_y, the Pythagorean theorem holds true:

$$|A| = \sqrt{A_x^2 + A_y^2}$$

and from the figures above:

$$A_x = A\sin\alpha$$
$$A_y = A\cos\alpha$$
$$\tan\alpha = \frac{A_y}{A_x}$$

We can write the same relationships for vector **B** by simply replacing ***A*** with ***B*** and the angles α with ϕ in each of the equations above.

Adding Vectors Using Components

Earlier we added vectors together graphically to find their resultant. Using the tip-to-tail method of adding vectors, we can find the resultant of **A** and **B**, which we will call **C**:

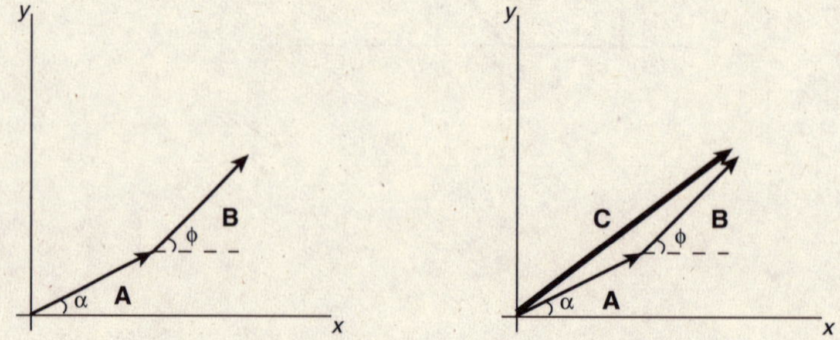

We can also use their components to find the resultant of vectors **A** and **B**. The *x*-components of the resultant vector **C** is the sum of the *x*-components of **A** and **B**. Similarly, the *y*-components of the resultant vector **C** is the sum of the *y*-components of **A** and **B**. So:

合成位移的第一种方法：

$C_x = A_x + B_x$ and $C_y = A_y + B_y$

and by the Pythagorean theorem:

$$|C| = \sqrt{C_x^2 + C_y^2}$$

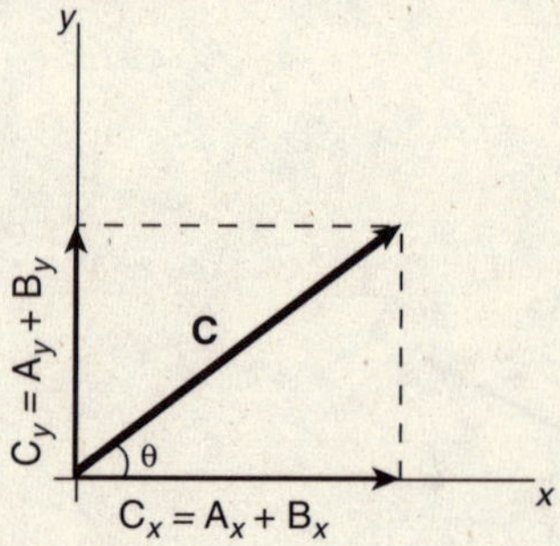

The algebraic sum of the *x*-components of two vectors equals the *x*-component of the resultant; the same is true of the *y*-components.

Example: Using the diagrams above, let **A** = 4 meters at 30° from the *x*-axis, and **B** = 3 meters at 45° from the *x*-axis. Find the magnitude and direction of the resultant vector **C**. (cos30 = 0.87, sin 30 = 0.50, cos45 = sin45 = 0.70)

Solution: First, we need to find A_x, A_y, B_x, and B_y.

$$A_x = A\cos 30 = (4\text{ m})\cos 30 = (4\text{ m})(0.87) = 3.5\text{ m}$$
$$A_y = A\sin 30 = (4\text{ m})\cos 30 = (4\text{ m})(0.50) = 2.0\text{ m}$$
$$B_x = B\cos 45 = (3\text{ m})\cos 45 = (3\text{ m})(0.70) = 2.1\text{ m}$$
$$B_y = B\sin 45 = (3\text{ m})\sin 45 = (3\text{ m})(0.70) = 2.1\text{ m}$$

Now we can find the *x*- and *y*-components of the resultant **C**:

$$C_x = A_x + B_x = 3.5 \text{ m} + 2.1 \text{ m} = 5.6 \text{ m}$$
$$C_y = A_y + B_y = 2.0 \text{ m} + 2.1 \text{ m} = 4.1 \text{ m}$$

The magnitude of the resultant **C** is

$$|\mathbf{C}| = \sqrt{C_x^2 + C_y^2} = \sqrt{(5.6 \text{ m})^2 + (4.1 \text{ m})^2} = 6.9 \text{ m}$$

and its angle from the *x*-axis can be found by

$$\tan \theta = \frac{C_y}{C_x}$$

$$\theta = \tan^{-1}\left[\frac{4.1}{5.6}\right] = 36.2°$$

The properties of vectors we've discussed here can be applied to any vector, including velocity, acceleration, force, and momentum. We will review the vector properties of these quantities in later chapters.

MEASUREMENT, SCALARS, AND VECTORS REVIEW QUESTIONS

1. The number 300,000,000 can be written as

   (A) 3×10^6
   (B) 3×10^7
   (C) 3×10^8
   (D) 3×10^9
   (E) 3×10^{-9}

2. The fundamental SI units for mass, length, and time, respectively, are

   (A) newton, meter, minute.
   (B) kilogram, meter, second.
   (C) pound, foot, hour.
   (D) pound, foot, second.
   (E) kilogram, centimeter, hour.

3. A millimeter is

   (A) 10^3 m
   (B) 10^2 m
   (C) 10^1 m
   (D) 10^{-3} m
   (E) 10^{-6} m

4. How many nanometers are in a kilometer?

   (A) 10^{-9}
   (B) 10^9
   (C) 10^{-12}
   (D) 10^{12}
   (E) 10^{27}

5. The data plotted on a graph of distance on the *y*-axis vs. time on the *x*-axis yields a linear graph. The slope of the graph is

   (A) $\frac{\Delta d}{\Delta t}$
   (B) $(\Delta d)(\Delta t)$
   (C) $\frac{\Delta t}{\Delta d}$
   (D) $(\Delta d) + (\Delta t)$
   (E) $(\Delta d) - (\Delta t)$

6. Consider the velocity vs. time graph shown.

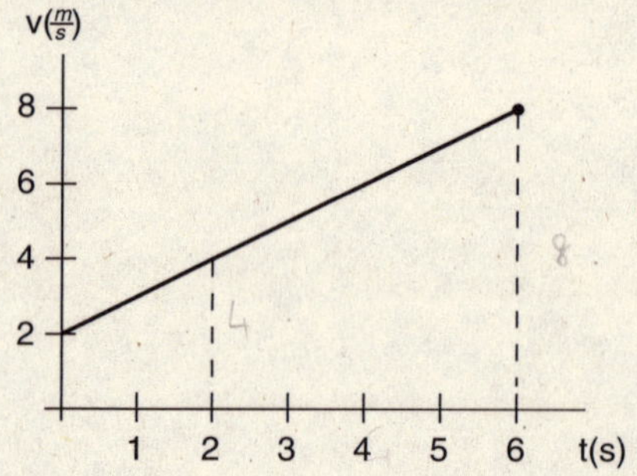

   The area under the graph from 2 s to 6 s is most nearly

   (A) 8
   (B) 16
   (C) 24
   (D) 32
   (E) 36

7. Data from an experiment shows that kinetic energy is proportional to the square of velocity. The plot of the graph of kinetic energy vs. velocity would be

(A) a straight horizontal line.
(B) a straight diagonal line sloping upward.
(C) a parabola.
(D) a hyperbola.
(E) a circle.

8. The cosine of an angle in a right triangle is equal to the

(A) sine of the angle.
(B) tangent of the angle.
(C) hypotenuse of the triangle.
(D) side adjacent to the angle.
(E) ratio of the side adjacent to the angle and the hypotenuse of the triangle.

9. Which of the following quantities is NOT a vector quantity?

(A) displacement
(B) mass
(C) resultant
(D) equilibrant
(E) 10 km at 30° north of east

10. The resultant of the two displacement vectors 3 m east and 4 m north is

(A) 5 m northeast
(B) 7 m northeast
(C) 1 m southwest
(D) 1 m northeast
(E) 12 m northeast

11. The x-component of the vector shown is most nearly (sin 30 = 0.5, cos 30 = 0.87, tan 30 = 0.58)

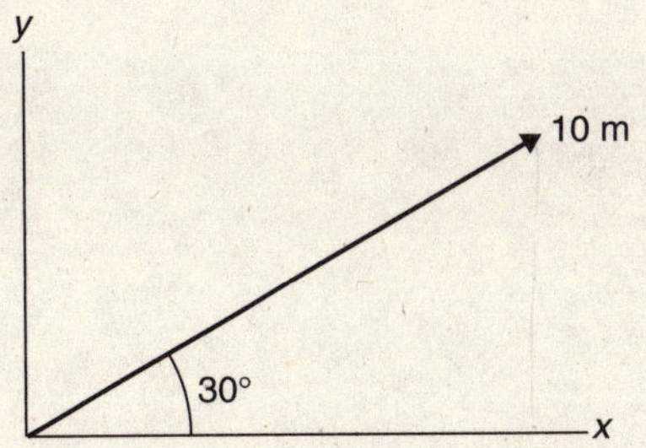

(A) 10 m
(B) 5 m
(C) 8.7 m
(D) 5.8 m
(E) 100 m

12. Two displacement vectors each having a y-component of 10 km are added together to form a resultant which forms an angle of 60° from the $+x$-axis. What is the magnitude of the resultant? ($\sin 60° = 0.87$, $\cos 60° = 0.5$)

(A) 23 m
(B) 40 m
(C) 12 m
(D) 20 m
(E) 30 m

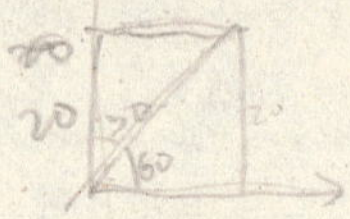

SOLUTIONS TO MEASUREMENT REVIEW QUESTIONS

1. C

A 3 followed by 8 zeros, or 3 multiplied by 10 eight times.

2. B

Mass is measured in kilograms, length in meters, and time in seconds.

3. D

The prefix *milli* means 10^{-3}.

4. D

$$1\text{ km}\left(\frac{10^3\text{m}}{\text{km}}\right)\left(\frac{10^9\text{nm}}{\text{m}}\right) = 10^{12}\text{nm}$$

5. A

Slope equals rise over run, which in this case would be $\frac{\Delta d}{\Delta t}$.

6. C

The total area can be calculated by adding the area bounded by $4(6 - 2) = 16$ and the area of the triangle bounded by $v = 4$ to $v = 8$ and $t = 6$ and $t = 2$: $\frac{1}{2}(8 - 4)(6 - 2) = 8$. Thus, the total area is $16 + 8 = 24$.

7. C

Since the kinetic energy is proportional to the square of the velocity, plotting *KE* vs. *v* would produce a parabola.

8. E

The cosine is defined as the ratio of the adjacent side to the hypotenuse.

9. B

Mass has no direction associated with it and therefore is a scalar, not a vector.

10. A

The two displacements represent the two legs of a right triangle. By the Pythagorean theorem, the hypotenuse is equal to:

$$\sqrt{3^2 + 4^2} = 5\text{m northeast}$$

11. C

The *x*-component of the vector is the side adjacent to the 30° angle and is found by $(10\text{ m})\cos 30° = 8.7\text{ m}$.

12. A

The figure below shows that the two y-components add up to 20 m, which is the y-component of the resultant. The y-component of the resultant (20 m) is the opposite side to the 60° angle, and the magnitude of the resultant can be found by:

$$\frac{20\text{ m}}{\sin 60} = 23\text{ m}$$

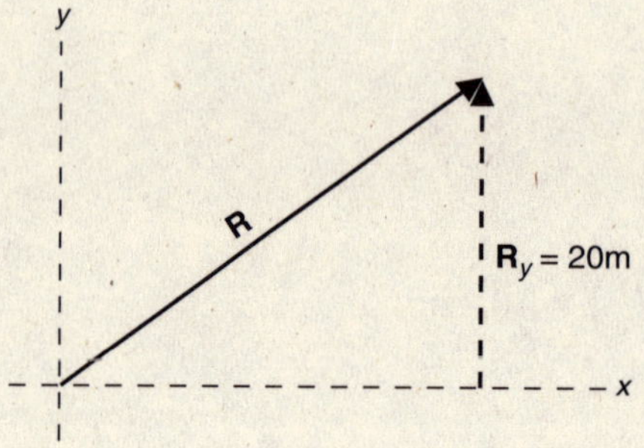

CHAPTER FOUR

Kinematics

We begin our review of motion with the study of *kinematics.* Kinematics is the study of *how* things move; that is, how far things move (distance and displacement), how fast they move (speed and velocity), how fast their velocity changes (acceleration), and how much time passes during the changes. Kinematics is the study of the relationships between all of these things without regard to *why* something is moving, a question we will take up in the next chapter.

Kinematics is the study of how things move.

DISTANCE AND DISPLACEMENT

Distance can be defined as total length moved. If you run around a circular track, you have covered a distance equal to the circumference of the track. Distance is a scalar, which means it has no direction associated with it. *Displacement,* however, is a vector, representing a change in position. Displacement is defined as the straight-line distance between two points, and is a vector which points from an object's initial position toward its final position. In our previous example, if you run around a circular track and end up at the same place you started, your displacement is zero, since there is no distance between your starting point and your ending point. For more discussion on displacement and other vectors, see chapter 3.

Distance is a scalar and *displacement* is a vector.

SPEED AND VELOCITY (速率和速度)

Average speed is defined as the total distance a moving object covers divided by the total time it takes to cover that distance:

$$\text{average speed} = \frac{\text{total distance}}{\text{total time}}$$

$$v = \frac{d}{t}$$

where v stands for speed, d is for distance, and t is time.

> **Example:** A sprinter can run 100 meters in 11 seconds. Find her average speed.
>
> **Solution:** $v = \frac{d}{t} = \frac{100 \text{ m}}{11 \text{ s}} = 9.1 \text{ m/s}$

Average velocity is defined a little differently than *average speed*. While average speed is the total distance divided by the total time, average velocity is the change in position divided by the change in time. Since velocity is a vector, we must define it in terms of another vector, displacement **s**. Often average speed and average velocity are interchangeable for the purposes of the SAT II: Physics Test. Speed is the magnitude of velocity; that is, speed is a scalar and velocity is a vector. For example, if you are driving west at 55 miles per hour, we say that your speed is 55 mph, and your velocity is 55 mph west. We will use the letter v for both speed and velocity in our calculations and will take the direction of velocity into account when necessary.

Speed is a scalar and *velocity* is a vector.

ACCELERATION

Acceleration $\boldsymbol{a}$ is defined as the rate of change of velocity $\boldsymbol{v}$, or the change in velocity $\boldsymbol{v}$ divided by the change in time $\boldsymbol{t}$:

$$\text{acceleration} = \frac{\text{change in velocity}}{\text{change in time}}$$

$$a = \frac{\Delta v}{\Delta t}$$

The Greek letter Δ stands for "change in," or final quantity (in this case, velocity) minus initial quantity.

Acceleration tells us how fast velocity is changing. For example, if you start from rest on the goal line of a football field, and begin walking up to a speed of 2 m/s for the first second, then up to 4 m/s, for the second second, then up to 6 m/s for the third second, you are speeding up with an average acceleration of 2 m/s for each second you are walking. We write:

$$a = \frac{\Delta v}{\Delta t} = \frac{2\text{m/s}}{1\text{s}} = 2\text{m/s/s} = 2\ \text{m/s}^2$$

In other words, you are changing your speed by 2 m/s for each second you walk. If you start with a high velocity and slow down, you are still accelerating, but your acceleration would be considered negative, compared to the positive acceleration discussed above.

Acceleration is the rate at which velocity is changing.

Example: A car moving initially at 10 m/s accelerates up to 30 m/s during at time of 4 seconds. Find the acceleration of the car.

Solution: The car's initial velocity $v_i = 10$ m/s, the final velocity $v_f = 30$ m/s, and the time interval $\Delta t = 4$ s. The acceleration can be found by:

$$a = \frac{\Delta v}{\Delta t} = \frac{v_f - v_i}{\Delta t} = \frac{30\ \text{m/s} - 10\ \text{m/s}}{4\ \text{s}} = 5\ \text{m/s}^2$$

Velocity is the rate of change of displacement, and acceleration is the rate of change of velocity.

Usually, the change in speed Δv is calculated by the final speed v_f minus the initial speed v_i. The initial and final speeds are called instantaneous speeds, since they each occur at a particular instant in time and are not average speeds.

THE KINEMATIC EQUATIONS FOR UNIFORMLY ACCELERATED MOTION

Kinematics is the study of the relationships between distance and displacement, speed and velocity, acceleration, and time. The kinematic equations are the equations of motion which relate these quantities to each other. These equations assume that the acceleration of an object is *uniform*, that is, constant for the time interval we are interested in. The kinematic equations listed below would not work for calculating velocities and displacements for an object that is accelerating erratically. Fortunately, the SAT II: Physics Test only deals with uniform acceleration, so the kinematic equations will be very helpful in solving problems on the test.

Quick Quiz:

A bicyclist decreases her speed by 3 m/s for each second until she stops after a time of 7 s. Which of the following quantities is constant?

a. speed
b. velocity
c. acceleration
d. displacement

Answer: c. acceleration, since she slows down at the same rate until she stops.

Kinematic Equations

Equation 1	$\bar{v} = \frac{d}{t}$
Equation 2	$d = \frac{1}{2}(v_f + v_i)t$
Equation 3	$\bar{a} = \frac{\Delta v}{\Delta t} = \frac{v_f - v_i}{\Delta t}$
Equation 4	$v_f = v_i + at$
Equation 5	$s = v_i t + \frac{1}{2}at^2$
Equation 6	$v_f^2 = v_i^2 + 2as$

where t = time, d = distance traveled, v_i = initial velocity, v_f = final velocity, or the velocity at the end of a time interval, a = acceleration, and s = displacement (or change in position, $\Delta x = x_f - x_i$).

The *kinematic equations* relate distance, position, displacement, velocity, acceleration, and time to each other.

Notice that in some of the equations above we have chosen to replace the time interval Δt with a t, representing total time. Also, these equations are the scalar form of the kinematic equations rather than the vector form. We need to remember that if we have a moving object with a displacement, velocity, or acceleration in the opposite direction of another vector quantity, we must use a negative sign to distinguish opposite direction.

When solving problems using the kinematic equations, we need to write down which quantities are given in the statement of the problem and the quantity we want to find, and use the equation that includes all of them.

Example: A car traveling initially at a speed of +12 m/s accelerates at a rate of +4 m/s^2 for a time of 5 s. What is the car's speed at the end of 5 s?

Solution: Let's write down the quantities we are given in the problem:

$v_i = +12$ m/s

$a = +4$ m/s^2

$t = 5$ s

We want to find the final velocity v_f, so we look at the list of equations and find the one that includes v_i, a, t, and v_f. Equation 4 contains all of these quantities.

$$v_f = v_i + at = 12 \text{ m/s} + (4 \text{ m/s}^2)(5 \text{ s}) = +32 \text{ m/s}$$

Example: The car in the above example now slows down at the rate of -2 m/s^2 for a distance of 60 meters. What is the car's velocity after covering this distance?

Solution: Organizing the new given quantities, we have:

$v_i = +32 \text{ m/s}$

$a = -2 \text{ m/s}^2$

$s = 60 \text{ m}$

and we are to find the final velocity v_f after the car travels 60 m. Notice that we are not given time in this example, so we need to match up the given quantities with an equation which does not include time, such as Equation 6:

$$v_f^{\,2} = v_i^{\,2} + 2as$$

$$v_f = \sqrt{v_i^2 + 2as} = \sqrt{(32 \text{ m/s})^2 + 2(-2 \text{ m/s}^2)(60 \text{ m})} = +28 \text{ m/s}$$

If you had difficulty working that last one without a calculator, don't worry; the numbers chosen for the SAT II: Physics Test will be simple enough for you to do the calculations without a calculator. You can find more examples using these equations in the review questions and explanations at the end of this chapter.

Free Fall

An object is in *free fall* if it is falling freely under the influence of gravity. Any object, regardless of its mass, falls near the surface of the Earth with an acceleration of 9.8 m/s^2, which we will denote with the letter g. We will round the free fall acceleration g to 10 m/s^2 for the purpose of the SAT II: Physics Test. This free fall acceleration assumes that there is no air resistance to impede the motion of the falling object, and this is a safe assumption on the SAT II: Physics Test unless you are told differently for a particular question on the exam.

Since the free fall acceleration is constant, we may use the kinematic equations to solve problems involving free fall. We simply need to replace the acceleration a with the specific free fall acceleration g in each equation.

All objects, regardless of mass, fall with an acceleration of 10 m/s^2 in the absence of air resistance.

Example: A ball is dropped from rest from a high window of a tall building and falls for 3 seconds.

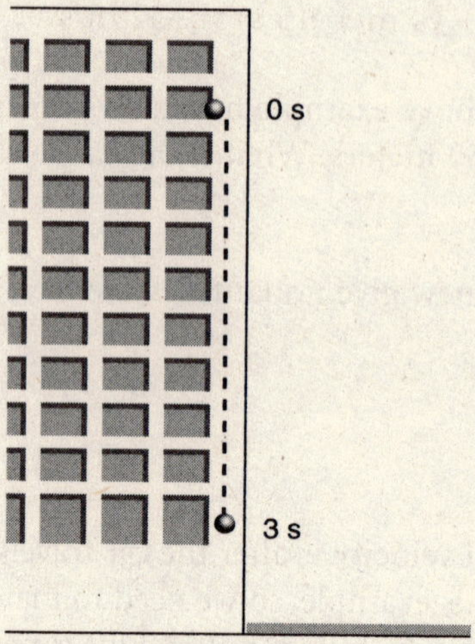

(忽略空气阻力)

Neglecting air resistance:

(A) What is the speed of the ball? $v = at = gt = +30$ m/s

(B) What is the distance the ball has fallen from where it was released? $S = \frac{1}{2}gt^2 = 5 \times 3 \times 3 = 45$ m

(C) What is the ball's acceleration? $a = g = 10$ m/s^2

Solution: Let's list the quantities given or implied in this problem:

$v_i = 0$, since the ball is dropped from rest

$t = 3$ s

$a = g = 10$ m/s^2, since the ball is falling freely under the influence of gravity

(A) We are looking for the speed v_f after 3 s, so we find an equation that includes v_f, v_i, a, and t. Equation 4 gives

$$v_f = v_i + gt = 0 + (10 \text{ m/s}^2)(3 \text{ s}) = 30 \text{ m/s}$$

(B) The distance fallen can be found by Equation 5, choosing the downward direction as positive:

$$s = v_i t + \tfrac{1}{2}gt^2 = (0)(3 \text{ s}) + \tfrac{1}{2}(10 \text{ m/s}^2)(3 \text{ s})^2 = 45 \text{ m}$$

(C) After 3 s of free fall, the ball is still accelerating at 10 m/s^2. The ball's *speed* is increasing, but the *rate* at which its speed is increasing (its *acceleration*) is constant.

Example: Suppose another ball is thrown upward with an initial velocity of 5 m/s from the window in the previous example. What is the ball's velocity (magnitude and direction) after 8 seconds?

Solution: In this case, the ball is thrown upward, and since gravity acts downward, we must assign a negative sign to either the acceleration of the ball, which is still 10 m/s^2 while it's in the air, or the initial velocity. Let's call the initial velocity positive since it is directed upward, and the acceleration negative since it is directed downward. Listing the quantities we are given:

If velocity and acceleration are in opposite directions in the same problem, we must give a negative sign to one of them and a positive sign to the other.

$v_i = +5$ m/s

$t = 8$ s

$a = g = -10$ m/s^2

5 = gt₁
t₁ = 0.5
Vf = gt₂ = 10×7.5 = 75 m/s

and we are trying to find v_f.

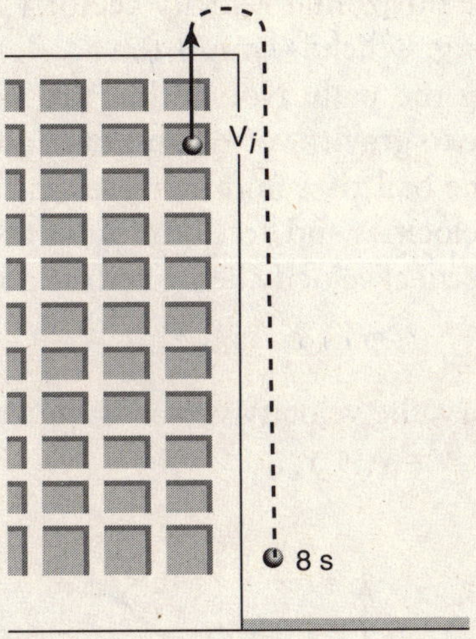

Equation 4 has everything we need:

$$v_f = v_i + at = (+5 \text{ m/s}) + (-10 \text{ m/s}^2)(8 \text{ s}) = -75 \text{ m/s}$$

Thus, after 8 seconds, the ball is moving downward with a speed of 75 m/s.

Projectile Motion (抛物线运动)

Projectile motion results when an object is thrown either horizontally through the air or at an angle relative to the ground. In both cases, the object moves through the air with a constant horizontal velocity (neglecting air friction), and at the same time is falling freely under the influence of gravity. In other words, the projected object is moving horizontally and vertically (垂直地) at the same time, and the resulting path of the projectile, called the *trajectory*, has a parabolic shape. For this reason, projectile motion is considered to be *two-dimensional* motion.

cannon 大炮
cannonball 炮弹

The motion of a projectile can be broken down into constant velocity and zero acceleration in the horizontal direction, and a changing vertical velocity due to the acceleration of gravity. Let's label any quantity in the horizontal direction with the subscript x, and any quantity in the vertical direction with the subscript y. If we fire a cannonball from a cannon on the ground pointing up at an angle θ, the ball will follow a parabolic path and we can draw the vectors associated with the motion at each point along the path:

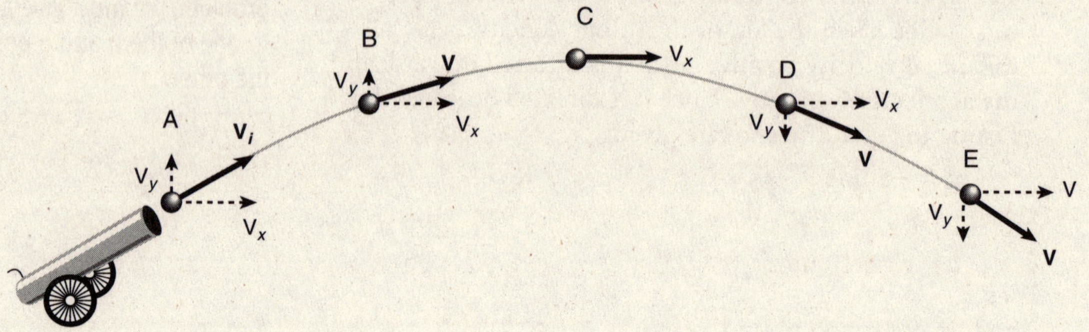

At each point, we can draw the horizontal velocity vector $\mathbf{v_x}$, the vertical velocity vector $\mathbf{v_y}$. The vertical acceleration vector **g**, which is simply the acceleration due to gravity, is directed downward at each point along the path. Notice that the length of the horizontal velocity vector and the acceleration due to gravity vector do not change, since they are constant. The vertical velocity decreases as the ball rises and increases as the ball falls. The motion of the ball is symmetric; that is, the velocities and acceleration of the ball on the way up is the same as on the way down, with the vertical velocity being zero at point C and reversing its direction at this point.

At any point along the trajectory, the velocity vector is the vector sum of the horizontal and vertical velocity vectors, that is, $\mathbf{v} = \mathbf{v_x} + \mathbf{v_y}$.

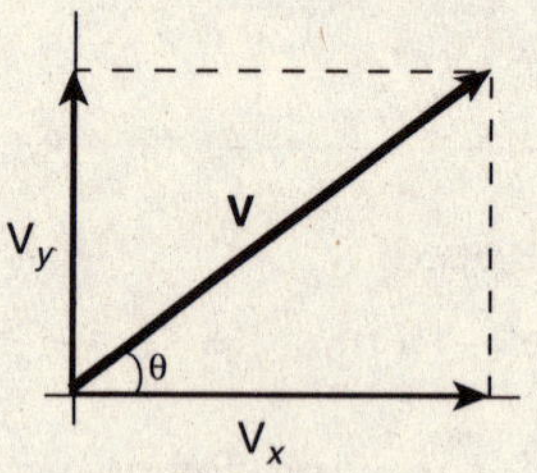

By the Pythagorean theorem:

$$v = \sqrt{v_x^{\,2} + v_y^{\,2}}$$

and

$$v_x = v\cos\theta$$

$$v_y = v\sin\theta$$

$$\theta = \tan^{-1}\left(\frac{v_y}{v_x}\right)$$

In both the horizontal and vertical cases, the acceleration is constant; that is, zero in the horizontal direction and 10 m/s^2 downward in the vertical direction. Therefore we can use the kinematic equations to describe the motion of a projectile.

Kinematic Equations for a Projectile

Horizontal motion (neglecting air friction)	Vertical motion
$a_x = 0$	$a_y = g = 10\ \text{m/s}^2$
$v_x = \dfrac{s_x}{t}$	$v_y = v_{iy} + gt$
$s_x = v_x t$	$s_y = v_{iy}t + \frac{1}{2}gt^2$

If the acceleration **g** and the initial vertical velocity $\mathbf{v_{iy}}$ are in opposite directions, we must give one of them a negative sign.

The horizontal velocity of a projectile is constant, but the vertical velocity is changed by gravity.

Example: A ping pong ball is launched from a level floor at an angle of 60º from the horizontal with an initial velocity of 6 m/s. (cos 60 = 0.5; sin 60 = 0.87; tan 60 = 1.73)

Neglecting air resistance:

(A) What is the maximum height reached by the ball?

(B) How much time does it take for the ball to land on the floor?

(C) How far horizontally does the ball travel before hitting the floor?

Solution: Let's write down the quantities we are given:

$\theta = 60º$

$v_i = 6$ m/s

$g = 10$ m/s^2

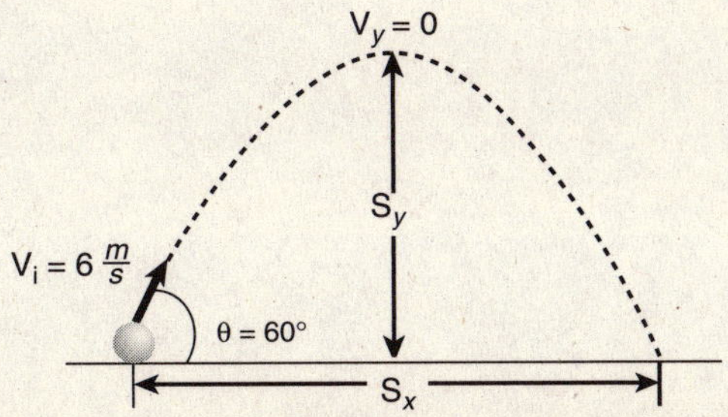

(A) At the ball's maximum height, $v_{fy} = 0$. We can use a vertical kinematic equation to find the time t_1 it takes to reach this maximum height, then find the maximum height.

$$v_{fy} = v_{iy} + gt_1$$

$$0 = (v \sin\theta) - gt_1$$

Note that g is negative, since the initial velocity is upward and positive.

$$t_1 = \frac{v\sin\theta}{g} = \frac{(6\text{ m/s})\sin 60}{10\text{ m/s}^2} = 0.5\text{ s}$$

Then the height of the ball from the ground at this time is

$$s_y = v_{iy}t_1 - \tfrac{1}{2}gt_1^2 = (6\text{ m/s})\sin 60(0.5\text{ s}) - \tfrac{1}{2}(10\text{ m/s}^2)(0.5\text{ s})^2 = 1.35\text{ m}$$

Thus, the ball is located 1.35 m above the ground after 0.5 s.

(B) By symmetry, the time t_2 it takes for the ball to land on the floor is twice the time it takes for the ball to reach maximum height. So, $t_2 = 2(0.5\text{ s}) = 1\text{ s}$.

(C) The horizontal distance traveled by the ball before hitting the ground is:

$$s_x = v_x t_2 = (v\cos\theta)t_2 = (6\text{ m/s})\cos 60(1\text{ s}) = 3\text{ m}$$

This maximum horizontal distance is also called the *range* of the projectile.

GRAPHS OF MOTION (运动图象)

Let's take some time to review how we interpret the motion of an object when we are given the information about it in graphical form. On the SAT II: Physics Test, you will need to be able to interpret three types of graphs: displacement vs. time, velocity vs. time, and acceleration vs. time.

Displacement vs. Time

s-t（位移－时间 图象）

Consider the displacement vs. time graph below:

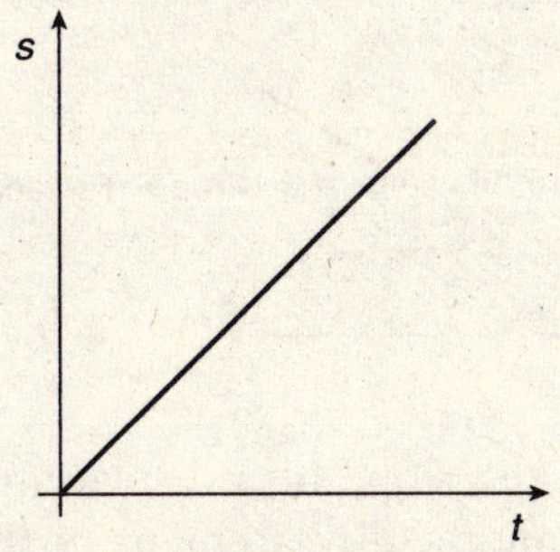

Since the graph is a straight, diagonal line, the slope is constant. As we discussed in chapter 3, the slope of a line is the rise divided by the run. The rise in this case is the displacement, or change in position, and the run is the change in time. Thus:

The slope of a displacement vs. time graph is velocity.

注：S-t图象的斜率是速度

$$\text{slope of an } s \text{ vs. } t \text{ graph} = \frac{rise}{run} = \frac{\Delta x}{\Delta t} = velocity$$

A constant slope in this case means a constant velocity. Remember, position and time are changing, but the velocity (slope) is constant.

Velocity vs. Time

v-t（速度－时间 图象）

Consider the velocity vs. time graph below:

注：v-t图象的斜率是加速度

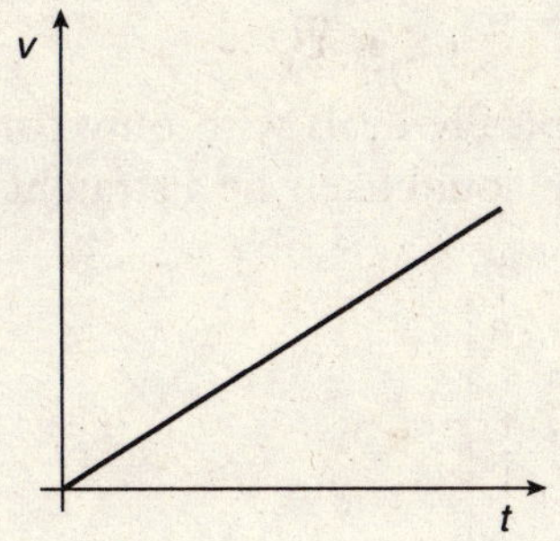

In this case, the straight diagonal line means that the velocity is changing with time, that is, the object is accelerating.

The slope of a *velocity vs. time* graph is acceleration, and the area under a *velocity vs. time* graph is the displacement.

$$\text{slope of a } v \text{ vs. } t \text{ graph} = \frac{rise}{run} = \frac{\Delta v}{\Delta t} = acceleration$$

The constant slope in this case means constant acceleration. Remember, the velocity is changing with time, but the acceleration is constant.

As we discussed in chapter 3, the area under a velocity vs. time graph is the displacement.

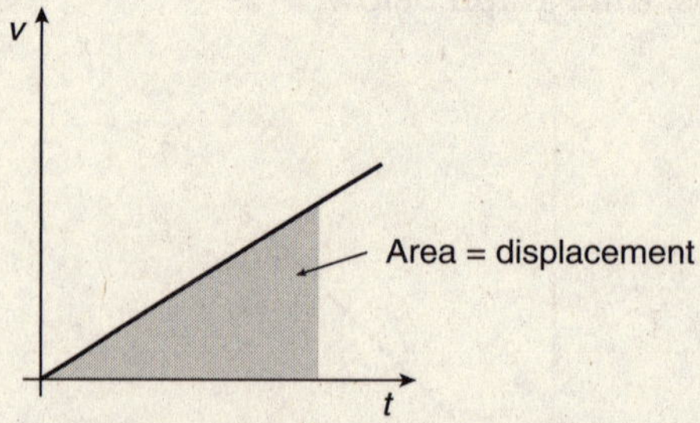

V-t图象的面积是位移

If the velocity is changing, then the slope of a displacement vs. time graph must also be changing. For constant acceleration, the displacement vs. time graph would be a parabola:

速度改变，即S-t图象的斜率改变，故一个恒定加速度的S-t图象是曲线图象

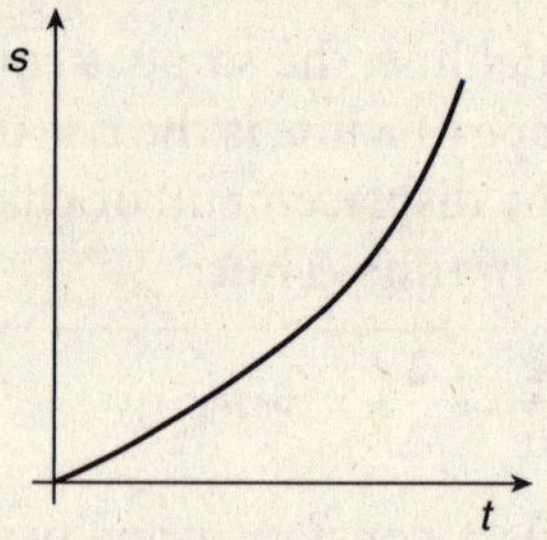

On this graph, the slope (velocity) is constantly increasing, indicating that the object is accelerating. The shape is parabolic because displacement is proportional to the square of time, as we've seen in the kinematic equations relating displacement and time for constant acceleration.

Acceleration vs. Time

a-t (加速度—时间图象)

Since the SAT II: Physics Test typically deals with only constant acceleration, any graph of acceleration vs. time on the exam would likely be a straight horizontal line:

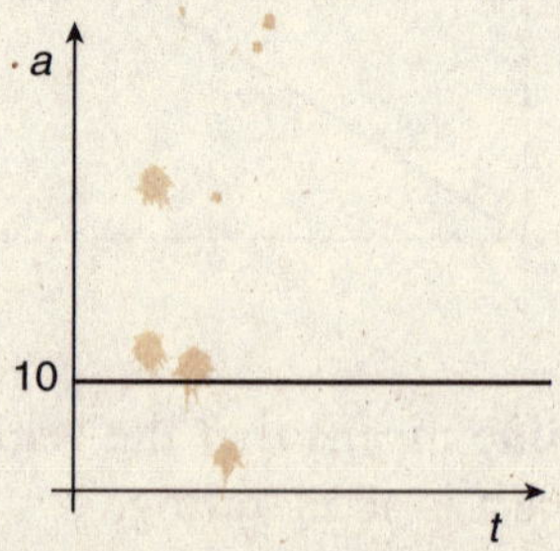

This graph tells us that the acceleration of this object is +10 m/s^2. If the object were accelerating negatively, the horizontal line would be below the time axis.

Example: A football player catches a ball at his goal line ($x = 0$) at $t = 0$, and his motion is then graphed on the displacement vs. time graph below.

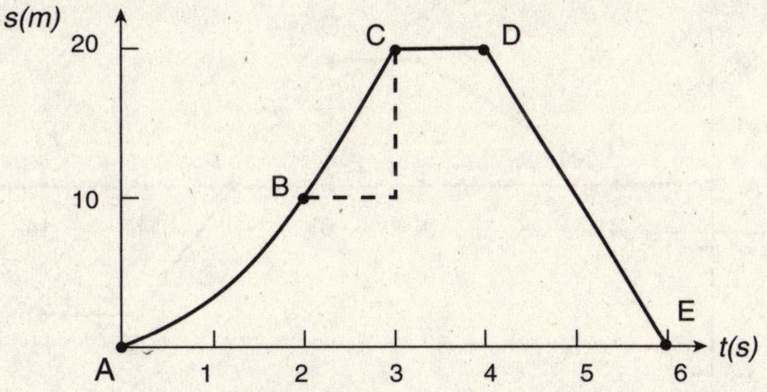

(A) During which interval is the player accelerating away from his goal line?

(B) What is his velocity between points B and C?

(C) Describe his motion between points C and D.

(D) What is his velocity (magnitude and direction) during interval from 4 s to 6 s?

Solution:

(A) The player is accelerating during the interval A to B, since the graph is parabolic. This tells us that his velocity is increasing as he runs away from the goal line.

(B) The player's velocity between points B and C is the slope of the line between these two points.

$$\text{slope} = \frac{rise}{run} = \frac{\Delta x}{\Delta t} = \frac{20\text{ m} - 10\text{ m}}{3\text{ s} - 2\text{ s}} = 10\text{ m/s}$$

(C) During the interval from C to D, the time changes from 3 s to 4 s, but the player's position is not changing. He remains at rest at a distance of 20 m from the goal line, and thus his velocity is zero during this time interval. We could also say that since the slope of the line is zero, his velocity is zero, and he is at rest.

(D) The player's velocity between 4 s and 6 s is found by the slope of the line in that interval:

$$\text{slope} = \frac{rise}{run} = \frac{\Delta x}{\Delta t} = \frac{0\text{ m} - 20\text{ m}}{6\text{ s} - 4\text{ s}} = -10\text{ m/s}$$

Since the slope is negative, the velocity is negative, which does *not* mean the player is slowing down. A negative velocity means the player is running back toward his own goal line, that is, he is moving backward compared to his motion in intervals AB and BC.

Example: The graph below represents the velocity vs. time graph for a car:

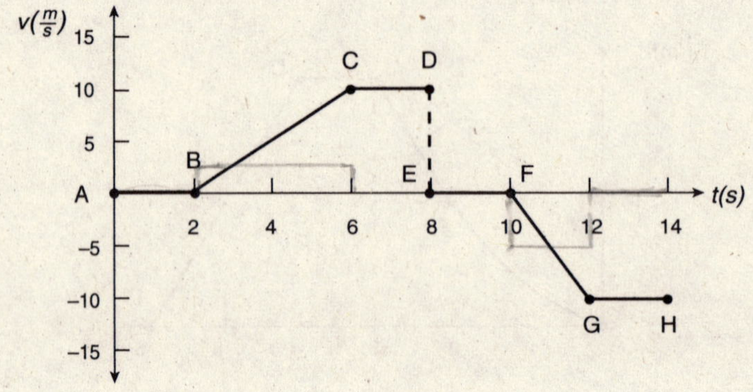

(A) During which intervals is the car at rest?

(B) During which intervals is the car moving at a constant nonzero velocity? State the velocity at each of these intervals.

(C) There are two intervals during which the car is accelerating. Find the acceleration in each of these intervals.

(D) Sketch the acceleration vs. time graph which corresponds to the motion of the car from 0 s to 14 s.

Solution:

(A) The car is at rest when the velocity is zero: A to B, and E to F.

(B) The velocity of the car is constant from C to D, when its velocity is +10 m/s, and from G to H, when its velocity is –10 m/s. The car is traveling backward during the interval G to H compared to the interval C to D.

(C) The car is accelerating during intervals B to C and F to G. During the interval B to C, the acceleration can be found by finding the slope of the v vs. t graph:

$$\text{slope of a } v \text{ vs. } t \text{ graph} = \frac{rise}{run} = \frac{\Delta v}{\Delta t} = \frac{10 \text{ m/s} - 0}{6 \text{ s} - 2 \text{ s}} = +2.5 \text{ m/s}^2$$

During the interval F to G:

$$\text{slope} = \frac{rise}{run} = \frac{\Delta v}{\Delta t} = \frac{-10 \text{ m/s} - 0}{12 \text{ s} - 10 \text{ s}} = -5 \text{ m/s}^2$$

In this last case, the car is still speeding up (from 0 to –10 m/s^2); it's just speeding up in a negative direction.

(D) The acceleration vs. time graph looks like this:

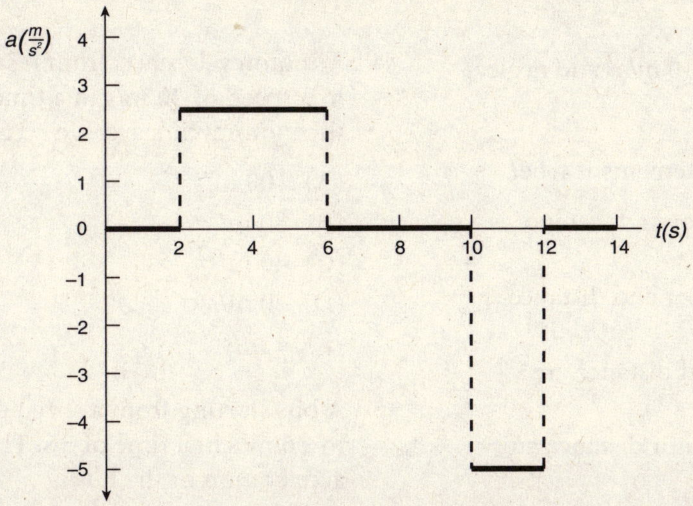

KINEMATICS REVIEW QUESTIONS

Unless otherwise noted, use $g = 10\ \text{m/s}^2$ and neglect air resistance.

1. Which of the following statements is true?
   (A) Displacement is a scalar and distance is a vector.
   (B) Displacement is a vector and distance is a scalar.
   (C) Both displacement and distance are vectors.
   (D) Neither displacement nor distance are vectors.
   (E) Displacement and distance are always equal.

2. Which of the following is the best statement for a velocity?
   (A) 60 miles per hour
   (B) 30 meters per second
   (C) 30 km at 45° north of east
   (D) 40 km/hr
   (E) 50 km/hr southwest

3. A jogger runs 3 km in 0.3 hr, then 7 km in 0.70 hr. What is the average speed of the jogger?
   (A) 10 km/hr
   (B) 3 km/hr
   (C) 1 km/hr
   (D) 0.1 km/hr
   (E) 100 km/hr

4. A motorcycle starts from rest and accelerates to a speed of 20 m/s in a time of 5 s. What is the motorcycle's average acceleration?
   (A) $100\ \text{m/s}^2$
   (B) $80\ \text{m/s}^2$
   (C) $40\ \text{m/s}^2$
   (D) $20\ \text{m/s}^2$
   (E) $4\ \text{m/s}^2$

5. A bus starting from a speed of +12 m/s slows to +6 m/s in a time of 3 s. The average acceleration of the bus is
   (A) $2\ \text{m/s}^2$
   (B) $4\ \text{m/s}^2$
   (C) $3\ \text{m/s}^2$
   (D) $-2\ \text{m/s}^2$
   (E) $-4\ \text{m/s}^2$

6. A train accelerates from rest with an acceleration of $4\ \text{m/s}^2$ for a time of 10 s. What is the train's speed at the end of 10 s?
   (A) 14 m/s
   (B) 6 m/s
   (C) 2.5 m/s
   (D) 0.4 m/s
   (E) 40 m/s

7. A football player starts from rest 20 meters from the goal line and accelerates away from the goal line at $2\ \text{m/s}^2$. How far from the goal line is the player after 4 s?
   (A) 8 m
   (B) 28 m
   (C) 32 m
   (D) 36 m
   (E) 52 m

8. A stone is dropped from rest. What is the acceleration of the stone immediately after it is dropped?

   (A) zero
   (B) 5 m/s^2
   (C) 10 m/s^2
   (D) 20 m/s^2
   (E) 30 m/s^2

Questions 9–11:

A ball is thrown straight upward with an initial velocity of +15 m/s.

9. What is the ball's acceleration just after it is thrown?

   (A) zero
   (B) 10 m/s^2 upward
   (C) 10 m/s^2 downward
   (D) 15 m/s^2 upward
   (E) 15 m/s^2 downward

10. How much time does it take for the ball to rise to its maximum height?

   (A) 25 s
   (B) 15 s
   (C) 10 s
   (D) 5 s
   (E) 1.5 s

11. What is the approximate maximum height reached by the ball?

   (A) 23 m
   (B) 15 m
   (C) 11 m
   (D) 8 m
   (E) 5 m

12. Which of the following is NOT true of a projectile launched from the ground at an angle (neglecting air resistance)?

   (A) The horizontal velocity is constant.
   (B) The vertical acceleration is upward during the first half of the flight, and downward during the second half of the flight.
   (C) The horizontal acceleration is zero.
   (D) The vertical acceleration is 10 m/s^2.
   (E) The time of flight can be found by horizontal distance divided by horizontal velocity.

13. A projectile is launched horizontally from the edge of a cliff 20 m high with an initial speed of 10 m/s. What is the horizontal distance the projectile travels before striking the level ground below the cliff?

   (A) 5 m
   (B) 10 m
   (C) 20 m
   (D) 40 m
   (E) 60 m

14. A projectile is launched from level ground with a velocity of 40 m/s at an angle of 30° from the ground. What will be the vertical component of the projectile's velocity just before it strikes the ground? (sin 30° = 0.5, cos 30 = 0.87)

   (A) 10 m/s
   (B) 20 m/s
   (C) 30 m/s
   (D) 35 m/s
   (E) 40 m/s

15. Each of the choices below shows a pair of graphs. Which choice shows the pair of graphs which are equivalent to each other?

(A) 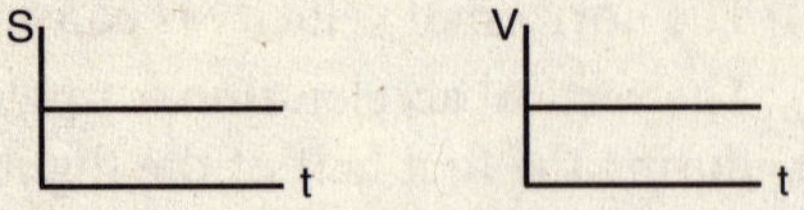

(B) 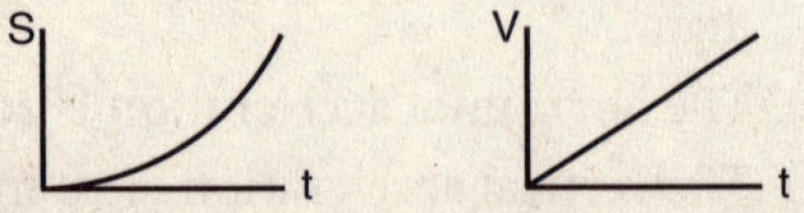

(C) 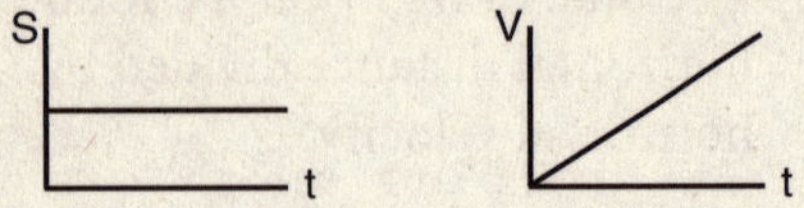

(D) 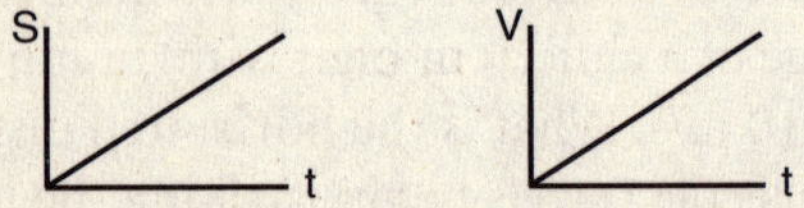

(E) 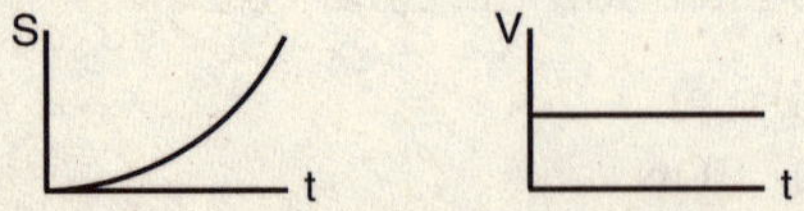

Questions 16–17:

Consider the velocity vs. time graph below:

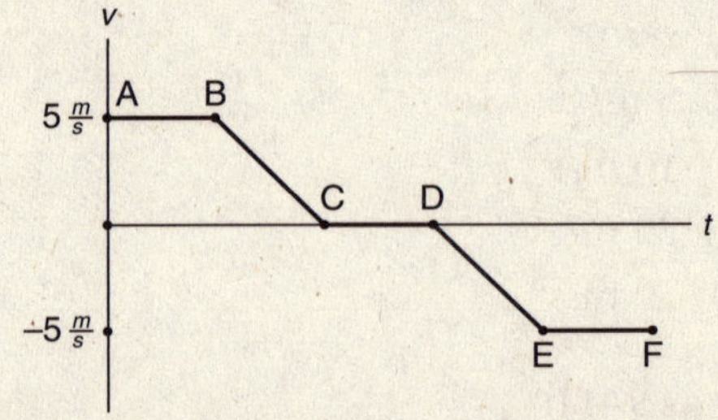

16. During which interval is the object at rest?

(A) AB
(B) BC
(C) CD
(D) DE
(E) EF

17. During which interval is the speed of the object increasing?

(A) AB
(B) BC
(C) CD
(D) DE
(E) EF

SOLUTIONS TO KINEMATICS REVIEW QUESTIONS

1. B

Displacement is the straight-line length from an origin to a final position and includes direction, whereas distance is simply length moved.

2. E

Velocity is a vector and therefore direction should be included.

3. A

Average speed is total distance divided by total time. The total distance covered by the jogger is 10 km and the total time is 1 hour, so the average speed is 10 km/hr.

4. E

$$a = \frac{\Delta v}{\Delta t} = \frac{20 \text{ m/s}}{5 \text{ s}} = 4 \text{ m/s}^2$$

5. D

$$a = \frac{v_f - v_i}{t} = \frac{6 \text{ m/s} - 12 \text{ m/s}}{3 \text{ s}} = -2 \text{ m/s}^2$$

6. E

$$v_f = v_i + at = 0 + (4 \text{ m/s}^2)(10 \text{ s}) = 40 \text{ m/s}$$

7. D

$$x_f = x_i + v_i t + \tfrac{1}{2}at^2 = (20 \text{ m}) + 0 + \tfrac{1}{2}(2 \text{ m/s}^2)(4 \text{ s})^2 = 36 \text{ m}$$

8. C

The acceleration due to gravity is 10 m/s^2 downward at all points during the stone's fall.

9. C

After the ball is thrown, the only acceleration it has is the acceleration due to gravity, 10 m/s^2 downward.

10. E

At the ball's maximum height, $v_f = 0$. Thus:

$$v_f = v_i + gt = 0$$

$$t = \frac{v_i}{g} = \frac{15 \text{ m/s}}{10 \text{ m/s}^2} = 1.5 \text{ s}$$

11. C

$$s = \tfrac{1}{2}gt^2 = \tfrac{1}{2}(10\text{ m/s}^2)(1.5\text{ s})^2 = 11.25\text{ m}$$

12. B

Since the vertical acceleration is due to gravity, it is always downward.

13. C

First we find the time of flight, which can be calculated from the height:

$$d = \tfrac{1}{2}gt^2,\text{ so } t = \sqrt{\frac{2d}{g}} = \sqrt{\frac{2(20\text{ m})}{10\text{ m/s}^2}} = 2\text{ s}$$

Then, $x = v_x t = (10\text{ m/s})(2\text{ s}) = 20\text{ m}$

14. B

Neglecting air resistance, the y-component of the velocity projectile just before it lands is equal to the y-component of the velocity when it is first fired:

$$v_y = (40\text{ m/s})\sin 30^\circ = 20\text{ m/s}$$

15. B

Both graphs imply the object is accelerating.

16. C

The object is at rest during the interval in which the velocity is zero.

17. D

The speed is increasing from zero at point D to –5 m/s at point E, implying the object is going backward but speeding up from D to E.

CHAPTER FIVE

Dynamics

In the last chapter, we discussed *kinematics*, the study of *how* motion occurs. In this chapter, we review *dynamics*, the study of the *causes* of motion, namely, *forces*. A force is a push or a pull. As we discussed in chapter 3, a force is a vector quantity, and therefore has magnitude (size) and direction (angle). The SI unit for force is the *newton*, named for the originator of the study of dynamics, Isaac Newton. In 1687, Newton published his three laws of motion.

NEWTON'S FIRST LAW OF MOTION

The first law of motion states that an object in a state of constant velocity (including zero velocity) will continue in that state unless acted upon by an unbalanced, or net, force. Thus, a book at rest on your desk will remain at rest until you or something else applies a force to it. The property of the book that causes it to follow Newton's first law of motion is its *inertia*. Inertia is the resistance of an object to changing its state of motion or state of rest. It is difficult to push a car to get it moving from a state of rest because of its inertia, just as it is difficult to stop a car which is moving because of its inertia. We measure inertia by measuring the mass of an object, or the amount of material it contains. Thus, the SI unit for inertia would be the kilogram. We often refer to Newton's first law as the *law of inertia*. Remember, you don't need an unbalanced force to keep an object at rest or moving at a constant velocity; its own inertia takes care of that.

Newton's first law is the *law of inertia*.

NEWTON'S SECOND LAW OF MOTION

The law of inertia tells us what happens to an object when there are no unbalanced forces acting on it. Newton's second law tells us what happens to an object that *does* have an unbalanced force acting on it: It accelerates in the direction of the unbalanced force. Another name for an unbalanced force is a *net force*, meaning a force which is not canceled by any other force acting on the object. Sometimes the net force acting on an object is called an *external force*.

Newton's second law can be written as force equals mass times acceleration.

Newton's second law can be stated like this: *A net force acting on a mass causes that mass to accelerate in the direction of the net force. The acceleration is proportional to the force* (if you double the force, you double the amount of acceleration), *and inversely proportional to the mass of the object being accelerated* (twice as big a mass will only be accelerated half as much by the same force). In equation form, we write Newton's second law as:

$$\boldsymbol{F_{net}} = m\boldsymbol{a}$$

where $\boldsymbol{F_{net}}$ and $\boldsymbol{a}$ are vectors pointing in the same direction. We see from this equation that the newton is defined as a $(kg)m/s^2$, and happens to be a little less than a quarter of a pound.

Example: Two ropes pull on a 50 kg cart as shown below. What is the acceleration of the cart?

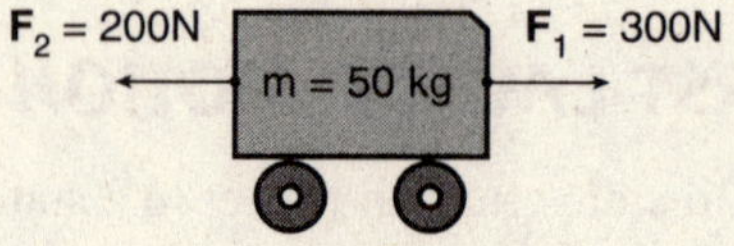

Solution: Before calculating the acceleration of the cart using Newton's second law, we must first find the net force. Since the rope on the right is pulling with a force which is greater than the force pulling to the left, the net force must be to the right and equal to the difference between the forces, since they are acting in opposite directions.

$$F_{net} = F_1 - F_2 = 300\text{ N} - 200\text{ N} = 100\text{ N to the right}$$

Then, $a = \dfrac{F_{net}}{m} = \dfrac{100\text{ N}}{50\text{ kg}} = 2\text{m/s}^2$ to the right.

But what if the forces are acting at an angle other than 0° or 180°?

Example: Two forces act on a 40 kg sled resting on ice as shown from the top view below. What is the magnitude and direction of the acceleration of the sled?

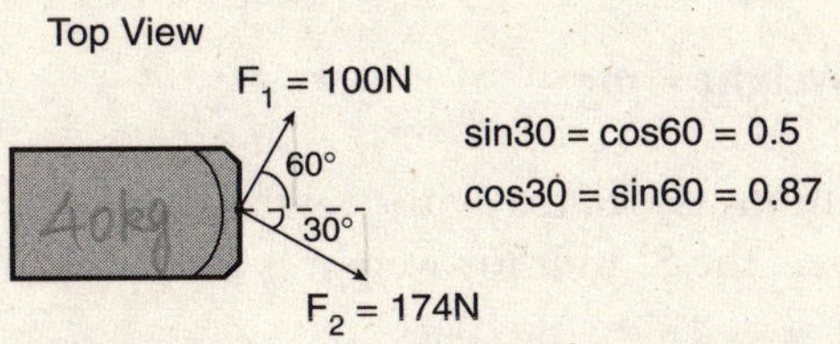

Solution: Once again we need to find the net force before we can find the acceleration of the sled. Since the weight of the sled or the normal force exerted by the ground on the sled do not contribute to the acceleration of the sled in this case, we will not include them in our diagram. Let's break the forces down into their components:

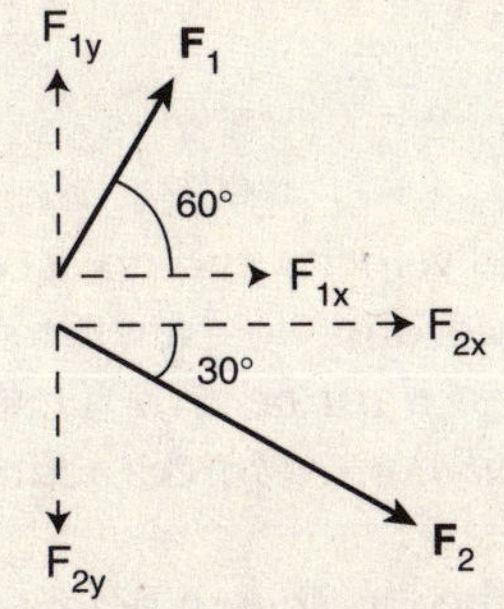

$$F_{1x} = F_1\cos60 = (100\text{ N})(0.5) = 50\text{ N to the right}$$

$$F_{2x} = F_2\cos30 = (174\text{ N})(0.87) = 150\text{ N}$$

$$F_{1y} = F_1\sin60 = (100\text{ N})(0.87) = 87\text{ N}$$

$$F_{2y} = -F_2\sin30 = (174\text{ N})(0.5) = -87\text{ N}$$

Notice that the *y*-components of the forces cancel out, so we are left with a net force to the right:

$$F_{net} = F_{1x} + F_{2x} = 50\text{ N} + 150\text{ N} = 200\text{ N to the right}$$

So the acceleration of the sled is:

$$a = \frac{F_{net}}{m} = \frac{200\text{ N}}{40\text{ kg}} = 5\text{ m/s}^2\text{ to the right}$$

Weight (重量)

As we mentioned earlier, mass is a measure of the inertia of an object. If you take a 1 kg mass to the moon, it will still have a mass of 1 kg. Mass does not depend on gravity, but weight does. The *weight* of an object is defined as the amount of gravitational force acting on its mass. Since weight is a force, we can calculate it using Newton's second law:

$$F_{net} = ma \text{ becomes } \textbf{Weight} = mg$$

where the specific acceleration associated with weight is, not surprisingly, the acceleration due to gravity. Like any force, the SI unit for weight is the newton.

Mass is the measure of the amount of substance in an object, and is measured in kilograms. *Weight* is the gravitational force pulling down on an object, and is measured in newtons.

Example: Find the weight in newtons of a girl having a mass of 40 kg.

Solution: $W = mg = (40 \text{ kg})(10 \text{ m/s}^2) = 400$ N. $1 N \approx \frac{1}{4}$ pound (lbs)

Since one newton is a little less than a quarter of a pound, this girl weighs about 100 lbs in U.S. Customary units.

Static Equilibrium (静止平衡)

A system is said to be *static* if it has no velocity and no acceleration. In other words, it's at rest and it's going to stay that way. According to Newton's first law, *if an object is in static equilibrium, the net force on the object must be zero.* That doesn't mean there are no forces acting on it; it means there are no *unbalanced* forces acting on it.

Example: Three ropes are attached as shown below. The tension forces in the ropes are $\mathbf{T_1}$, $\mathbf{T_2}$, and $\mathbf{T_3}$, and the mass of the hanging ball is 50 kg. What is the magnitude of the tension in each of the three ropes? (sin 30 = cos 60 = 0.5; sin 60 = cos 30 = 0.87)

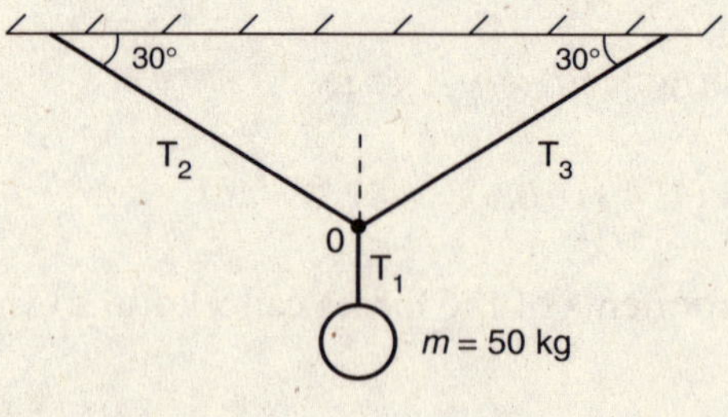

$T_1 = mg = 50 \times 10 = 500$ N

$T_2 \cdot \sin 30° + T_3 \cdot \sin 30° = T_1$

$T_2 + T_3 = 1000$ N

$T_2 \cdot \cos 30° = T_3 \cdot \cos 30°$

$T_2 = T_3$

$\therefore T_1 = T_2 = T_3 = 500$ N

Solution: Since the system is in equilibrium, the net force on the system must be zero. We start by drawing a *free-body force diagram* of all the forces acting on point O. A free-body force diagram is a vector diagram of all of the forces acting on a mass.

Always draw all of the forces acting on an object before answering a question about the forces or net force.

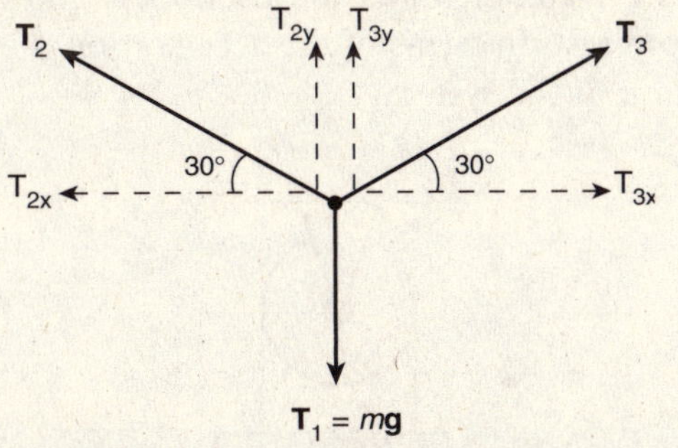

Tension T_1 is simply the weight of the hanging ball:

$$T_1 = W = mg = (50 \text{ kg})(10 \text{ m/s}^2) = 500 \text{ N}$$

To find T_2 and T_3, we can break them down into their vector components:

$$T_{2x} = T_2\cos30 \text{ and } T_{2y} = T_2\sin30$$

$$T_{3x} = T_3\cos30 \text{ and } T_{3y} = T_3\sin30$$

Since the forces are in equilibrium, the vector sum of the forces in the x-direction must equal zero. Therefore:

$$T_{2x} = T_{3x}$$

$$T_2\cos30 = T_3\cos30$$

$$T_2 = T_3$$

The sum of the forces in the y-direction must also be zero:

$$T_{2y} + T_{3y} = W$$

$$T_2\sin30 + T_3\sin30 = W$$

$$T_2(0.5) + T_3(0.5) = 500 \text{ N}$$

Since $T_2 = T_3$, it follows that each tension is also equal to 500 N.

Blocks and Pulley (滑轮组)

A typical example used to illustrate weight and Newton's second law is a blocks and pulley system. Let's look at a couple of examples of the blocks and pulley system.

Example: Two blocks of mass *m* and *3m* are connected by a string which is passed over a pulley of negligible mass and friction, as shown below. The system is released from rest. In terms of the acceleration due to gravity *g*, what is the acceleration of the system?

$T - mg = ma$

$3mg - T = 3ma$

$2mg = 4ma$

$a = \frac{1}{2}g$

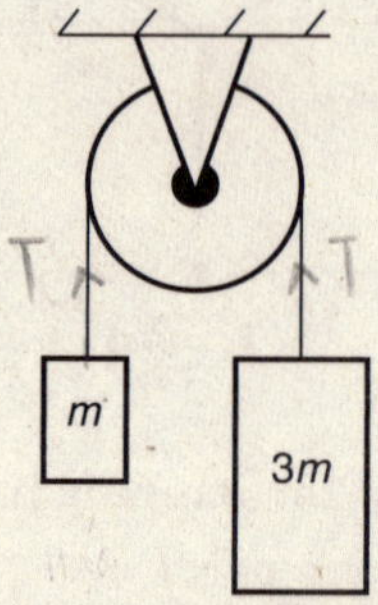

Solution: First, we should draw a free-body force diagram for each block. There are two forces acting on each of the masses: weight downward and the tension in the string upward. Our free-body force diagrams should look like this:

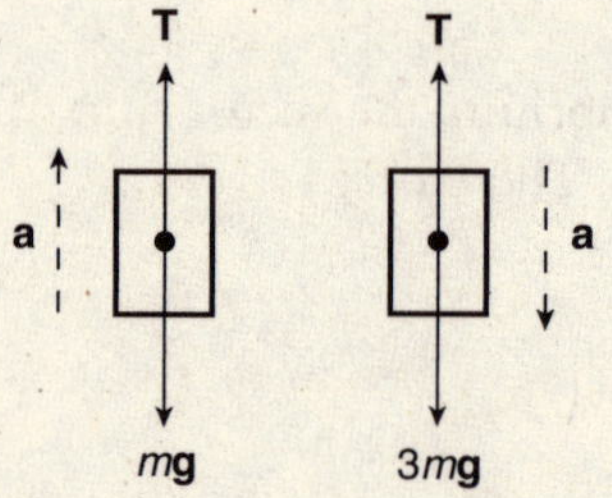

Writing Newton's second law for each of the blocks:

$F_{net} = ma$

$(T - mg) = ma$

$T = mg + ma$

$F_{net} = 3ma$

$(3mg - T) = 3ma$

$T = 3mg - 3ma$

Notice that the tension T acting on block m is greater than block m's weight, but block $3m$ has a greater weight than the tension T. This is, of course, the reason block $3m$ accelerates downward and block m accelerates upward. The magnitude of the tension acting on each block is the same, and the magnitudes of their accelerations are the same. Setting their tensions equal to each other, we get:

$$mg + ma = 3mg - 3ma$$

Canceling the masses and solving for a, we get:

$$a = \tfrac{1}{2}g$$

Of course, the block of mass m accelerates upward, while the block $3m$ accelerates downward.

Example: A block of mass m rests on a horizontal table of negligible friction. A string is tied to the block, passed over a pulley, and another block of mass $4m$ is hung on the other end of the string, as shown in the figure below. Find the acceleration of the system.

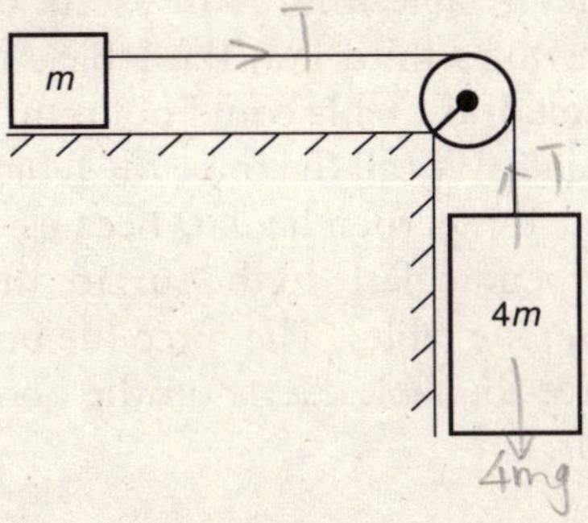

$$ma = T$$
$$4ma = 4mg - T$$
$$4mg - ma = 4ma$$
$$5ma = 4mg$$
$$a = \frac{4}{5}g$$

Solution: Once again, let's draw a free-body force diagram for each of the blocks, and then apply Newton's second law.

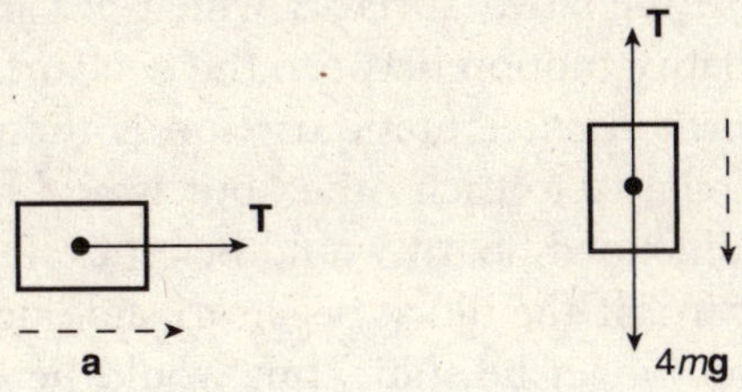

$F_{net} = ma$ $\qquad\qquad$ $F_{net} = 4ma$

$T = ma$ $\qquad\qquad$ $(4mg - T) = 4ma$

$\qquad\qquad\qquad\qquad$ $T = 4mg - 4ma$

Setting the two equations for T equal to each other:

$$ma = 4mg - 4ma$$

$$a = \frac{4}{5}g$$

See if you can make an educated guess at the answer to a question before working it out, and eliminate impossible or unreasonable answers.

Could you have guessed at this solution before we worked it out? If the hanging mass $4m$ were not connected to the mass m on the table, it would fall freely at g. Since it is connected to the mass, the weight of mass $4m$ must accelerate $4m + m = 5m$ of mass, resulting in an acceleration only 4/5 that of free-fall acceleration.

NEWTON'S THIRD LAW OF MOTION （牛顿第三定律）

Newton's third law is the *law of action and reaction.*

Newton's third law is sometimes called the *law of action and reaction*. It states that for every action force, there is an equal and opposite reaction force. For example, let's say your physics book weighs 6 N. If you set it on a level table, the book exerts 6 N of force on the table. By Newton's third law, the table must exert 6 N back up on the book. If the table could not return the 6 N of force on the book, the book would sink into the table. We call the force the table exerts on the book the *normal force*. *Normal* is another word for perpendicular, because the normal force always acts perpendicularly to the surface that is applying the force (in this case, the table). The force the book exerts on the table and the force the table exerts on the book are called an *action-reaction pair*. 牛三律：两个物体之间的作用力和反作用力总是大小相等，方向相反，作用在一条直线上。

支持力 N

The normal force N is the reaction force that a surface exerts on an object and is perpendicular to the surface.

Friction （摩擦）

Friction is a resistive force between two surfaces which are in contact with each other. Friction always opposes the relative motion between the two surfaces. There are two types of friction: *static friction* and *kinetic friction*. Static friction is the resistive force between two surfaces that are not moving relative to each other, but would be moving if there were no friction. A block at rest on an inclined board would be an example of static friction acting between the block and the board. If the block began to slide down the board, the friction between the surfaces would no longer be static, but would be kinetic, or sliding, friction. Sliding friction is typically less than static friction for the same two surfaces in contact.

We can define a ratio which gives an indication of the roughness between two surfaces in contact with each other. This ratio is the frictional force between the surfaces divided by the normal force acting on the surfaces, and is called the *coefficient of friction*. The coefficient of friction is represented by the Greek letter μ (mu). Equations for the coefficients of static and kinetic friction are:

The frictional force between two surfaces is actually due to the interaction between the electrons in the atoms contained in the two surfaces.

$$\mu_s = \frac{f_s}{N} \text{ and } \mu_k = \frac{f_k}{N}$$

where f_s is the static frictional force and f_k is the kinetic frictional force. (Note: There are times when these equations are not quite valid, but we don't have to worry about those cases for the SAT II: Physics Test.)

Example: A 2 kg block rests on a rough horizontal table as shown below. A rope is attached to the block and is pulled with a force of 8 N. As a result, the block accelerates at 2.5 m/s^2. What is the coefficient of kinetic friction μ_k between the block and the table?

Quick Quiz:

True or False: The frictional force must equal the weight of the object on which it is acting.

Answer: False, although it can be proportional to the weight, since sometimes a heavier object creates more friction.

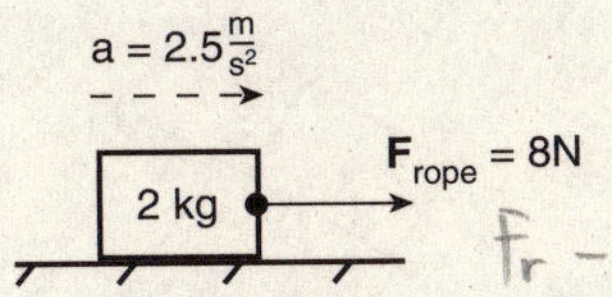

Solution: First, let's draw all of the forces acting on the block:

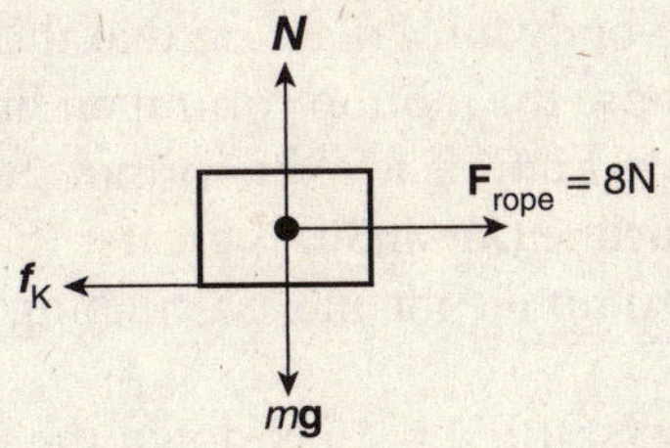

In the vertical direction, the normal force N must be equal and opposite to the weight W down.

$$N = W = mg = (2 \text{ kg})(10 \text{ m/s}^2) = 20 \text{ N}$$

In the horizontal direction, the net force is the difference between the force F_{rope} the rope exerts on the block and the force of friction f_k on the block:

$$F_{net} = (F_{rope} - f_k) = ma$$

Then $f_k = F_{rope} - ma = 8 \text{ N} - (2 \text{ kg})(2.5 \text{ m/s}^2) = 3 \text{ N}$

The coefficient of friction is:

$$\mu_k = \frac{f_k}{N} = \frac{3\ N}{20\ N} = 0.15$$

Notice that the coefficient of friction doesn't have any units, since it is the ratio of two forces. This is the coefficient of kinetic friction, since the block is moving.

Inclined Plane

If the surface on which the block rests is not horizontal but inclined at an angle, the normal force is not equal and opposite the weight of the block. Let's look at an example.

Example: A 3 kg block is placed on a rough inclined plane. The coefficient of static friction between the block and the plane is 0.2, and the angle of incline is 30°. Will the block slide down the plane, or remain at rest? (sin30 = cos60 = 0.5; cos30 = sin60 = 0.87)

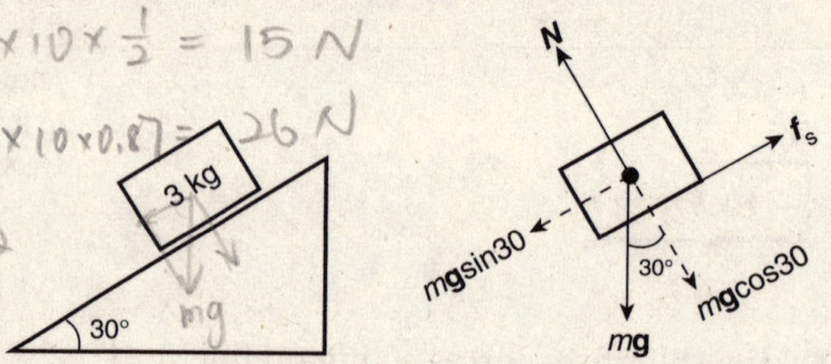

Solution: We see in the free-body force diagram that the frictional force points up the incline to oppose the motion (or impending motion) of the block. What we really want to know is whether or not the component of the weight directed down the incline (*mg*sin30) is greater than the magnitude of the frictional force f_s directed up the incline. Calculating each of those forces:

$$mg\sin 30 = (3\text{ kg})(10\text{ m/s}^2)(0.5) = 15\text{ N down the incline}$$

The normal force *N* is equal and opposite to the component of the weight *mg*cos30. So:

$$f_s = \mu_s N = \mu_s(\text{mgcos}30) = (0.2)(3\text{ kg})(10\text{ m/s}^2)(0.87)$$

$$= 5.2\text{ N up the incline.}$$

Thus, the force down the incline is greater than the force up the incline, and the block will overcome friction to slide down the incline.

DYNAMICS REVIEW QUESTIONS

1. The amount of force needed to keep a 0.1 kg hockey puck moving at a constant speed of 5 m/s on frictionless ice is

   (A) zero
   (B) 0.1 N
   (C) 0.5 N
   (D) 5 N
   (E) 50 N

2. A force of 20 N is needed to overcome a frictional force of 5 N and accelerate a 3 kg mass across a floor. What is the acceleration of the mass?

   (A) 4 m/s^2
   (B) 5 m/s^2
   (C) 7 m/s^2
   (D) 20 m/s^2
   (E) 60 m/s^2

3. A force of 50 N directed at an angle of 45° from the horizontal pulls a 70 kg sled across a frictionless pond. The acceleration of the sled is most nearly (sin 45° = cos 45° = 0.7)

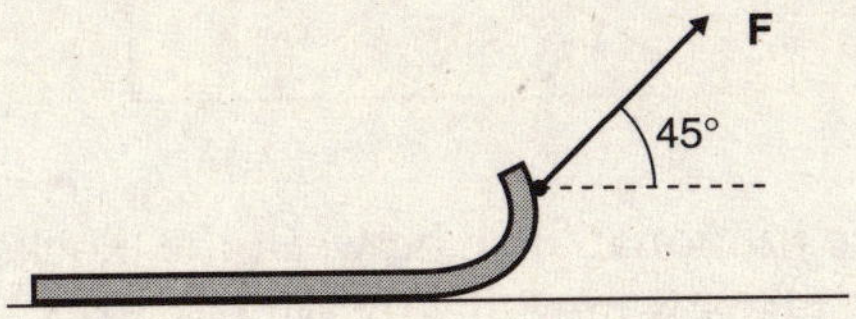

   (A) 0.5 m/s^2
   (B) 0.7 m/s^2
   (C) 5 m/s^2
   (D) 35 m/s^2
   (E) 50 m/s^2

4. Which of the following is true of the magnitudes of tensions T_1, T_2, and T_3 in the ropes in the diagram shown?

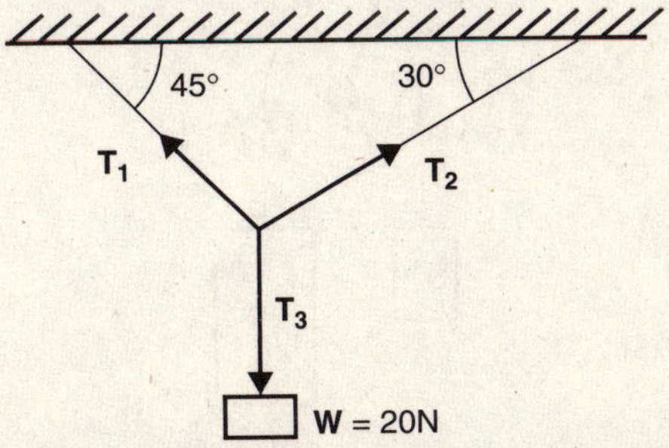

   (A) The magnitude of tension T_3 must be greater than 20 N.
   (B) The magnitude of the tension T_2 is greater than T_1.
   (C) The sum of the *y*-components of T_1 and T_2 is equal to 20 N.
   (D) The sum of the magnitudes of T_1 and T_2 is equal to T_3.
   (E) The sum of the magnitudes of T_2 and T_3 is equal to T_1.

5. Two blocks of mass m and $5m$ are connected by a light string which passes over a pulley of negligible mass and friction. What is the acceleration of the masses in terms of the acceleration due to gravity, g?

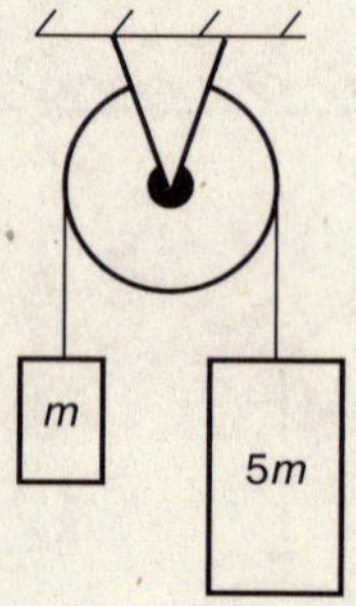

(A) $4g$
(B) $5g$
(C) $6g$
(D) $\frac{4}{5}g$
(E) $\frac{2}{3}g$

6. A 1 kg block rests on a frictionless table and is connected by a light string to another block of mass 2 kg. The string is passed over a pulley of negligible mass and friction. What is the acceleration of the masses?

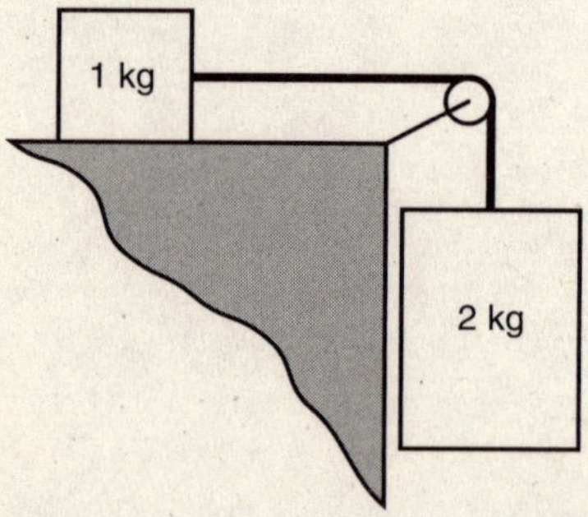

(A) 5 m/s^2
(B) 6.7 m/s^2
(C) 10 m/s^2
(D) 20 m/s^2
(E) 30 m/s^2

7. A ball falls freely toward the earth. If the action force is the earth pulling down on the ball, the reaction force is

(A) the ball pulling up on the earth.
(B) air resistance acting on the ball.
(C) the ball striking the earth when it lands.
(D) the inertia of the ball.
(E) There is no reaction force in this case.

8. Friction

(A) can occur only between two surfaces that are moving relative to one another.
(B) is equal to the normal force divided by the coefficient of friction.
(C) opposes the relative motion between the two surfaces in contact.
(D) only depends on one of the surfaces in contact.
(E) is always equal to the applied force.

9. A 2 kg wooden block rests on an inclined plane as shown.

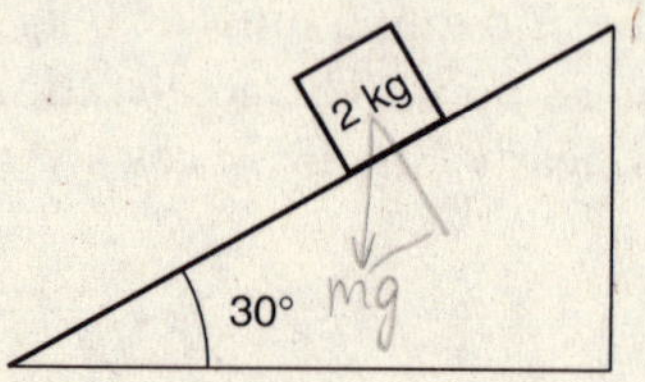

The frictional force between the block and the plane is most nearly (sin 30 = 0.5, cos 30 = 0.87, tan 30 = 0.58)

(A) 2 N
(B) 10 N
(C) 12 N
(D) 17 N
(E) 20 N

SOLUTIONS TO DYNAMICS REVIEW QUESTIONS

1. A

The law of inertia states that no force is needed to keep an object moving with a constant velocity.

2. B

The net force is 20 N – 5 N = 15 N, and the acceleration is

$$a = \frac{F_{net}}{m} = \frac{15\text{ N}}{3\text{ kg}} = 5\text{ m/s}^2$$

3. A

Only the x-component of the force accelerates the sled: $F_x = (50\text{ N})\cos 45° = 35$ N, and

$$a = \frac{F_x}{m} = \frac{35\text{ N}}{70\text{ kg}} = 0.5\text{ m/s}^2$$

4. C

$T_{1y} + T_{2y}$ must equal the weight, since the system is in static equilibrium.

5. E

The net force acting on the system is $5mg - mg = 4mg$. Thus, according to Newton's second law, $4mg = 6ma$, and $a = \frac{2}{3}g$.

6. B

The weight of the 2 kg block is the net force accelerating the entire system.

$$2mg = 3ma$$
$$a = \tfrac{2}{3}g = 6.7\text{ m/s}^2$$

7. A

Newton's third law states that if the earth is pulling on the ball, the ball is pulling on the earth.

8. C

Friction acts on each of the surfaces in contact that are moving or have the potential for moving relative to each other.

9. B

Since the block is in static equilibrium, the frictional force must be equal and opposite to the component of the weight pointing down the incline.

$$f = mg\sin 30° = (2\text{ kg})(10\text{ m/s}^2)\sin 30° = 10\text{ N}$$

CHAPTER SIX

Momentum and Impulse

MOMENTUM

The *momentum* of an object is defined as the product of its *mass* and the *velocity* at which it is moving. The usual symbol for momentum is p, and thus the equation for momentum is:

$$p = mv$$

Since mass is a scalar and velocity is a vector, momentum is also a vector and is always in the same direction as the velocity. Momentum can be thought of as inertia in motion, and can be measured by how difficult it is to move a mass from rest or stop an object that is already moving. Since the momentum of an object includes both its mass and velocity, two different masses can have the same momentum depending on their respective velocities.

***Momentum* is mass times velocity.**

Example: How fast must a small car of mass 800 kg be moving in order to have the same momentum as a large truck of mass 2,000 kg moving at 20 m/s?

Solution: The momentum of the truck is:

$$p = mv = (2{,}000 \text{ kg})(20 \text{ m/s}) = 40{,}000 \text{ kg m/s}$$

Then the small car's speed can be found by:

$$v = \frac{p}{m} = \frac{40{,}000 \text{ kg(m/s)}}{800 \text{ kg}} = 50 \text{ m/s}$$

IMPULSE (冲量)

Newton's second law states that an unbalanced (net) force acting on a mass will accelerate the mass in the direction of the force. Another way of saying this is that a net force acting on a mass will cause the mass to change its momentum. We can rearrange the equation for Newton's second law to emphasize the change in momentum:

$$F_{net} = ma = m\left(\frac{\Delta v}{\Delta t}\right)$$

or

$$F\Delta t = m\Delta v = mv_f - mv_i$$

Impulse is the product of the force acting and the time during which it acts.

The left side of the equation ($F\Delta t$) is called the *impulse*, and the right side ($m\Delta v$) is the *change in momentum*. This equation reflects the *impulse-momentum theorem*, and in words can be stated: *A force acting on a mass during a time causes the mass to change its momentum.*

Example: Consider the force vs. time graph below:

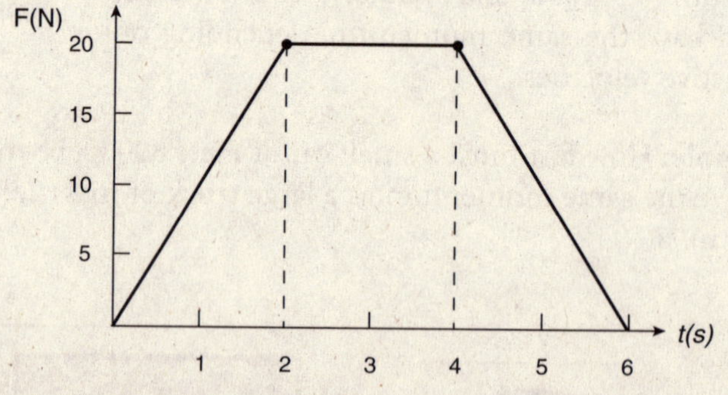

(A) What is the impulse acting on the object between 2 s and 4 s? $F\Delta t = 20 \times 2 = 40$ Ns

(B) What is the change in momentum between 2 s and 4 s? $\Delta p = 40$ kg·m/s

(C) If the mass of the object is 10 kg and it has a velocity of 2 m/s at 2 s, what is its velocity at 4 s?

$10v - 10 \times 2 = 40$

$10v = 60$

$v = 6$ m/s

Solution:

(A) The impulse $F\Delta t$ is the area under the graph from 2 s to 4 s:

$$Area = F\Delta t = (20\text{ N})(4\text{ s} - 2\text{ s}) = 40\text{ N s}$$

(B) The impulse-momentum theorem states that the change in momentum of an object is equal to the impulse applied. Thus:

$$\text{change in momentum} = F\Delta t = 40\text{ N s}$$

(C) The impulse-momentum theorem states that

$$F\Delta t = m(v_f - v_i)$$

$$40\text{ N s} = (10\text{ kg})(v_f - 2\text{ m/s})$$

where v_f is the velocity at a time of 4 s.

Solving for v_f, we get $v_f = 6$ m/s.

CONSERVATION OF MOMENTUM IN ONE DIMENSION (动量守恒)

We've seen that if you want to change the momentum of an object or a system of objects, Newton's second law says that you have to apply an unbalanced force. This implies that if there are no unbalanced forces acting on a system, the total momentum of the system must remain constant. This is another way of stating Newton's first law, the law of inertia, discussed in chapter 5. If the total momentum of a system remains constant during a process, such as an explosion or collision, we say that the momentum is *conserved*. There are three typical examples of conservation of momentum questions on the SAT II: Physics Test: recoil, inelastic collision, and elastic collision.

Recoil (反冲)

0.03× 300 = 3 V
V = 3 m/s to the west

Example: A 3 kg rifle which is initially at rest contains a bullet with a mass of 0.03 kg. The bullet is fired from the rifle at a speed of 300 m/s toward the east. What is the recoil velocity of the rifle?

Solution: Since the rifle and bullet were motionless before the rifle was fired, the total momentum of the system is zero both before and after the rifle was fired, since all forces in this case are internal, and don't change the total momentum of the system. We write the momentum for each mass before and after the rifle is fired, and solve for the recoil velocity of the rifle. We will denote any velocity or momentum that occurs after a process with a *prime* (′).

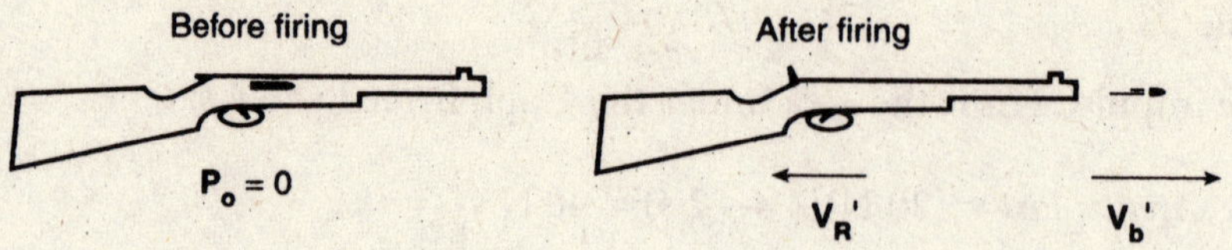

Momentum before firing = Momentum after firing

$$p_o = p_b' + p_R'$$

$$0 = m_b v_b' + m_R v_R'$$

$$0 = (0.03 \text{ kg})(300 \text{ m/s}) + (3\text{kg})v_R'$$

$$v_R' = -3 \text{ m/s}$$

Note that the recoil velocity of the rifle is negative, indicating that it moves in the opposite direction (west) of the bullet, as we would expect. Also note that the velocity of the rifle is one-hundredth of the velocity of the bullet, since the rifle is one hundred times more massive than the bullet. In other words, *the momentum of the bullet must be equal and opposite to the momentum of the rifle.*

We can generalize this result to an object which explodes and sends its fragments in several directions, as shown below. Again, the total vector momentum before the explosion must equal the total vector momentum after the explosion.

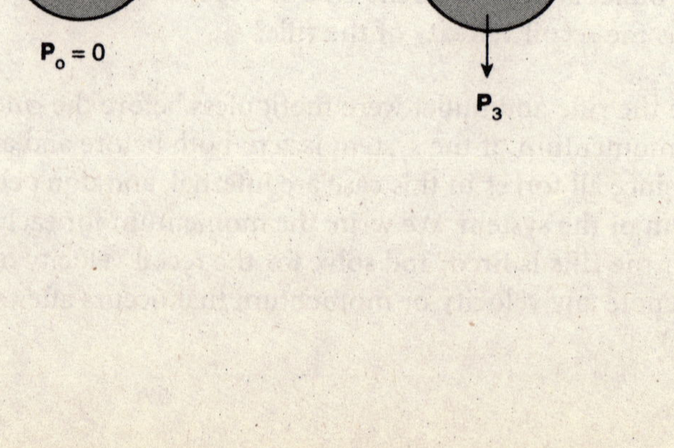

Inelastic Collision (非弹性碰撞)

An inelastic collision occurs when two colliding objects stick together after impact. Momentum is conserved in an inelastic collision, but kinetic energy is not.

There are three types of conservation of momentum problems: recoil, inelastic collision, and elastic collision.

Example: A car of mass 700 kg and moving at a speed of 30 m/s collides with a stationary truck of mass 1,400 kg, and the two vehicles lock together on impact. What is the combined velocity of the car and truck after the collision?

Solution: Once again, momentum is conserved in the collision.

Before the Inelastic Collision

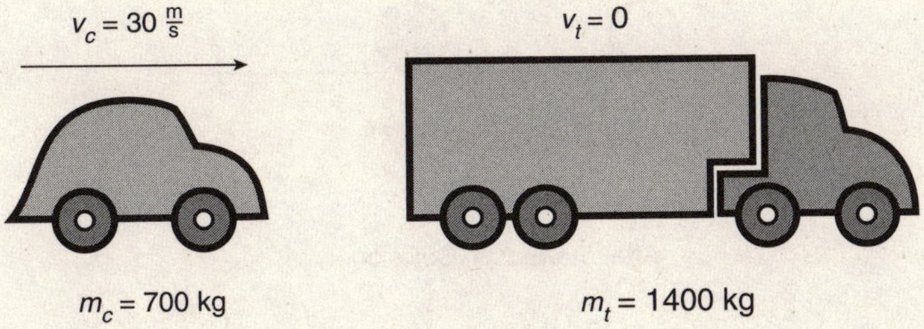

$700 \times 30 + 0 = (700 + 1400)v$

$v = 10$ m/s

After the Inelastic Collision

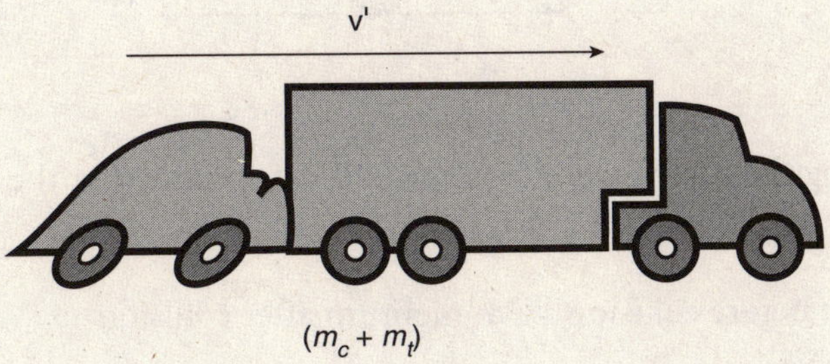

Momentum before collision = Momentum after collision

$p_c = p_{c+t}'$

$m_c v_c = (m_c + m_t)v'$

$(700 \text{ kg})(30 \text{ m/s}) = (700 \text{ kg} + 1{,}400 \text{ kg})v'$

$v' = 10$ m/s

Elastic Collision

In an elastic collision, the colliding objects bounce off each other, and momentum is conserved. As we shall see in the next chapter, kinetic energy is also conserved in an elastic collision.

> **Example:** A white pool ball of mass 0.5 kg moving at 10 m/s collides head-on with a 0.4 kg red pool ball initially at rest. After the collision, the white pool ball continues in the same direction with a velocity of 1.1 m/s. Neglecting friction, what is the velocity of the red ball after the collision?

Before the Elastic Collision

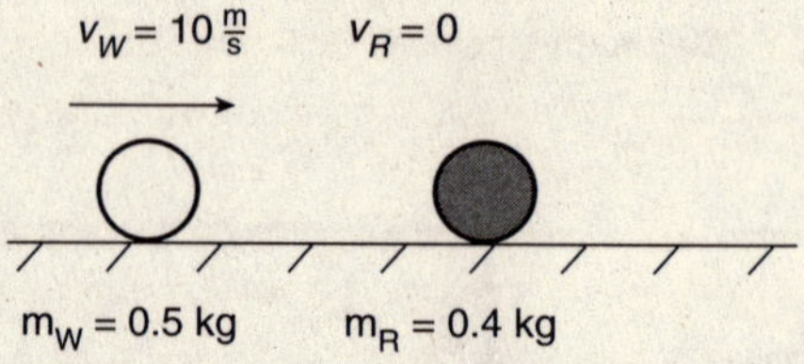

After the Elastic Collision

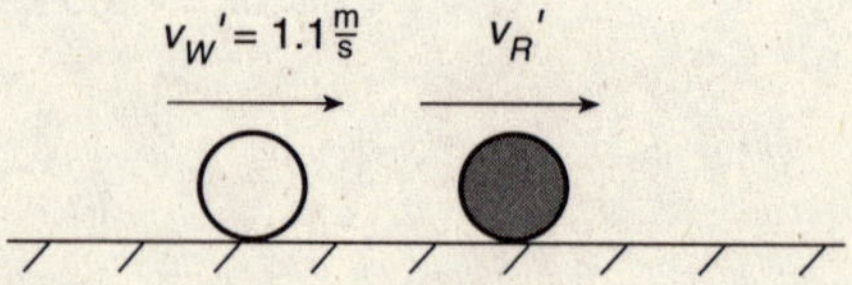

> **Solution:** Writing the momentum for each ball before and after the collision, we have:
>
> Momentum before collision = Momentum after collision
>
> $m_w v_w + m_r v_r = m_w v_w' + m_r v_r'$
>
> $(0.5\text{ kg})(10\text{ m/s}) + (0) = (0.5\text{ kg})(1.1\text{ m/s}) + (0.4\text{ kg})v_r'$
>
> $v_r' = 11.1$ m/s to the right

MOMENTUM AND IMPULSE REVIEW QUESTIONS

1. A hockey puck is sliding on the ice with a momentum of 5 kg m/s when it is struck by a hockey stick giving it an impulse of 100 N s in the direction of motion of the puck. Afterward, the momentum of the puck is

    (A) 500 kg m/s
    (B) 105 kg m/s
    (C) 100 kg m/s
    (D) 50 kg m/s
    (E) 20 kg m/s

2. A force acts on a mass as shown in the force vs. time graph.

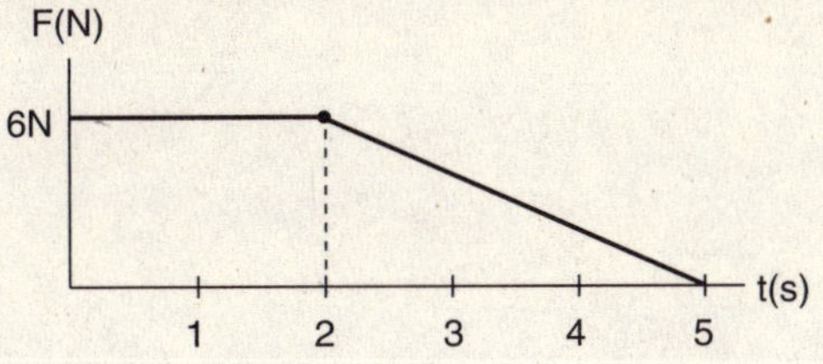

The change in momentum of the mass between 2 s and 5 s is

    (A) 9 kg m/s
    (B) 18 kg m/s
    (C) 21 kg m/s
    (D) 30 kg m/s
    (E) 42 kg m/s

3. A cannonball is fired from a cannon so that the cannon recoils backward as the ball is fired forward. Which of the following statements is true?

    (A) The velocity of the cannonball is equal and opposite to the velocity of the cannon.
    (B) The mass of the cannonball and the cannon must be equal.
    (C) The momentum of the cannonball must be greater than the magnitude of the momentum of the cannon.
    (D) The momentum of the cannon must be equal to the magnitude of the momentum of the cannonball.
    (E) The momentum of the cannon must be greater than the magnitude of the momentum of the cannonball.

4. A 5,000 kg railroad car collides and sticks to a stationary railroad car of mass 7,000 kg and they move off together with a speed of 5 m/s. What was the speed of the 5,000 kg car before the collision?

    (A) 5 m/s
    (B) 12 m/s
    (C) 14 m/s
    (D) 35 m/s
    (E) 60 m/s

5. Which of the following quantities are conserved in an elastic collision?

    (A) momentum only
    (B) momentum and potential energy
    (C) kinetic energy only
    (D) momentum and kinetic energy
    (E) momentum and velocity

6. A 1 kg pool ball moving at +10 m/s strikes a 2 kg pool ball which is initially at rest. Which of the following statements is true immediately after the elastic collision?

    (A) The 1 kg pool ball is at rest.
    (B) The 2 kg ball must have a negative velocity.
    (C) The 2 kg ball's speed must be less than 10 m/s.
    (D) The two pool balls stick together.
    (E) The total kinetic energy of the two pool balls must be equal to 100 J.

SOLUTIONS TO MOMENTUM AND IMPULSE REVIEW QUESTIONS

1. B

The amount of impulse added is equal to the change in momentum. Thus:

$$5\frac{kgm}{s} + 100 \text{ N s} = 105\frac{kgm}{s}$$

2. A

The area under the graph from 2 s to 5 s is equal to the impulse and therefore the change in momentum. The area under the triangle is

$$\tfrac{1}{2}(6 \text{ N})(3 \text{ s}) = 9\frac{kgm}{s}$$

3. D

Conservation of momentum tells us that the momentum of the cannonball must be equal and opposite to the momentum of the cannon.

4. B

For the inelastic collision, momentum before the collision equals the momentum after the collision:

$$m_1v_1 = (m_1 + m_2)v'$$

$$(5{,}000 \text{ kg})v_1 = (12{,}000 \text{ kg})(5 \text{ m/s}) \text{ yields } v_1 = 12 \text{ m/s}$$

5. D

By definition, both momentum and kinetic energy are conserved in an elastic collision.

6. C

Since the 1 kg ball retains some of its momentum after the collision and transfers some of its momentum to the 2 kg ball, the speed of the 2 kg ball must be less than the original speed of the 1 kg ball.

CHAPTER SEVEN

Work and Energy

WORK

As you know by now, there are many words in physics that may be used quite differently outside the context of a physics course. The concept of *work* is certainly one of these words. In physics, work is defined as the scalar product of force **F** and displacement **s**, that is,

$$W = \mathbf{F} \cdot \mathbf{s}$$

Here, the force and displacement vectors are multiplied together in such a way that the product yields a scalar. Thus, work is not a vector, and has no direction associated with it. Since work is the product of force and displacement, it has units of newton-meters, or *joules* (J). A joule is the work done by applying a force of one newton through a displacement of one meter. One joule is about the amount of work you do in lifting your calculator to a height of one meter.

> ***Work*** **is the scalar product of force and displacement.**

In the equation for work above, the scalar product of force and displacement can also be written as:

$$W = Fs \cos\theta$$

where θ is the angle between the applied force and the displacement. Let's look at a couple of examples.

Example: A box is pulled across a horizontal floor of negligible friction with a horizontal force of 40 N through a displacement of 10 m. How much work is done on the box?

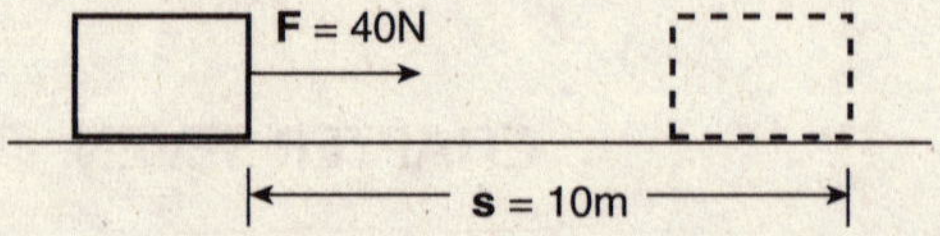

Solution: Since the force and the displacement are in the same direction, we can write:

$$W = Fs = (40\text{ N})(10\text{ m}) = 400\text{ J}$$

Example: Suppose the force applied to the box in the previous example were directed at 60° above the horizontal as shown in the figure below. If the box is pulled by this force through a displacement of 10 m, how much work is done on the box? (sin60 = cos30 = 0.87; sin30 = cos60 = 0.5)

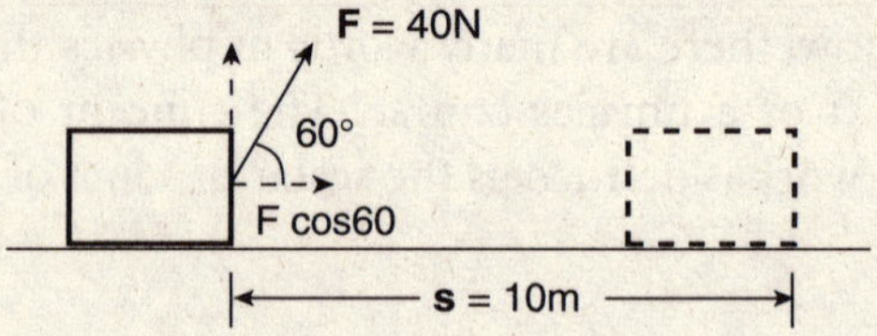

Solution: For work to be done, we must only consider the component of the force which lies in the direction of the displacement, namely $F\cos\theta$. Then the work done is:

$$W = Fs\cos 60 = (40\text{ N})(10\text{ m})(0.5) = 200\text{ J}$$

Quick Quiz:

A boy pushes on a stationary brick wall with a force of 50 N for a time of 15 seconds. How much work did he do on the wall?

a. 750 J b. 50 J c. 15 J d. none

Answer: d. none. If the wall is not displaced, there is no work done on it.

Thus, the person pulling the box across the floor only accomplishes half as much work by pulling up at an angle of 60°. This does not, however, mean that the person is only half as tired when he finishes the job; work is not a measure of how tired you are after performing the work. It is a measure of the product of the force that was applied in the direction of the displacement. Work is also a measure of the energy that was transferred while the force was being applied.

ENERGY (能)

Energy is the ability to do work.

Energy is the ability to do work, and when work is done, there is always a transfer of energy. Energy can take on many forms, such as potential energy, kinetic energy, and heat energy. The unit for energy is the same as the unit for work, the joule. This is because *the amount of work done on a system is exactly equal to the change in energy of the system.* This is called the *work-energy theorem.* Let's look at two examples of the work-energy theorem.

Potential Energy (势能) · [PE]

Potential energy is the energy a system has because of its position or configuration. When you stretch a rubber band, you store energy in the rubber band as elastic potential energy. When you lift a mass upward against gravity, you do work on the mass and therefore change its energy. The work you do on the mass gives the mass potential energy relative to the ground. To lift it, you must apply a force equal to the weight mg of the mass through a displacement height h.

Potential energy **is the potential for doing work.**

The work done in lifting the mass is:

$$W = Fs = (mg)h$$

which must also equal its potential energy:

$$PE = mgh$$

Example: A 5 kg bowling ball is lifted straight up to a shelf 2 meters above the floor. What is the work done on the bowling ball, and how much potential energy does the ball have when it is on this shelf?

PE = W = mgh = 5x10x2 = 100 J

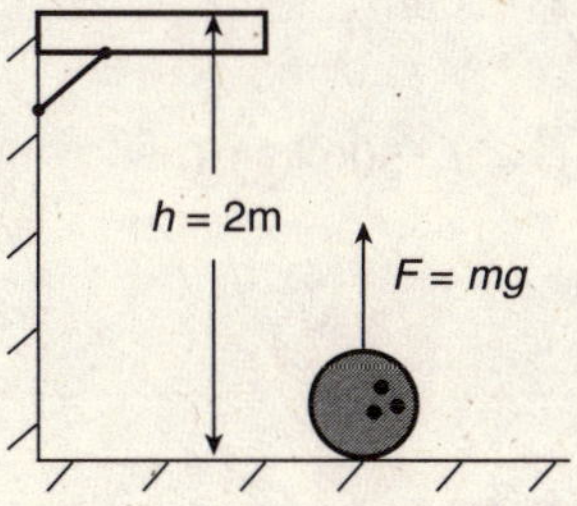

Solution: The work done in lifting the bowling ball to a height of 2 m is:

$$W = mgh = (5\text{ kg})(10\text{ m/s}^2)(2\text{ m}) = 100\text{ J}$$

The potential energy of the ball must also be 100 J relative to the floor, since all of the work done on the ball went into increasing its potential energy.

Kinetic Energy

Kinetic energy is energy of motion.

Kinetic energy is the energy an object has because it is moving. The kinetic energy of a moving object depends on its mass and the square of its velocity:

$$KE = \tfrac{1}{2}mv^2$$

But in order for a mass to gain kinetic energy, work must be done on the mass to push it up to a certain speed or to slow it down. The work-energy theorem states that the change in kinetic energy of an object is exactly equal to the work done on it, assuming there is no change in the object's potential energy.

Example: A 500 kg car is accelerated from rest by a force of 2,000 N for a displacement of 400 m. What is the speed of the car at the end of 400 m?

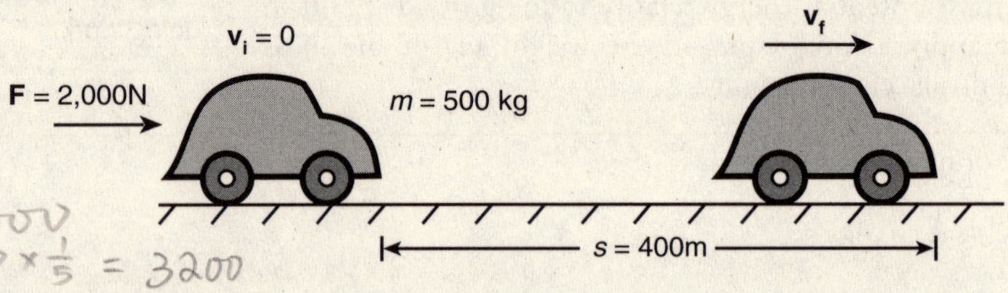

Solution: The work done on the car is equal to its change in kinetic energy. The work done is:

$$W = Fs = (2{,}000\ \text{N})(400\ \text{m}) = 800{,}000\ \text{J}$$

Thus, the 800,000 J is also equal to the change in kinetic energy of the car:

$$800{,}000\ J = \Delta KE = \tfrac{1}{2}mv_f^2 - \tfrac{1}{2}mv_i^2$$

$$800{,}000\ \text{J} = \tfrac{1}{2}(500\ \text{kg})\ v_f^2 - \tfrac{1}{2}(500\ \text{kg})(0)^2$$

Solving for v_f, we get 56.6 m/s.

CONSERVATION OF ENERGY

When work is done on a system, the energy of that system changes from one form to another, but the *total amount* of energy remains the same. We say that total energy is *conserved*, that is, remains constant during any process. This is also called the *law of conservation of energy*.

As an example, let's consider a 10-kg stone that sits just on the edge of a cliff 10 meters above the ground.

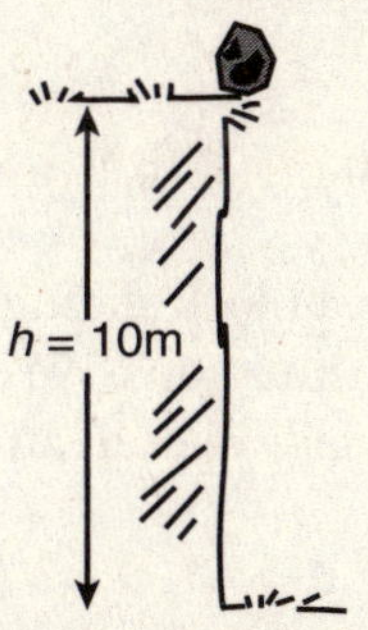

The potential energy of the stone relative to the ground is: PE=10 x 10 x 10 = 1000J

$$PE = mgh = (10 \text{ kg})(10 \text{ m/s}^2)(10 \text{ m}) = 1{,}000 \text{ J}$$

If the stone slips off the edge of the cliff and falls, the potential energy of the stone begins changing into kinetic energy as the stone gains speed while losing height above the ground. Both the stone's potential and kinetic energies change, but the total amount of energy is conserved. In other words, the sum of the potential energy and kinetic energy at any time is equal to the total energy, which remains constant throughout the fall. The chart below summarizes the changes in potential and kinetic energies as the stone falls.

***Conservation of energy* means that the total energy of a system remains constant.**

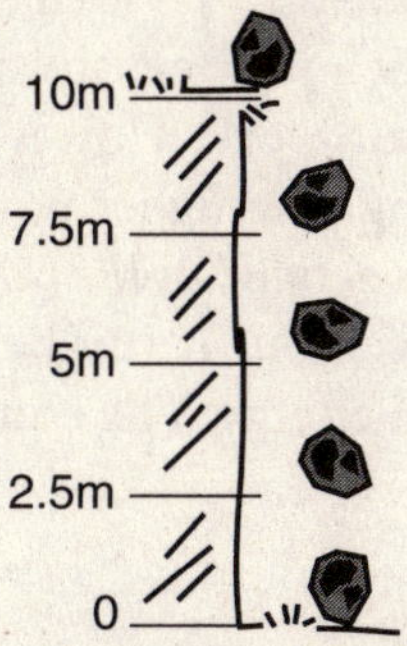

PE	*KE*	Total *E*
1,000J	0	1,000J
750J	250J	1,000J
500J	500J	1,000J
250J	750J	1,000J
0	1,000J	1,000J

We can use the fact that the sum of the potential and kinetic energies remains constant during free fall to solve for quantities such as speed.

Example: What is the speed of the stone in the example above just before it strikes the ground?

PE = 1000J   KE = 1000   ½mv² = 1000
v² = 200
v = 14.1 m/s

Solution: Conservation of energy states that the total energy at the top of the cliff equals the total energy at the bottom of the cliff.

$$PE_{top} + KE_{top} = PE_{bottom} + KE_{bottom}$$

But the KE at the top and the PE at the bottom are zero, so we are left with:

$$PE_{top} = KE_{bottom}$$

$$mgh = \tfrac{1}{2} m v_{bottom}^{2}$$

$$v = \sqrt{2gh} = \sqrt{2(10\ \text{m/s}^2)(10\ \text{m})} = 14.1\ \text{m/s}$$

These same principles can be applied to a block sliding down a frictionless ramp, a pendulum swinging from a height, and many other situations. We could use Newton's laws and kinematics to solve these types of problems, but usually conservation of energy is easier to apply.

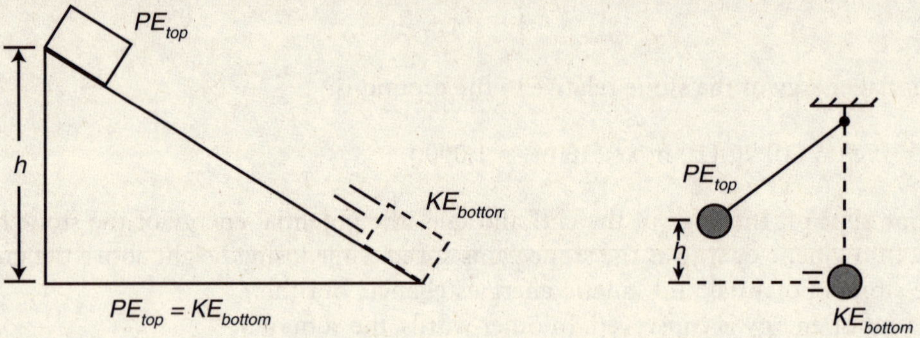

POWER (功率)

Work can be done slowly or quickly, but the time taken to perform the work doesn't affect the amount of work which is done, since there is no element of time in the definition for work. However, if you do the work quickly, you are operating at a higher power level than if you do the work slowly. *Power is defined as the rate at which work is done.* Often we think of electricity when we think of power, but it can be applied to mechanical work and energy as easily as it is applied to electrical energy. The equation for power is:

Power is the rate at which work is done.

$$P = \frac{Work}{time}$$

and has units of joules/second or watts (W). A machine is producing one watt of power if it is doing one joule of work every second. A 100 watt light bulb uses 100 joules of energy each second.

Example: A motor lifts a 100 kg crate onto a deck which is 5 meters high in a time of 10 seconds. How much power does the machine produce?

$P = \frac{100 \times 10 \times 5}{10} = 500\ W$

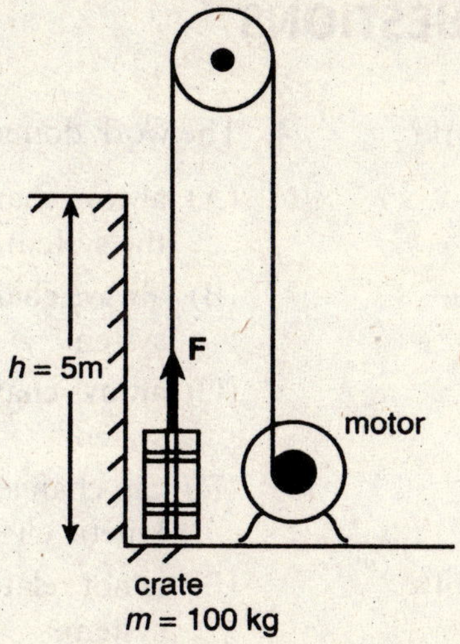

Solution: Power is work divided by time:

$$P = \frac{W}{t} = \frac{mgh}{t} = \frac{(100 \text{ kg})(10 \text{ m/s}^2)(5 \text{ m})}{10 \text{ s}} = 500 \text{ watts}$$

Any motor could do this same amount of work, but not necessarily as quickly, and therefore not at this 500 watt power level. For example, a small motor could lift the 100 kg crate if it were first broken into 1 kg pieces. Then the small motor could lift one piece at a time to the height if 5 meters and accomplish the same amount of work at a lower power level.

The letter W can stand for weight, work, or watt. Read the test questions carefully to be sure you know which one is being asked about in the question.

WORK AND ENERGY REVIEW QUESTIONS

1. Which of the following is NOT true of work?
   (A) It is the scalar product of force and displacement.
   (B) It is a vector that is always in the same direction as the force.
   (C) It is measured in joules.
   (D) It has the same units as energy.
   (E) It takes energy to perform work.

2. A 5 kg box is pushed across a level floor with a force of 50 N for a displacement of 10 m, then lifted to a height of 3 m. What is the total work done on the box?
   (A) 250 J
   (B) 500 J
   (C) 650 J
   (D) 2,500 J
   (E) 7,500 J

3. A 30 kg cart is pushed up the inclined plane shown by a force **F** to a height of 5 m.

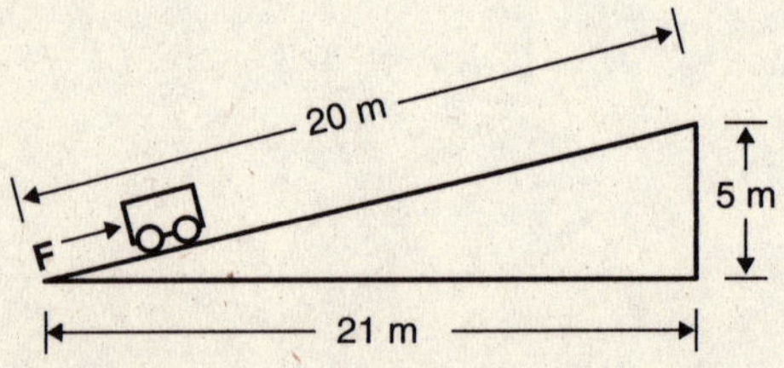

   What is the potential energy of the cart when it reaches the top of the inclined plane?
   (A) 1,500 J
   (B) 630 J
   (C) 600 J
   (D) 300 J
   (E) 150 J

4. The work done on a system
   (A) always changes the potential energy of the system.
   (B) always changes the kinetic energy of the system.
   (C) always changes the momentum of a system.
   (D) can change either the potential energy or kinetic energy of the system.
   (E) is not related to the energy of the system.

5. A ball falls from a height *h* from a tower. Which of the following statements is true?
   (A) The potential energy of the ball is conserved as it falls.
   (B) The kinetic energy of the ball is conserved as it falls.
   (C) The difference between the potential energy and kinetic energy is a constant as the ball falls.
   (D) The sum of the kinetic and potential energies of the ball is a constant.
   (E) The momentum of the ball is constant as the ball falls.

6. A 0.5 kg ball is dropped from a third story window that is 10 m above the sidewalk. What is the speed of the ball just before it strikes the sidewalk?
   (A) 5 m/s
   (B) 10 m/s
   (C) 14 m/s
   (D) 28 m/s
   (E) 200 m/s

7. Drew lifts a 50-kg crate onto a truck bed 1 meter high in 2 seconds. Perry lifts fifty 1 kg boxes onto the same truck in a time of 2 minutes. Which of the following statements is true?

   (A) Drew does more work than Perry does.
   (B) Perry does more work than Drew does.
   (C) They do the same amount of work, but Drew operates at a higher power level.
   (D) They do the same amount of work, but Perry operates at a higher power level.
   (E) Drew and Perry do the same amount of work and operate at the same power level.

8. A crane can lift a 600 kg mass to height of 20 m in 2 minutes. The power at which the crane is operating is

   (A) 1,000 watts
   (B) 6,000 watts
   (C) 60,000 watts
   (D) 240,000 watts
   (E) 1,440,000 watts

SOLUTIONS TO WORK AND ENERGY REVIEW QUESTIONS

1. B

Work is a scalar, not a vector.

2. C

$$\text{Total Work} = Fd + mgh = (50\text{N})(10\text{ m}) + (5\text{ kg})(10\text{ m/s}^2)(3\text{ m}) = 650\text{ J}$$

3. A

The potential energy is only equal to the work done against gravity.

$$PE = mgh = (30\text{ kg})(10\text{ m/s}^2)(5\text{ m}) = 1{,}500\text{ J}$$

4. D

Work done equals either a change in potential or kinetic energy, or sometimes both.

5. D

The total energy is the sum of the potential and kinetic energies, and must remain constant.

6. C

Potential energy at the top equals the kinetic energy at the bottom:

$$mgh = \tfrac{1}{2} mv^2$$

$$(0.5\text{kg})(10\text{ m/s}^2)(10\text{ m}) = \tfrac{1}{2}(0.5\text{ kg})v^2$$

$$v = 14\text{ m/s}$$

7. C

Drew and Perry do the same amount of work, but Drew does it faster and therefore operates at a higher power level.

8. A

$$P = \frac{Work}{time} = \frac{mgh}{t} = \frac{(600\text{ kg})(10\text{ m/s}^2)(20\text{ m})}{120\text{ s}} = 1{,}000\text{ watts}$$

CHAPTER EIGHT

Circular Motion and Rotation

UNIFORM CIRCULAR MOTION (匀速圆周运动) [UCM]

Up until this chapter, we've been reviewing motion in a straight line. The law of inertia states that if an object is moving, it will continue moving in a straight line at a constant velocity until a net force causes it to speed up, slow down, or change direction. Imagine a ball moving in a straight line in space, but tied to a string as shown in the figure below.

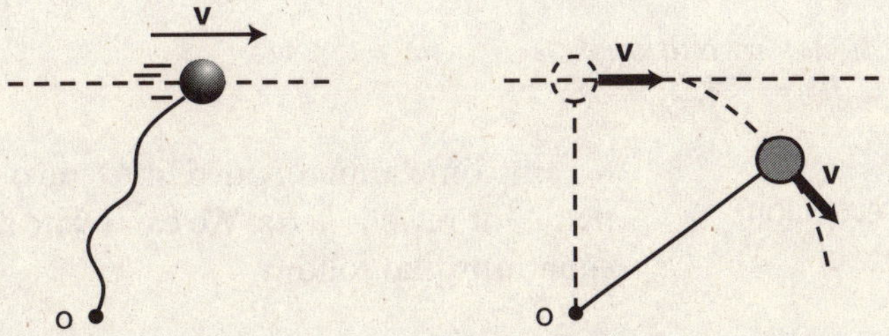

The ball will continue to move in a straight line until it reaches the end of the string, at which time it will be pulled to a central point O and begin moving in a circle. As long as the string is pulled toward this central point, the ball will continue moving in a circle at a constant speed. *An object moving in a circle at a constant speed is said to be in uniform circular motion (UCM).* Notice that even though the speed is constant, the velocity vector is not constant, since it is always changing direction due to the central force that the string applies to the ball, which we will call the *centripetal force* $\mathbf{F_c}$. *Centripetal* means "center-seeking."

向心力

向心力

Centripetal force is the center-seeking force that causes an object to follow a circular path.

There are three vectors associated with uniform circular motion: *velocity* (**v**), *centripetal force* ($\mathbf{F_c}$), and *centripetal acceleration* ($\boldsymbol{a_c}$). These vectors are drawn in the diagram below.

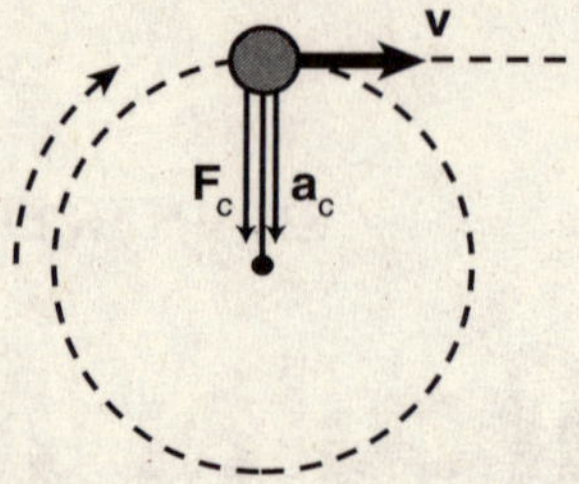

Notice that the velocity vector is tangent to the path of the ball and points in the direction the ball would move if the string were to break at that instant. The centripetal force and acceleration are both pointing toward the center. Although the centripetal force does not change the speed of the ball, it is always pulling the ball away from its inertial straight-line path and toward the center of the circle. Thus, the centripetal force accelerates the ball toward the center of the circle, constantly changing its direction.

> The three vectors associated with UCM are velocity, centripetal force, and centripetal acceleration.

The time it takes for the ball to complete one revolution is called the *period* (T). Since period is a time, we will measure it in seconds, minutes, hours, or even years. On the other hand, *frequency* (f) is the number of revolutions the ball makes per unit time. Units for frequency would include:

$$\frac{\textit{revolutions}}{\textit{second}},\ \frac{\textit{revolutions}}{\textit{hour}},$$

or any time unit divided into revolutions or cycles. Another name for rev/s is *hertz*. We can relate all of these quantities in the equations that follow.

> *Period* is the time for one revolution, and *frequency* is the number of revolutions per unit time.

The constant speed of any object can be found by:

$$v = \frac{d}{t}$$

For an object moving in a circle and completing one revolution, the speed of the object can be found by

$$v = \frac{\textit{circumference}}{\textit{period}} = \frac{2\pi r}{T},$$ where r is the radius of the circle,

and the centripetal acceleration is:

$$a_c = \frac{v^2}{r}$$

Newton's second law states that the net force and acceleration must be in the same direction and are related by the equation:

$$\boldsymbol{F}_{net} = m\boldsymbol{a} = \frac{mv^2}{r}$$

Some other examples of centripetal force are the gravitational force keeping a satellite in orbit, and friction between a car's tires and the road that causes the car to turn in a circle.

Example: A 100 kg racecar moves around a flat circular track of radius 50 m with a constant speed. The car makes one revolution around the track every 10 seconds. Find:

(A) the speed of the car

(B) the acceleration of the car

(C) the force between the tires and the road that keeps the car moving in a circle.

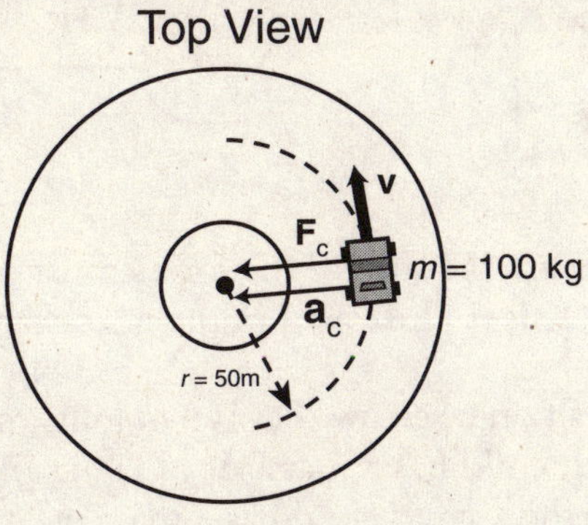

Solution:

(A) The speed of the car is:

$$v = \frac{circumference}{period} = \frac{2\pi r}{T} = \frac{2(3.14)(50\text{ m})}{10\text{ s}} = 31.4\text{ m/s}$$

(B) $a_c = \dfrac{v^2}{r} = \dfrac{(31.4\text{ m/s})^2}{50\text{ m}} = 19.7\text{ m/s}^2$

(C) The frictional force between the tires and the road provides the centripetal force:

$$F_c = ma_c = (100\text{ kg})(19.7\text{ m/s}^2) = 1{,}970\text{ N}$$

Quick Quiz:

The centripetal force acting on a ball in UCM pulls the ball inward toward the center. What keeps the ball from moving all the way to the center of the circle?

a. inertia

b. centrifugal force

c. centripetal force

d. gravity

Answer: a. inertia. The ball wants to follow a straight line because of its inertia, but the tension in the string pulls it inward away from its straight-line path.

Centripetal force pulls an object toward the center of its circular path; *centrifugal* force means "center-fleeing," and is not a real force, but the perceived effect of inertia.

TORQUE [扭矩]

Torque is the result of a force acting at a distance from a rotational axis, and may cause a rotation about the axis only if there is a component of the force perpendicular to the radius. Torque is the product of the perpendicular component of the force ($F_{\perp}$) and the distance (r) (as in radius) from the rotational axis:

$$\text{Torque} = F_{\perp}r$$

The unit for torque is the *newton-meter*. To open a hinged door, you apply a force perpendicular to the door at the doorknob, which is mounted a certain distance from the hinges, and create a torque that causes the door to rotate. When you use a wrench to tighten a bolt, you apply a force at the end of the wrench and perpendicular to it to get the most torque to turn the bolt.

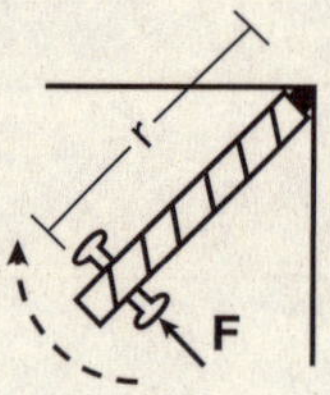

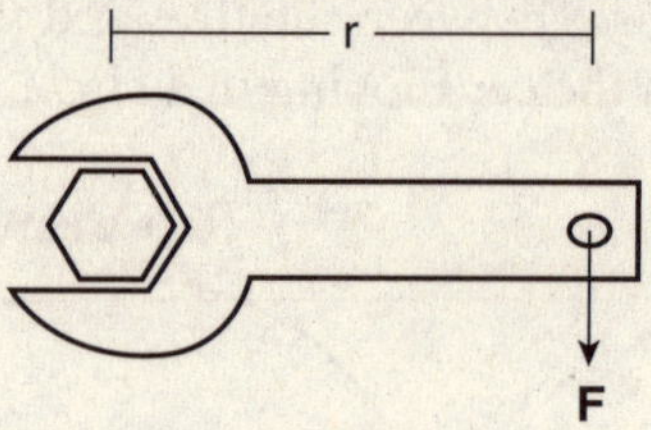

For a system in static equilibrium, the sum of the forces must equal zero, and the sum of the torques must also equal zero. This is illustrated in the next example.

Example: Two children sit on a seesaw that is 9 m long and pivoted on an axis at its center. The first child has a mass m_1 of 20 kg and sits at the left end of the seesaw, while the second child has a mass m_2 of 40 kg and sits somewhere on the seesaw to the right of the axis. At what distance r_2 from the axis should the second child sit to keep the seesaw horizontal?

$20 \times 10 \times 4.5 = 40 \times 10 \times r_2$

$r_2 = 2.25\text{m}$

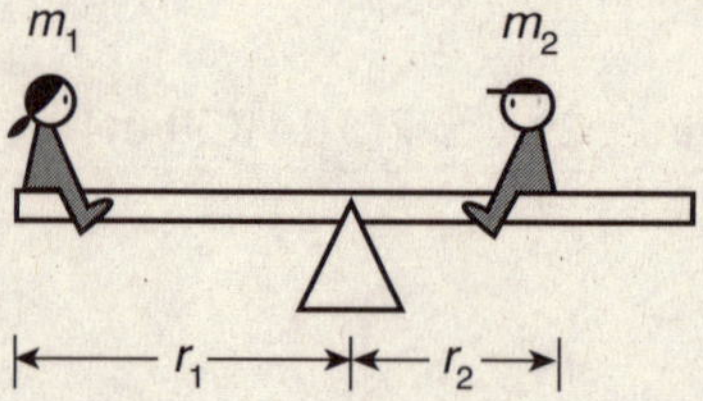

Solution: For the seesaw to remain horizontal, the torque on the left must equal the torque on the right. The forces acting on the seesaw on either side are just the weight mg of each child. So:

torque on the left = torque on the right

$F_1r_1 = F_2r_2$,

where each force is applied perpendicular to the seesaw board,

$(m_1g)r_1 = (m_2g)r_2$

Canceling the g's and solving for r_2 we get:

$$r_2 = \frac{m_1 r_1}{m_2} = \frac{(20\text{ kg})(4.5\text{ m})}{40\text{ kg}} = 2.25\text{ m to the right of the axis.}$$

Could you have guessed this answer before we worked it out? Since the child on the right is twice as heavy as the child on the left, he should sit half as far from the axis on the right side to balance the torque the lighter child is producing on the left side.

ANGULAR MOMENTUM (角动量) [L]

In an earlier chapter, we studied linear momentum (p), which is the product of the mass of an object and its velocity. For an object moving in a curved path such as a circle, we can also define its *angular momentum*. Consider a ball on the end of a string which is being swung in a circular path. Its angular momentum (L) is defined as the product of its mass, velocity, and radius of orbit:

$$L = mvr$$

We are interested in angular momentum because, as long as there are no external torques, the quantity mvr is conserved. For example, consider a ball on the end of a string which is passed through a glass tube and swung in a horizontal circle at a large radius R. If we pull the string through the tube to shorten the radius to r, the speed must increase to make up for the loss in radius to conserve angular momentum:

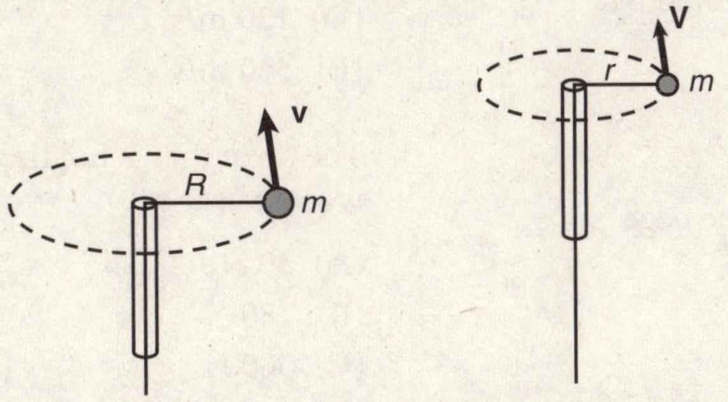

angular momentum before = angular momentum after

$$mvR = mVr$$

We will revisit the law of conservation of angular momentum when we review gravitation and satellite orbits in chapter 10.

CIRCULAR MOTION AND ROTATION REVIEW QUESTIONS

Questions 1–2:

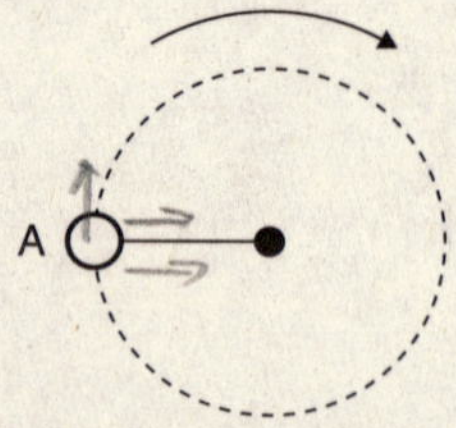

1. A ball on the end of a string is swung in a horizontal circle, rotating clockwise as shown. When the ball is at point A, the direction of the velocity, centripetal force, and centripetal acceleration vectors, respectively, are

   (A) → ← ←
   (B) ← → →
   (C) ↑ → ←
   (D) ↑ → →
   (E) ↓ → →

2. If the string were suddenly cut when the ball is at point A in the figure above, the subsequent motion of the ball would be

   (A) to move to the right.
   (B) to move to the left.
   (C) to move to the top of the page.
   (D) to move to the bottom of the page.
   (E) to move up and to the left.

Questions 3–5:
A 30 kg child sits on the edge of a carnival ride at a radius of 2 m. The ride makes 2 revolutions in 4 s.

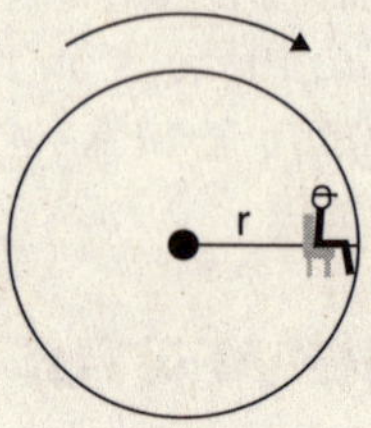

3. The period of revolution for this ride is

   (A) ½ rev/s
   (B) ½ s
   (C) 2 rev/s
   (D) 2 s
   (E) 4 s

4. The speed of the child is most nearly

   (A) 4 m/s
   (B) 6 m/s
   (C) 24 m/s
   (D) 120 m/s
   (E) 360 m/s

5. The force that is holding the child on the ride is most nearly

   (A) 30 N
   (B) 180 N
   (C) 300 N
   (D) 540 N
   (E) 4,320 N

6. Torque
   - (A) is the vector product of force and lever arm displacement.
   - (B) is a scalar and has no direction associated with it.
   - (C) is always equal to force.
   - (D) is always greater for shorter lever arms.
   - (E) must always equal zero.

7. Two blocks of mass 3 kg and 4 kg hang from the ends of a rod of negligible mass marked in seven equal parts.

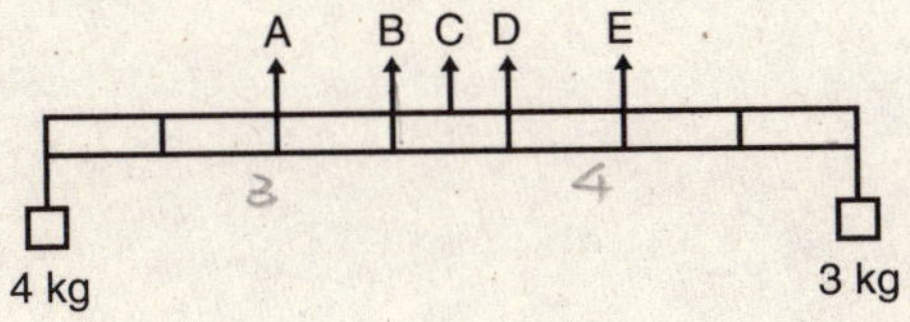

At which of the points indicated should a string be attached if the rod is to remain horizontal when suspended from the string?

- (A) A
- (B) B
- (C) C
- (D) D
- (E) E

8. A ball is tied to a string that passes through a glass tube and is attached to a hanging mass.

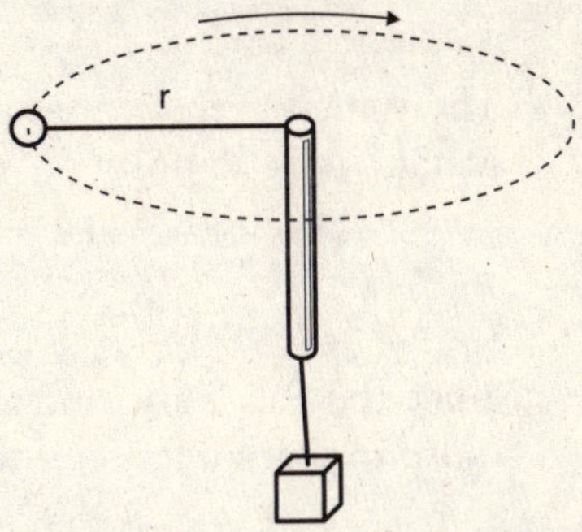

The tube is held vertically as the ball is swung in a horizontal circle at a radius *r* and speed *v*. If the string is pulled through the tube by hanging a larger mass on the string so that the radius becomes 1/2 *r*, the speed of the ball will become

- (A) $\frac{1}{2} v$
- (B) $2v$
- (C) $\frac{1}{4} v$
- (D) $4v$
- (E) v (unchanged)

SOLUTIONS TO CIRCULAR MOTION AND ROTATION REVIEW QUESTIONS

1. D

The velocity vector points tangent to the circle in the direction that the ball is moving the instant it is at point A, which is toward the top of the page. The centripetal force and acceleration both point toward the center of the circle.

2. C

Since the ball is moving toward the top of the page at the instant it is at point A, it would move in that direction if the string were to be cut there.

3. D

The period is $\frac{4\text{s}}{2\text{rev}} = 2$ s.

4. B

$$v = \frac{2\pi r}{T} = \frac{2\pi(2\text{ m})}{2\text{ s}} = 6\text{ m/s}$$

5. D

The force holding the child on is the centripetal force.

$$F = \frac{mv^2}{r} = \frac{(30\text{ kg})(6\text{ m/s})^2}{2\text{ m}} = 540\text{ N}$$

6. A

Torque is a vector and it is the product of force and lever arm displacement.

7. B

The string must be hung a little closer to the 4 kg mass than the 3 kg mass so that the torque to the left of the string will equal the torque to the right of the string. We can write (3 kg)(4 length units) = (4 kg)(3 length units).

8. B

According to conservation of angular momentum, if the radius is halved the speed will double.

CHAPTER NINE

Vibrations

Another type of motion commonly found in nature is *vibrational motion*, which we sometimes call *harmonic motion*. An object is in harmonic motion if it follows a repeated path at regular time intervals. Two common examples of harmonic motion often studied in physics are a mass on a spring and a pendulum.

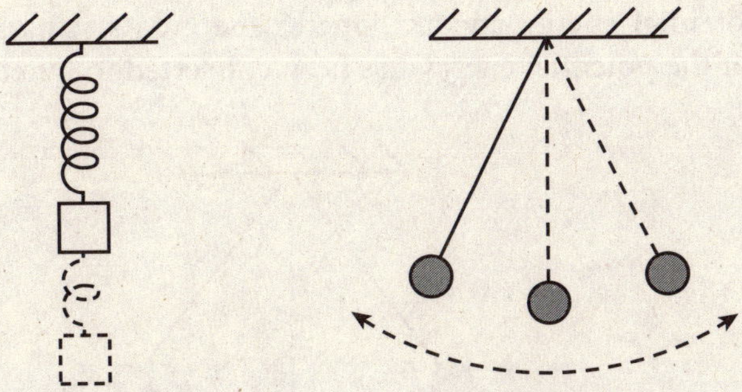

Mass on a spring   Pendulum

THE DYNAMICS OF HARMONIC MOTION

As an object vibrates, it has both a *period* and a *frequency*. Recall that the period of vibration is the time it takes for one complete cycle of motion (i.e., the time it takes for the object to return to its original position). For a pendulum, this would be the time for one complete swing, up and back. The frequency is the number of cycles per unit time, such as cycles per second, or hertz. The period and frequency of a pendulum depend only on the length of the pendulum and the acceleration due

> An object is moving with *harmonic motion* if it follows a repeated path at constant time intervals.

to gravity. The lowest point in the swing of a pendulum is called the *equilibrium position*, and the maximum displacement from equilibrium is called the *amplitude*. The amplitude can be measured by the maximum angle or by the linear horizontal distance from equilibrium, as shown below.

amplitude 振幅

restoring force 回复力

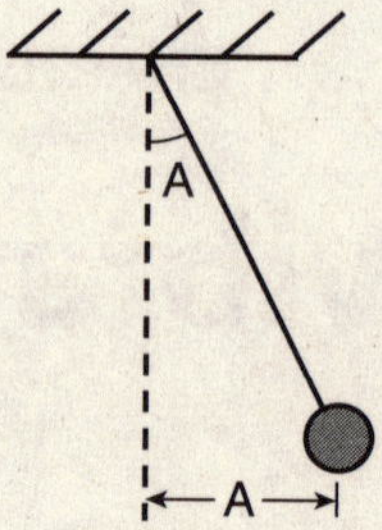

Since the pendulum vibrates about the equilibrium position, there must be a force that is trying to restore it back toward the center of the swing. This force is called the *restoring force*, and it is greatest at the amplitude and zero as the pendulum passes through the equilibrium position. Newton's second law tells us that if there is a net force, there must be an acceleration, and if the force is maximum at the amplitude, the acceleration must be maximum at the amplitude as well. The velocity, however, is zero at the amplitude and maximum as it passes through the equilibrium position. It might be helpful to think of the swing of a pendulum in terms of energy. At maximum amplitude, the pendulum mass has maximum potential energy and no kinetic energy. As it passes through the equilibrium position, all of the potential energy has been converted to kinetic energy.

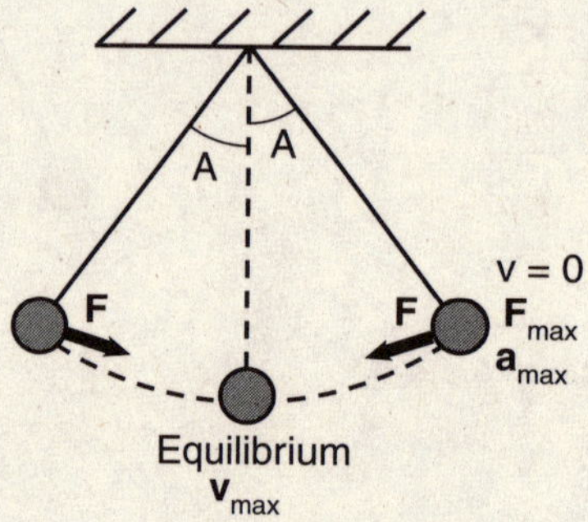

We can apply these same concepts to a mass vibrating at the end of a spring. On the figure below, we've labeled the equilibrium position, the amplitude, maximum force and acceleration, and maximum velocity.

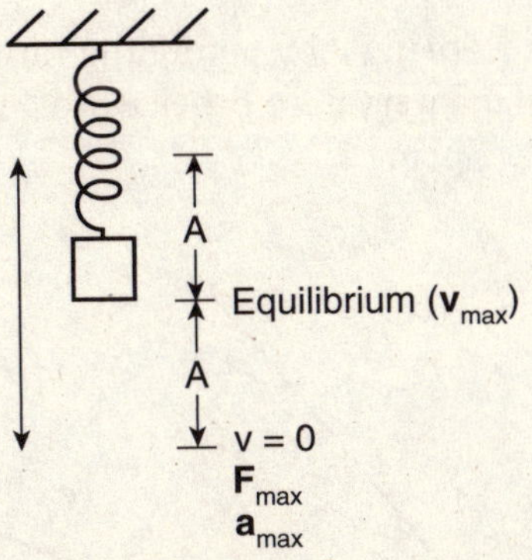

> **Quick Quiz:**
> Which of the following occurs at the same time for a swinging pendulum?
>
> maximum force, maximum velocity, maximum acceleration, amplitude, equilibrium position
>
> Answer: The maximum force and acceleration occur when the pendulum is at its amplitude, and the maximum velocity occurs at the equilibrium position.

The period and frequency of a mass vibrating on a spring depend on the stiffness in the spring. For a stiffer spring, it takes more force to stretch the spring to a particular length. The amount of force needed per unit length is called the *spring constant* (k), measured in newtons per meter. The relationship between force, stretched length, and k for an ideal (or linear) spring is called *Hooke's law*:

$$F_{spring} = kx$$

where x is the stretched length of the spring. For an ideal spring, the stretch is proportional to the force. If we pull with twice the force, the spring will stretch twice as far.

Example: The graph below represents the force F vs. stretched length x for two springs:

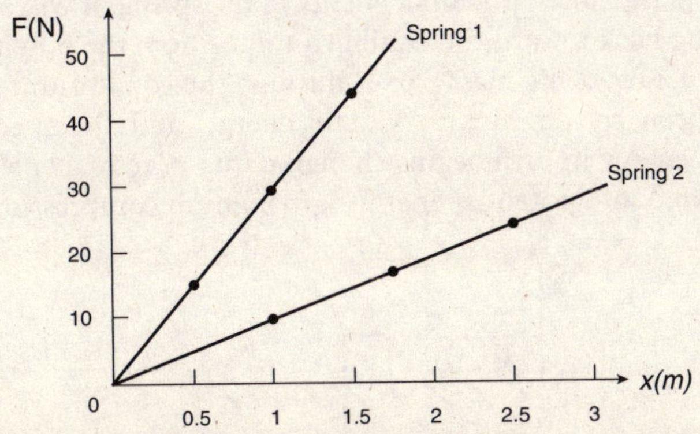

(A) Which is the stiffer spring? Explain.

(B) What is the spring constant for Spring 2?

Solution:

(A) Spring 1 is the stiffer spring, since it takes more force to stretch it the same length as Spring 2.

(B) We can find the spring constant k for Spring 2 by taking the ratio of the force to the stretch for a particular interval. In other words, we can find the slope of the F vs. x graph for Spring 2.

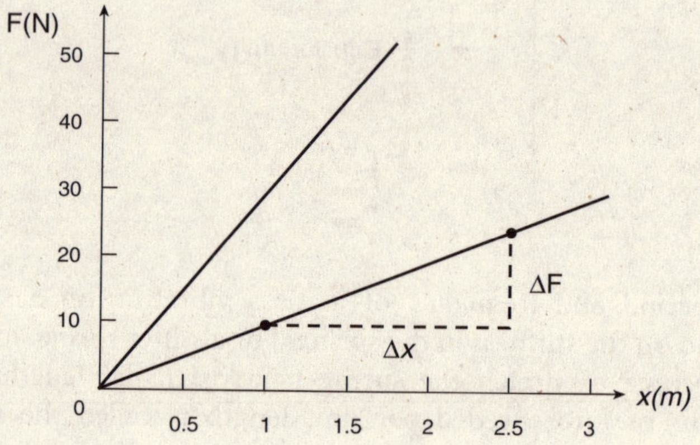

A linear spring follows *Hooke's law*: The force is proportional to the distance the spring is stretched.

$$k = \frac{25\text{ N} - 10\text{ N}}{2.5\text{ m} - 1\text{ m}} = 10\frac{\text{N}}{\text{m}}$$

ENERGY CONSIDERATIONS IN HARMONIC MOTION

As an object vibrates in harmonic motion, energy is transferred between potential energy and kinetic energy. Consider a mass sitting on a surface of negligible friction and attached to a linear spring. If we stretch a spring from its equilibrium (unstretched) position to a certain displacement, we do work on the mass against the spring force. By the work-energy theorem, the work done is equal to the stored potential energy in the spring. If we release the mass and allow it to begin moving back toward the equilibrium position, the potential energy begins changing into kinetic energy. As the mass passes through the equilibrium position, all of the potential energy has been converted into kinetic energy, and the speed of the mass is maximum. The kinetic energy in turn begins changing into potential energy, until all of the kinetic energy is converted into potential energy at maximum compression.

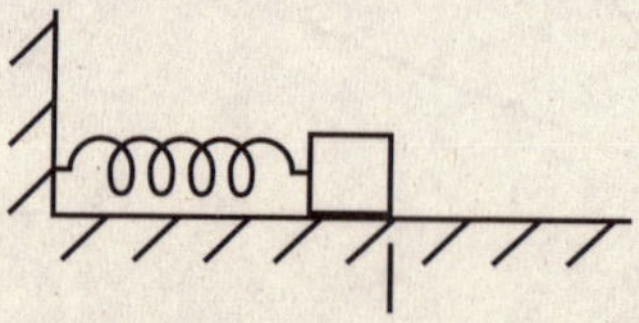

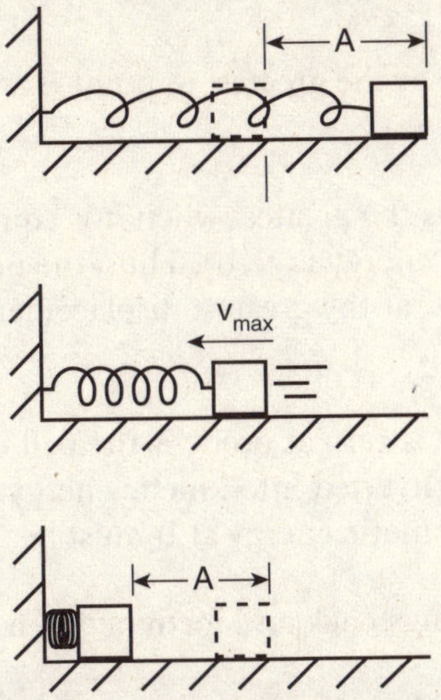

The compressed spring then accelerates the mass back through the equilibrium to the original starting position, and the entire process repeats itself. If we neglect friction on the surface and in the spring, the total energy of the system remains constant; that is:

total energy = potential energy + kinetic energy = a constant

Thus, whatever potential energy is lost must be gained by kinetic energy, and vice versa.

> **Example:** Consider the mass suspended on a spring. The mass vibrates between levels A and C, and level B is halfway between A and C. Assume there is no loss of energy due to friction as the mass oscillates, and the potential energy at point B is zero.

The *total energy* of a mass on a spring or a pendulum is the sum of the potential and kinetic energies.

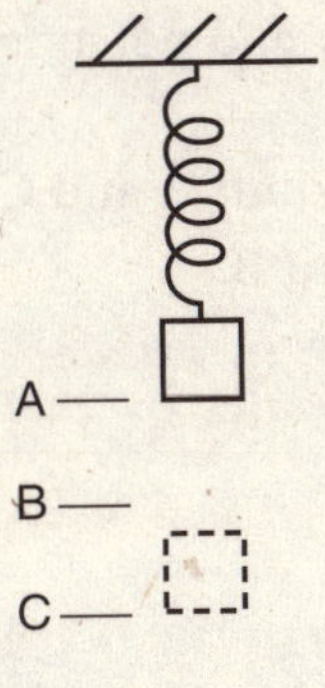

(A) At which point(s) is the speed of the mass the greatest?
(B) At which point(s) is the potential energy the greatest?
(C) If the potential energy at point A is 20 joules, what is the kinetic energy at point B?

Solution:

(A) The speed of the mass is the greatest at point B as it passes through the equilibrium position.

(B) The potential energy is the greatest when it is equal to the total energy (i.e., when the kinetic energy is zero). Thus, the potential energy is zero at points A and C, at the greatest displacement from the equilibrium position.

(C) If the potential energy is zero at point B, then all of the potential energy at point A is converted into kinetic energy as the mass passes through B. Thus, the kinetic energy at B must be 20 joules.

Example: A pendulum swings back and forth between points A and C as shown below.

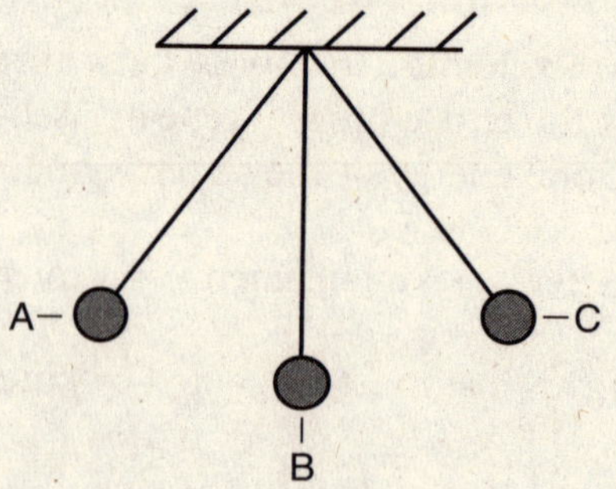

Which of the following statements are true?

(A) The potential energy at A is equal to the kinetic energy at C.

(B) The kinetic energy at B is equal to the total energy of the pendulum.

(C) The pendulum has both kinetic energy and potential energy at point A.

(D) The potential energy is maximum at points A and C.

(E) The kinetic energy is maximum at point B.

Solution:

(A) False. Since the mass is at rest at points A and C, there is no kinetic energy at A and C.

(B) True. Since there is no potential energy at point B, all of the energy of the pendulum must be kinetic at that point.

(C) False. The pendulum has no kinetic energy at point A, since it is momentarily at rest at that point.

(D) True. Since points A and C are at the amplitude of the swing of the pendulum, the potential energy of the pendulum is maximum at these points.

(E) True. The pendulum accelerates as it falls from point A to point C, and reaches its maximum speed at point C, and therefore its maximum kinetic energy.

VIBRATIONS REVIEW QUESTIONS

1. According to Hooke's law for a mass vibrating on an ideal spring, doubling the stretch distance will

   (A) double the velocity of the mass.
   (B) double the force that the spring exerts on the mass.
   (C) quadruple the force the spring exerts on the mass.
   (D) double the period.
   (E) double the frequency.

2. For an ideal spring, the slope of a force vs. displacement graph is equal to

   (A) the work done by the spring.
   (B) the amplitude.
   (C) the period.
   (D) the frequency.
   (E) the spring constant.

3. Consider the force vs. displacement graph shown for an ideal spring.

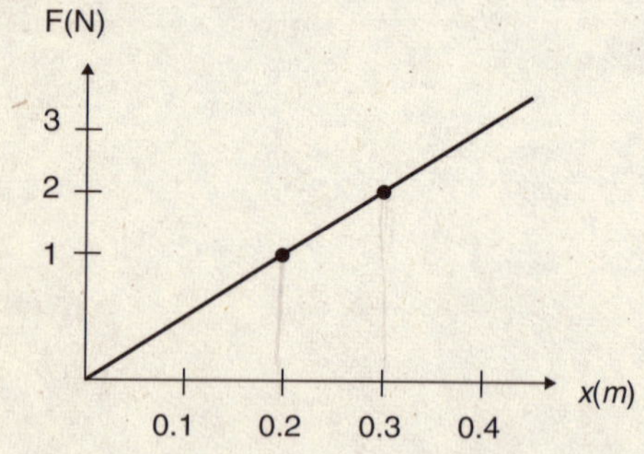

   The work done in stretching the spring from 0.2 m to 0.3 m is

   (A) 1 J
   (B) 10 J
   (C) 0.15 J
   (D) 1.5 J
   (E) 15 J

4. A pendulum swings with an amplitude θ as shown.

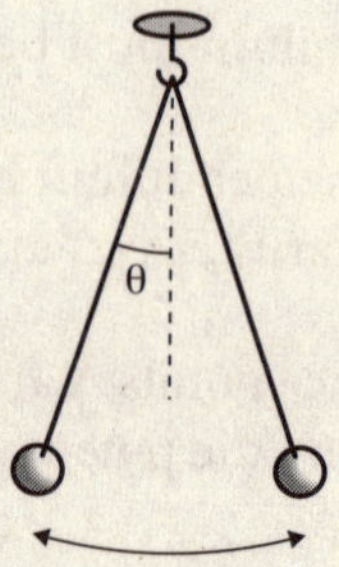

   If the amplitude is increased and the pendulum is released from a greater angle,

   (A) the period will decrease.
   (B) the period will increase.
   (C) the period will not change.
   (D) the frequency will increase.
   (E) the frequency will decrease.

5. A mass on an ideal spring vibrates between points A and E as shown.

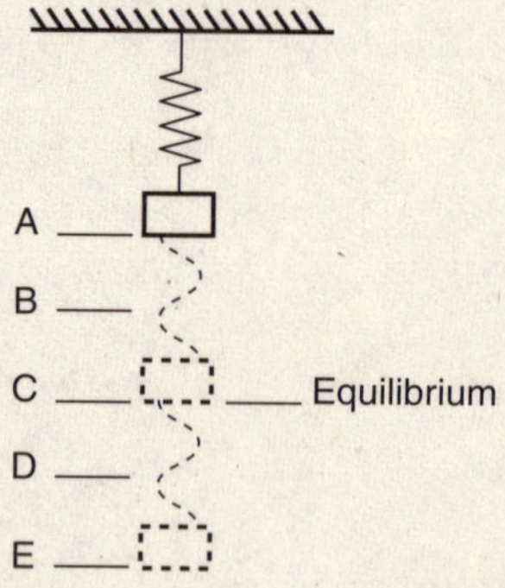

   At which point is the acceleration of the mass the greatest?

   (A) A
   (B) B
   (C) C
   (D) D
   (E) The acceleration is the same at all points.

D 6. A mass vibrates on an ideal spring as shown.

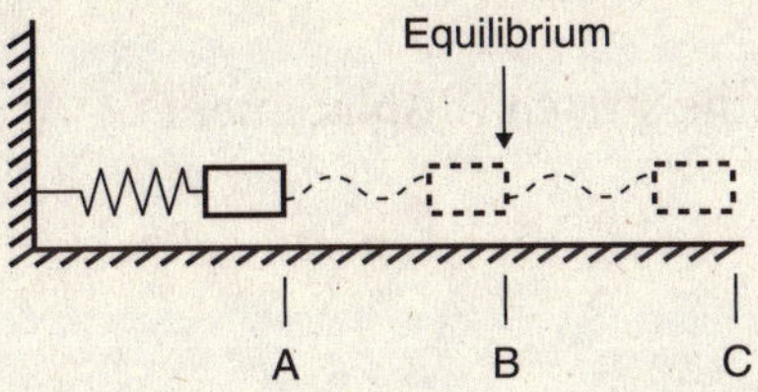

The total energy of the spring is 100 J. What is the kinetic energy of the mass at point B?

(A) 25 J
(B) 50 J
(C) 75 J
(D) 100 J
(E) 200 J

SOLUTIONS TO VIBRATIONS REVIEW QUESTIONS

1. B

Hooke's law states that the force is proportional to the stretch (displacement).

2. E

$$slope = \frac{\Delta F}{\Delta x} = k$$

3. C

The work done by the spring is equal to the area under the *F* vs. *x* graph from 0.2 m to 0.3 m, which is:

$$(1)(0.1) + \tfrac{1}{2}(1)(0.1) = 0.15 \text{ J}$$

4. C

The period (and thus the frequency) is not affected by a change in amplitude, only a change in length or gravity.

5. A

The acceleration is the greatest at the same point that the restoring force is the greatest, at the amplitude.

6. D

At point B, the equilibrium position, all of the energy is kinetic (100 J).

CHAPTER TEN

Gravity

NEWTON'S LAW OF UNIVERSAL GRAVITATION

Gravitational Force

Newton's law of universal gravitation states that every mass in the universe attracts every other mass in the universe. The gravitational force between two masses is proportional to the product of the masses and inversely proportional to the square of the distance between their centers. The equation describing the gravitational force is:

$$F_G = \frac{Gm_1m_2}{r^2}$$

Gravitational force is proportional to the product of the masses and inversely proportional to the square of the distance between their centers.

where F_G is the gravitational force, m_1 and m_2 are the masses in kilograms, and r is the distance between their centers. The constant G simply links the units for gravitational force to the other quantities, and in the metric system happens to be equal to 6.67×10^{-11} Nm^2/kg^2. The questions on the SAT II: Physics Test will typically not use this constant, but will instead ask you about the proportionalities involved in the relationship between force, mass, and separation distance.

Example: Two planets each of mass m and separated by a distance r are attracted to each other by a gravitational force F_1. If the mass of one of the planets is doubled, and the distance between the planets is also doubled, what will be the new force between the masses in terms of F_1?

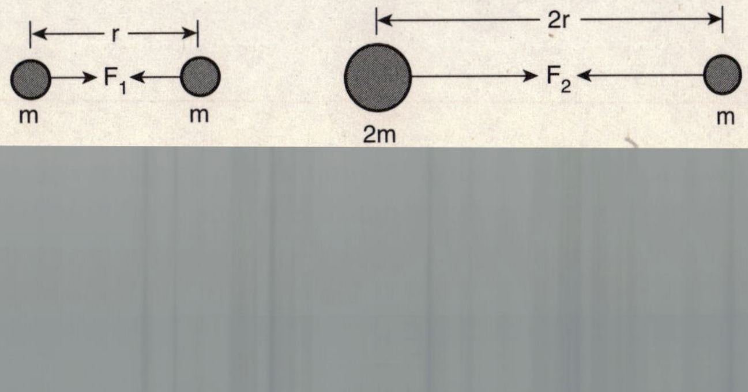

Solution: According to the equation for Newton's law of gravitation:

$$F_1 = \frac{Gmm}{r^2}$$

So the new force between the masses is:

$$F_2 = \frac{G(2m)(m)}{(2r)^2} = \frac{2Gmm}{4r^2}$$

So the new force F_2 is only half as much as the original force F_1. Note that regardless of the masses of the two objects, they always exert equal and opposite forces on each other.

Gravitational Acceleration

We've been treating the acceleration due to gravity as if it were a constant as long as we stay near the surface of the Earth. But if we travel an appreciable distance from the center of the Earth, such as twice the Earth's radius, the acceleration due to gravity is changed significantly. Like the gravitational force, the *gravitational acceleration* is inversely proportional to the square of the distance between the centers of the two masses.

Gravitational acceleration is proportional to gravitational force (Newton's second law).

Example: A ball is dropped near the surface of the Earth and accelerates toward the Earth with an acceleration of 10 m/s^2. With what initial acceleration will the ball fall if it is dropped from a height of three Earth radii from the center of the Earth?

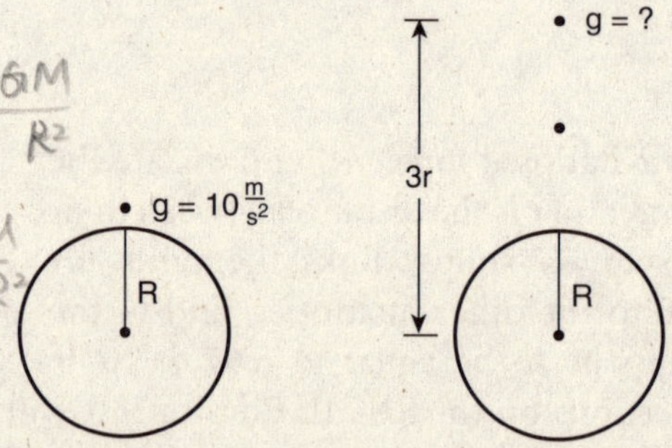

Solution: Since the acceleration due to gravity is inversely proportional to the square of the distance from the center of the Earth, then we can write:

$$g \propto \frac{1}{r^2}$$

where the ∝ sign means "proportional to." If the ball is dropped at a height of three earth radii, we have:

$$g \propto \frac{1}{(3r)^2} = \frac{1}{9r^2}$$

Thus, the acceleration due to gravity at three earth radii is 1/9 as much as it is near the earth's surface (one earth radius). The value for g is then

$$\frac{1}{9}(10 \text{ m/s}^2) = 1.1 \text{ m/s}^2$$

at this height.

ORBITING SATELLITES (卫星轨道)

Both the total energy and angular momentum of a satellite remain constant.

As a satellite such as the moon orbits the earth, it is pulled toward the earth with a gravitational force that is acting as a centripetal force. As described in chapter 8, the inertia of the satellite causes it to tend to follow a straight-line path, but the centripetal gravitational force pulls it toward the center of the orbit. Although we often approximate the orbits of satellites around the earth (or sun, in the case of the orbits of the planets) as circular, they are actually elliptical, with the earth (or sun) located at one focus of the ellipse.

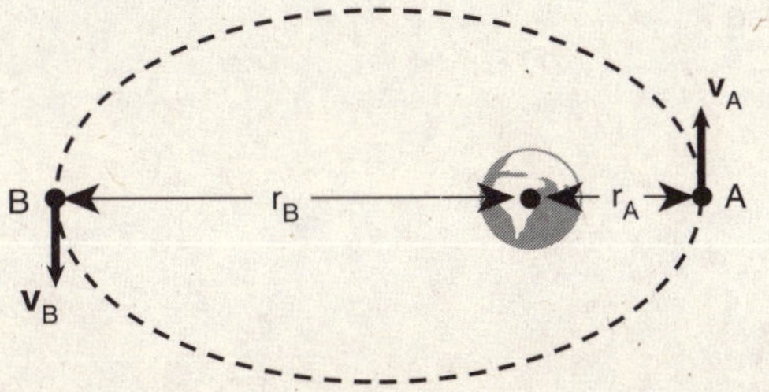

Comets follow highly elliptical orbits around the sun. Halley's comet takes only a few months to get around the sun on its closest approach, but takes more than seven decades to reach its farthest point from the sun and then return.

This means that there are times when the satellite is closer to the earth than other times. If the satellite is at point A in the figure above, the gravitational force acting on the satellite is greater than at point B, and the speed at A is greater than at B. This is consistent with the law of conservation of angular momentum discussed in chapter 8:

angular momentum at A = angular momentum at B

$$mv_A r_A = mv_B r_B$$

This equation shows that the closer the satellite gets to the planet (smaller r), the faster it goes (larger v). As the satellite orbits, potential and kinetic energy each change, but total energy (potential plus kinetic) remains constant.

Example: A satellite orbiting the earth moves from point A to point B in an elliptical orbit as shown below.

Fill in the table below with *increases*, *decreases*, or *constant* to describe the quantities related to the orbit.

Quantity	**Increases, decreases, constant**
Speed	decreases
Angular momentum	constant
Kinetic energy	decreases
Total energy	constant
Gravitational force	decreases

Solution:

Quantity	**Increases, decreases, constant**	**Explanation**
Speed	Decreases	The satellite is moving farther away from the planet and slowing down.
Angular momentum	Constant	Conservation of angular momentum
Kinetic energy	Decreases	Since the speed is decreasing, the kinetic energy must also be decreasing.
Total energy	Constant	Conservation of total energy
Gravitational force	Decreases	The satellite is getting farther away, thus the force is getting weaker by the inverse square law.

GRAVITY REVIEW QUESTIONS

1. Which of the following diagrams of two planets would represent the largest gravitational force between the masses?

   (A) 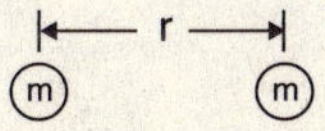

   (B) 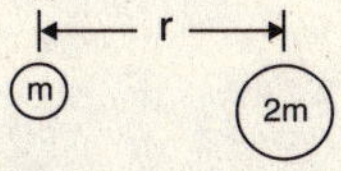

   (C) 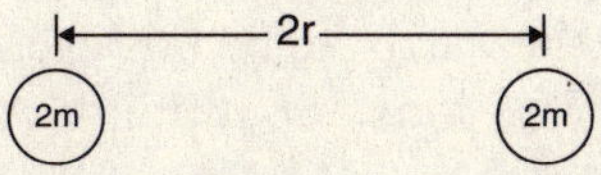

   (D) 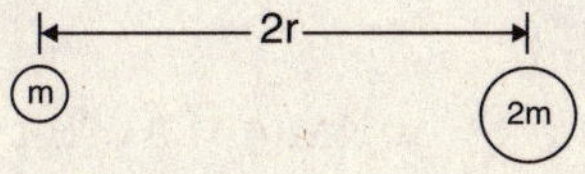

   (E) 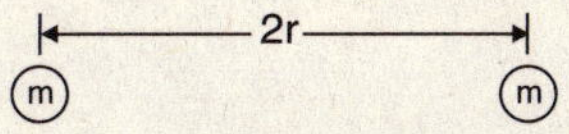

2. A satellite is in orbit around the earth. Consider the following quantities:

   I. distance from the center of the earth
   II. mass of the earth
   III. mass of the satellite

   The gravitational acceleration *g* depends on which of the above?

   (A) I only
   (B) I and II only
   (C) III only
   (D) I and III only
   (E) I, II, and III

3. As a satellite orbits the earth, it passes points A and B. The satellite is traveling faster at point B than at point A. Consider the following statements:

   I. The gravitational force is greater at B than at A.
   II. The speed is greater at B than at A.
   III. The kinetic energy is greater at B than at A.
   IV. The angular momentum is greater at B than at A.

   Which of the above statements are true?

   (A) I and II only
   (B) I only
   (C) II only
   (D) I, II, and III only
   (E) I, II, III, IV

4. As a a planet orbits the sun, which of the following must remain constant?

   (A) velocity
   (B) gravitational force
   (C) radius of orbit
   (D) angular momentum
   (E) the product of mass and velocity

SOLUTIONS TO GRAVITY REVIEW QUESTIONS

1. B

Gravitational force is proportional to the product of the masses and inversely proportional to the square of the distance between the masses. Answers A and C are equivalent, while B is larger than either. Answers D and E are smaller than C.

2. B

The acceleration due to gravity does not depend on the mass of the orbiting (falling) object, but only on the mass of the earth and the distance between the masses.

3. D

If the satellite is moving faster at B, then it is closer to the earth at B; however, angular momentum is conserved as the satellite orbits, and is therefore constant at points A and B.

4. D

Angular momentum is conserved and must remain constant throughout the orbit.

CHAPTER ELEVEN

Electric Fields, Forces, and Potentials

CHARGE

Charge is the fundamental quantity that underlies all electrical phenomena. The symbol for charge is q, and the SI unit for charge is the *Coulomb* (C). The fundamental carrier of negative charge is the electron, with a charge of -1.6×10^{-19} C. The proton, found in the nucleus of any atom, carries exactly the same charge as the electron, but is positive. The neutron, also found in the nucleus of the atom, has no charge. When charge is transferred, only electrons move from one atom to another. Thus, the transfer of charge is really just the transfer of electrons. We say that an object with a surplus of electrons is negatively charged, and an object having a deficiency of electrons is positively charged. Charge is conserved during any process, and so any charge lost by one object must be gained by another object.

The electron is the fundamental carrier of negative charge.

The Law of Charges

The law of charges states that like charges repel each other and unlike charges attract each other. This law is fundamental to understanding all electrical phenomena.

The law of charges: Like charges repel, unlike charges attract.

Example: Four charges, A, B, C, and D, exist in a region of space. Charge A attracts B, but B repels C. Charge C repels D, and D is positively charged. What is the sign of charge A?

Solution: If D is positive and it repels C, C must also be positive. Since C repels B, B must also be positive. A attracts B, so A must be negatively charged.

Example: Two charged spheres of equal size carry a charge of +6 C and –4 C, respectively. The spheres are brought in contact with one another for a time sufficient to allow them to reach an equilibrium charge. They are then separated. What is the final charge on each sphere?

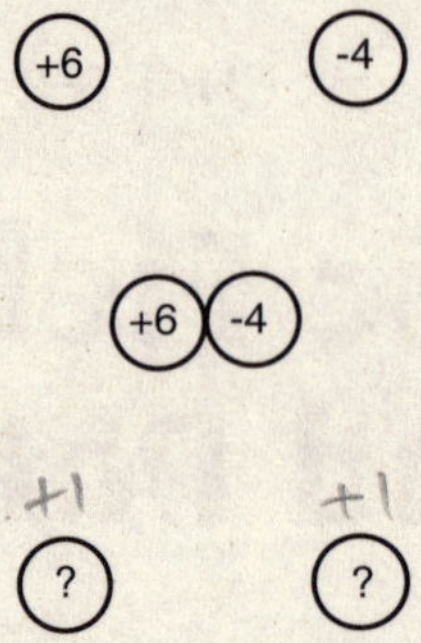

Charge is conserved during any process.

Solution: The total charge on the two spheres is +6 C + –4 C = +2 C, and this is the magnitude of the equilibrium charge. When they are separated, they divide the charge evenly, each keeping a charge of +1 C.

COULOMB'S LAW (库仑定律)

We know that two charges exert either an attractive or repulsive force on each other, but what is the nature of this force? It turns out that the force between any two charges follows the same basic form as Newton's law of universal gravitation; that is, the electric force is proportional to the magnitude of the charges and inversely proportional to the square of the distance between the charges. The equation for Coulomb's law is:

$$F_E = \frac{Kq_1q_2}{r^2}$$

$K = 9 \times 10^9\ Nm^2/C^2$

where F_E is the electric force, q_1 and q_2 are the charges, r is the distance between their centers, and K is a constant which happens to equal $9 \times 10^9\ \text{Nm}^2/\text{C}^2$. Usually, you will be asked questions about the proportionality between the quantities in Coulomb's law, rather than the equation.

Example: Two point charges q_1 and q_2 are separated by a distance r.

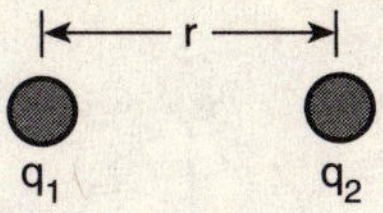

The following choices refer to the electric force on these two charges.

(a) It is quadrupled.
(b) It is doubled.
(c) It remains the same.
(d) It is halved.
(e) It is quartered.

(A) What happens to the force on q_2 if the charge on q_1 is doubled?
(B) What happens to the force on q_2 if the charge on q_1 is doubled and the distance between the charges is also doubled?

Solution:

(A) If q_1 is doubled, then:

$$F = \frac{K(2q_1)(q_2)}{r^2} = 2F_E$$

Thus the new force between the charges is doubled, answer (b).

(B) If q_1 is doubled and r is doubled, then:

$$F = \frac{K(2q_1)(q_2)}{(2r)^2} = \frac{2}{4}F_E = \frac{1}{2}F_E$$

Thus, the new force is half as much as the original force, answer (d).

SEPARATION AND TRANSFER OF CHARGE IN AN ELECTROSCOPE

An electroscope is a device that consists of a metal ball or plate connected to a metal rod with two thin metal leaves attached at the bottom. The rod and leaves are insulated so as not to pick up any extra charges from the air. Remember, even neutral objects have charges in them; they just have an equal number of positive and negative charges. We can use other charged objects to redistribute the charges in a neutral object without actually changing the amount of charge.

We can use an electroscope to study how charges in the ball, rod, and leaves are separated when we bring another charged object near the electroscope or touch the ball of the electroscope with the charged object. For example, your physics teacher may have charged a hard rubber rod negatively by rubbing it on a piece of fur. The rod becomes negatively charged because it strips electrons off the fur. If we bring the negatively charged rubber rod near the ball of the electroscope, the charges on the electroscope are separated. Use the diagram of the rod and electroscope below to draw how the positive and negative charges are distributed on the electroscope, then check your answer below.

An electroscope can be used to study how charges distribute themselves.

Did you draw the ball as positive and the leaves as negative and repelling each other? The free electrons in the ball are repelled by the nearby negatively charged rod and therefore move down to the leaves of the electroscope. Since both leaves are then negative, they repel each other. We say that the ball of the electroscope is positively charged since it now has a lack of electrons. Your drawing should look like this:

But what if we touch the ball of the electroscope with the negatively charged rod? Draw the distribution of charges below:

Conduction is the transfer of charge by actual contact.

The negative charges (electrons) flow into the electroscope, and we say that the electroscope is now negatively charged. This time, we didn't just separate charges in the metal, we actually transferred charge from one object to another by *conduction*, bringing the two objects in contact with each other. Your drawing should look like this:

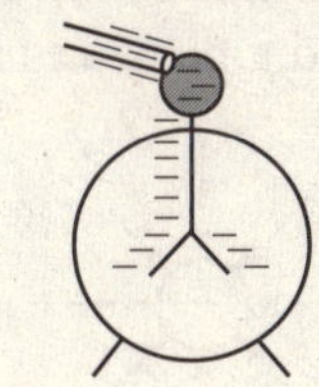

ELECTRIC FIELD

An *electric field* is the condition of space around a charge (or distribution of charges) in which another charge will feel a force. Electric field lines always point in the direction that a positive charge would feel a force. For example, if we take a charge Q to be the source of an electric field E, and we bring a very small positive "test" charge q nearby to test the strength and direction of the electric field, then q will feel a force that is directed radially away from Q.

Conductors, like metals, have electrons that are loosely bound to the outskirts of their atoms, and can therefore easily move from one atom to another. An *insulator*, like wood or glass, does not have many loosely bound electrons, and therefore cannot pass charge easily.

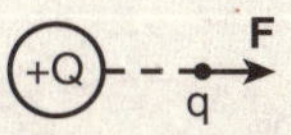

The magnitude of the electric field is given by the equation:

$$E = \frac{F}{q}$$

where electric field E is measured in Newtons per coulomb, and F is the force acting on the charge q which is feeling the force in the electric field. The test charge q would feel a force radially outward anywhere around the source charge Q, so we would draw the electric field lines around the positive charge Q like this:

Electric field is the force per unit charge in a region of space.

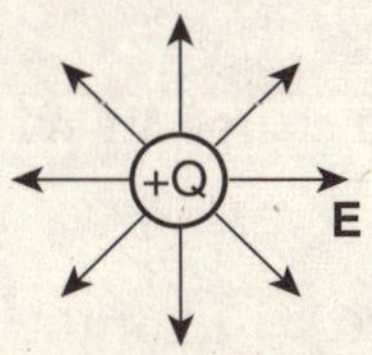

Electric field lines are always drawn in the direction a *positive* charge will feel a force.

Remember, electric field lines in a region are always drawn in the direction a positive charge would feel a force in that region. They can also represent the path a positive charge would follow in that region. The diagrams below show how we would draw the electric field lines around a negative charge, two positives, two negatives, and a positive and a negative charge.

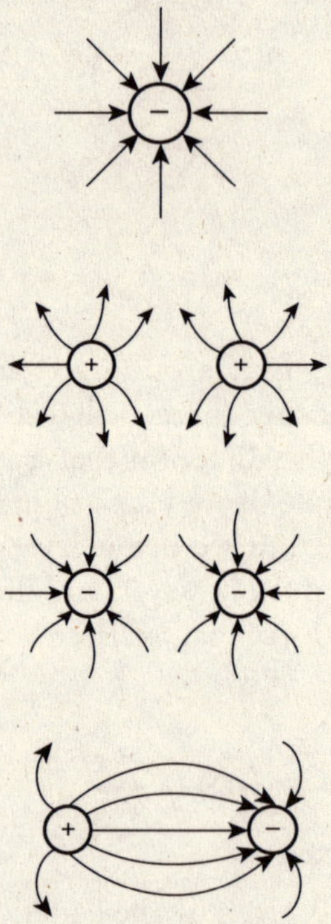

Electrons (*negative* charges) are moved when charge is transferred, but electric field lines are drawn in the direction a *positive* charge would move.

Example: A charge of $+6 \times 10^{-6}$ C is brought near a negative charge in a region where the electric field strength due to the negative charge is 20 N/C. What are the magnitude and direction of the force acting on the positive charge?

F = Eq
= 20 × 6 × 10⁻⁶
= 1.2 × 10⁻⁴ N to the left

Solution: Electric field, force, and charge are related by the equation, so the force can be found by:

$$F = qE = (6 \times 10^{-6}\ \text{C})(20\ \text{N/C}) = 1.2 \times 10^{-5}\ \text{N}$$

The force is directed radially inward toward the negative charge, since the positive charge is attracted to the negative charge.

Electric Potential (电势)

The *electric potential V* is defined in terms of the work we would have to do on a charge to move it against an electric field. For example, if we wanted to move a positive charge from point A to point B in the electric field shown below, we would have to do work on the charge, since the electric field would push against us.

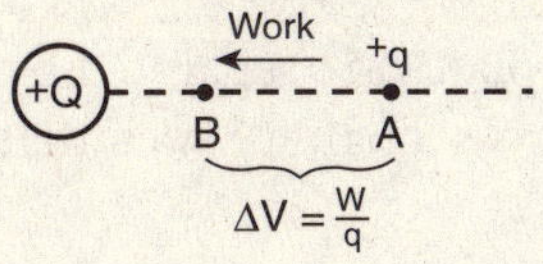

We say that there is a potential difference ΔV between points A and B, and the equation for potential difference between two points is:

$$\Delta V = \frac{Work}{q}$$

and is measured in joules/coulomb, or volts. When we apply potential difference to circuits in the next chapter, we will often call it *voltage*. If we place the charge q at point B and let it go, it would "fall" toward point A. We say that positive charges naturally want to move from a point of high potential (B) to low potential (A), and we refer to the movement of the positive charges as *current*. We will return to voltage and current in chapter 12.

Quick Quiz:

True or False: A single positive charge creates a uniform electric field around itself.

Answer: False. The electric field around a single positive charge gets weaker as you go farther from the charge, and therefore cannot be uniform.

Uniform Electric Field (匀强电场)

We can create a uniform electric field in a region of space by taking two metal plates, setting them parallel to each other and separating them by a distance d, and placing a voltage V (as from a battery) across the plates so that one of the plates will be positive and the other negative.

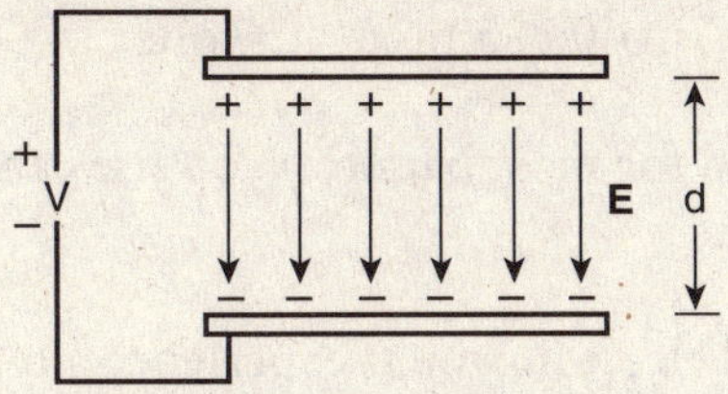

The positive charges on the top plate will line up uniformly with the negative charges on the bottom plate so that each positive charge lines up with a negative charge directly across from it. This arrangement of charges creates electric field lines that run from the positive charges to the negative charges and are uniformly spaced to produce a uniform (constant) electric field everywhere between the plates. Conducting plates which are connected this way are called a *capacitor*. Capacitors are used to store charge and electric field in a circuit that can be used at a later time. We will discuss capacitors further in chapter 12.

Electric *field* is the force per unit charge, and electric *potential* is the work per unit charge.

Capacitor 电容器

The electric field, voltage, and distance between the plates are related by the equation:

$$E = \frac{V}{d}$$

It follows from this equation that the unit for electric field is volts/meter, which is equivalent to newtons/coulomb.

Example: Two charged parallel conducting plates are separated by a distance d of 0.005 m and are connected across a battery such that the electric field between the plates is 300 V/m.

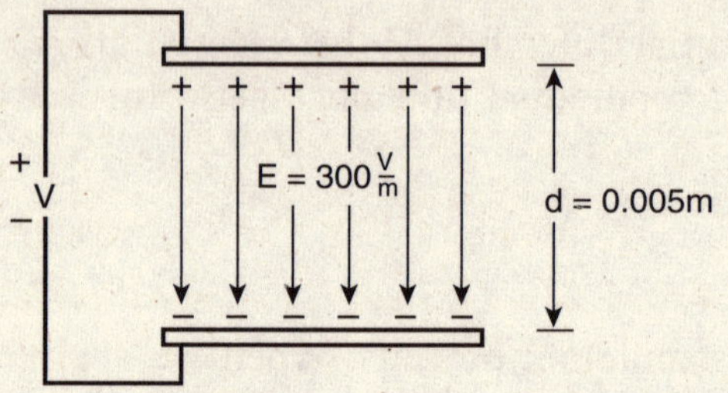

(A) What is the potential difference (voltage) of the battery that is connected across the plates? V = Ed = 1.5 ~~voltage~~ volts

(B) An electron is placed halfway between the plates. Describe the subsequent motion of the electron.

(C) A neutron is placed between the plates and near the top plate. Describe the force acting on the neutron at this location.

Solution:

(A) We are given the equation $E = \frac{V}{d}$, so it follows that:

$$V = Ed = (300 \text{ V/m})(0.005 \text{ m}) = 1.5 \text{ volts}$$

This voltage is typical for most batteries you use each day, such as size AA, C, or D.

(B) A negatively charged electron would experience a force upward toward the positive plate, and therefore would accelerate upward. We could also say that the electron will move in a direction opposite to the direction of the electric field, since electric field lines are always drawn in the direction a positive charge would move.

(C) Since a neutron has no charge, it will not experience a force from either the top or bottom plate, and consequently will not move.

ELECTRIC FIELDS, FORCES, AND POTENTIAL REVIEW QUESTIONS

1. When charge is transferred from one object to another, which of the following are actually transferred?

   (A) electrons
   (B) protons
   (C) neutrons
   (D) quarks
   (E) photons

2. Two conducting spheres of equal size have a charge of –3 C and +1 C, respectively. A conducting wire is connected from the first sphere to the second. What is the new charge on each sphere?

   (A) –4 C
   (B) +4 C
   (C) –1 C
   (D) +1 C
   (E) zero

3. According to Coulomb's law, if the electric force between two charges is positive, which of the following must be true?

   (A) One charge is positive and the other charge is negative.
   (B) The force between the charges is repulsive.
   (C) The force between the charges is attractive.
   (D) The two charges must be equal in magnitude.
   (E) The force must be directed toward the larger charge.

4. Two charges q_1 and q_2 are separated by a distance r and apply a force F to each other. If both charges are doubled, and the distance between them is halved, the new force between them is

   (A) ½F
   (B) ¼F
   (C) 4F
   (D) 8F
   (E) 16F

5. Two uncharged spheres A and B are near each other. A negatively charged rod is brought near one of the spheres.

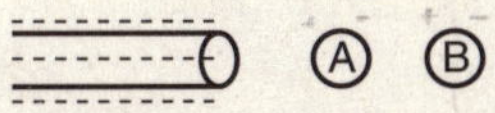

   The far right side of sphere B is

   (A) uncharged.
   (B) neutral.
   (C) positive.
   (D) negative.
   (E) equally positive and negative.

6. Two charges A and B are near each other, producing the electric field lines shown.

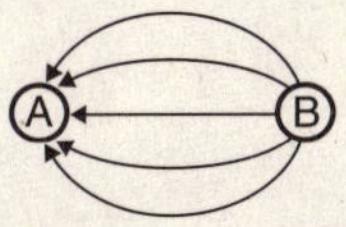

   What are the two charges A and B, respectively?

   (A) positive, positive
   (B) negative, negative
   (C) positive, negative
   (D) negative, positive
   (E) neutral, neutral

7. A force of 40 N acts on a charge of 0.25 C in a region of space. The electric field at the point of the charge is

(A) 10 N/C
(B) 100 N/C
(C) 160 N/C
(D) 40 N/C
(E) 0.00625 N/C

8. Electric potential

(A) is a vector quantity.
(B) is proportional to the work done in an electric field.
(C) is always equal to the electric field.
(D) is zero when a charge is in an electric field.
(E) is measured in N/C.

9. Two conducting plates are separated by a distance of 0.001 m. A 9 V battery is connected across the plates. The electric field between the plates is

(A) 9,000 V/m
(B) 900 V/m
(C) 9 V/m
(D) 0.009 V/m
(E) 0.00011 V/m

Questions 10–11:
Two charged parallel plates are oriented as shown.

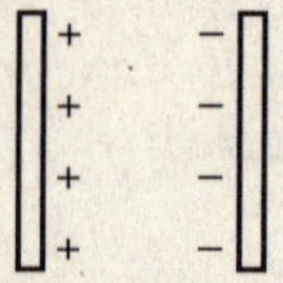

The following particles are placed between the plates, one at a time:

I. electron
II. proton
III. neutron

10. Which of the particles would move to the right between the plates?

(A) I and II only
(B) I and III only
(C) II and III only
(D) II only
(E) I only

11. Which of the particles would not experience a force while between the plates?

(A) I and II only
(B) II and III only
(C) I only
(D) III only
(E) I, II, and III

SOLUTIONS TO ELECTRIC FIELDS, FORCES, AND POTENTIAL QUESTIONS

1. A

Electrons are often free to move from one atom to another.

2. C

The total charge is –2 C, so when a wire connects the spheres, each sphere gets half of the –2 C, or –1 C each.

3. B

In the equation for electric force, two positive or two negative charges multiplied by each other yields a positive force, indicating repulsion.

4. E

$$F = \frac{K(2q_1)(2q_2)}{(\frac{1}{2})r^2} = 16F$$

5. D

The far right side of sphere B is negative, since the negative charges in the sphere are pushed as far away as possible by the negative charges on the rod.

6. D

Electric field lines begin on positive charges and end on negative charges; thus A is negative and B is positive.

7. C

$$E = \frac{F}{q} = \frac{40N}{0.25C} = 160\frac{N}{C}$$

8. B

Electric potential is defined as the work per unit charge in moving the charge to a particular place in an electric field.

9. A

$$E = \frac{V}{d} = \frac{9V}{0.001m} = 9{,}000\frac{V}{m}$$

10. D

Only the positively charged proton would move to the right, toward the negatively charged plate.

11. D

Since the neutron has no charge, it would not experience a force in an electric field.

Chapter 11 公式总结

① $F_E = \frac{K q_1 q_2}{r^2}$ $K = 9 \times 10^9 \ Nm^2/C^2$

② $E = \frac{F}{q}$

③ $\Delta V = \frac{Work}{q}$

④ $E = \frac{V}{d}$ (匀强电场)

CHAPTER TWELVE

Circuits

CURRENT, VOLTAGE, AND RESISTANCE

When we connect a battery, wires, and a light bulb in the circuit shown below, the bulb lights up. But what is actually happening in the circuit?

Recall from the previous chapter that the battery has a potential difference, or voltage, across its ends. One end of the battery is positive, and the other end is negative. When we connect the wires and light bulb to the battery in a complete circuit, charge begins to flow from one end of the battery, through the wires and the bulb, to the other end of the battery, causing the bulb to light. We say that the movement of positive charge from the positive end of the battery through the circuit to the negative end of the battery is called *conventional current*, or simply *current*. Current is the amount of charge moving through a conductor per second, and the unit for current is the coulomb/second, or *ampere*. We use the symbol *I* for current.

Current *I* is defined as the flow of positive charge through a conductor, and is measured in *amperes*, or *amps*.

Technically speaking, positive charges do not "flow" through a circuit, although it is common to speak of them as if they do. When the battery is connected to the circuit, the electrons on the negative side of the battery shift away from the negative end of the battery, bumping into electrons in the wire, creating a domino effect through the wire toward the positive end of the battery. This "wave" of bumping electrons heats the light bulb filament that glows as a result, giving off light and heat. The SAT II: Physics Test generally uses the conventional current model of current flow.

> *Resistance* is the opposition to the flow of current, and is measured in *ohms* (Ω).

As charge moves through the circuit, it encounters *resistance*, or opposition to the flow of current. Resistance is the electrical equivalent of friction. In our circuit above, the wires and the light bulb would be considered resistances, although usually the resistance of the wires is neglected. When a larger voltage is used, more charge moves through the wire, and thus more current flows through the circuit. In other words, the current is directly proportional to the voltage. This implies that the ratio of voltage to current is a constant. This constant is defined as the *resistance*, and is measured in *ohms* (Ω). An ohm is a volt/amp. The relationship between voltage, current, and resistance is called *Ohm's law*:

> *Ohm's law* relates voltage, current, and resistance in the equation $R = V/I$.

$$R = \frac{V}{I}$$

This relationship typically holds true for the purposes of the SAT II: Physics Test.

Schematic Diagram Symbols for Circuits

The table below summarizes the quantities discussed so far.

Quantity	**Symbol**	**Unit**	**Schematic Symbol**
Battery Voltage	V	Volts (V)	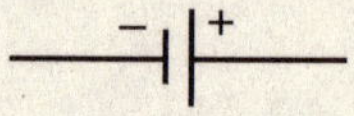
Current	I	Amps (A)	
Resistance	R	Ohms (Ω)	

Usually, when we draw circuit diagrams, we use the symbols in the table above. For example, the simple light bulb circuit would look like this:

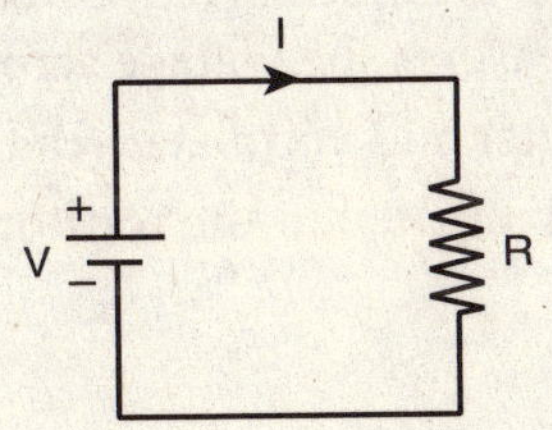

If we wanted to actually measure the current through the resistor and the voltage across the resistor, we would connect an *ammeter* (to measure current) in series with the resistor, and a *voltmeter* (to measure voltage) in parallel with the resistor, as shown below:

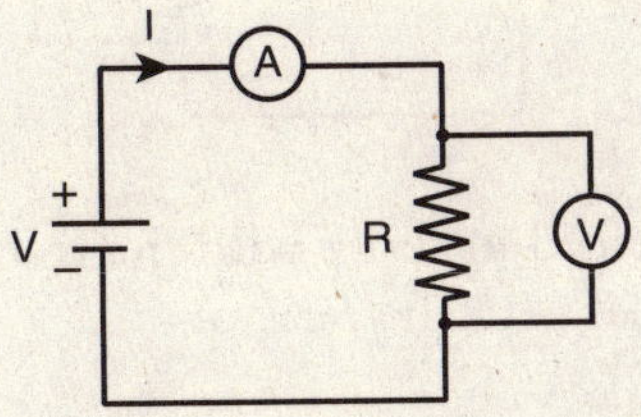

We place the ammeter in *series* with the resistor so that the same current will pass through the ammeter and the resistor, and we place the voltmeter in *parallel* with the resistor so that the voltage will be the same across the voltmeter and the resistor. Ammeters have a low resistance so as not to add to the total resistance of the circuit (and thus decrease the current in the circuit), and voltmeters have a high resistance so that current will not want to flow through them and bypass the resistor.

POWER AND JOULE'S LAW OF HEATING

In chapter 7 we defined *power* as the rate at which work is done, or the rate at which energy is transferred. When current flows through a resistor, heat is produced, and the amount of heat produced in joules per second is equal to the power in the resistor. The heating in the resistor is called *joule heating*, and the power dissipated in the resistor follows *Joule's law of heating*. The equation that relates power to the current, voltage, and resistance in a circuit is:

$$P = IV = I^2R = \frac{V^2}{R}$$

The unit for power is the joule/second, or *watt*.

Quick Quiz:

Ammeters

(A) have a low resistance and are always placed in parallel with a resistor.

(B) have a low resistance and are always placed in series with a resistor.

(C) have a high resistance and are always placed in parallel with a resistor.

(D) have a high resistance and are always placed in series with a resistor.

Answer: (B). Ammeters have a low resistance so as not to add to the resistance of the circuit, and must be connected in series with the resistor to get the same current as the resistor. If the question began with "Voltmeters," the answer would be (C).

Example: A simple circuit consists of a 12 volt battery and a 6 Ω resistor.

(A) Draw a schematic diagram of the circuit, and include an ammeter, which measures the current through the resistor, and a voltmeter, which measures the voltage across the resistor.

(B) What will the ammeter and voltmeter read? Ⓐ: 2 A Ⓥ: 12 V

(C) What is the power dissipated in the resistor? P = IV = 2×12 = 24 W

Solution:

(A) The schematic diagram would look like this:

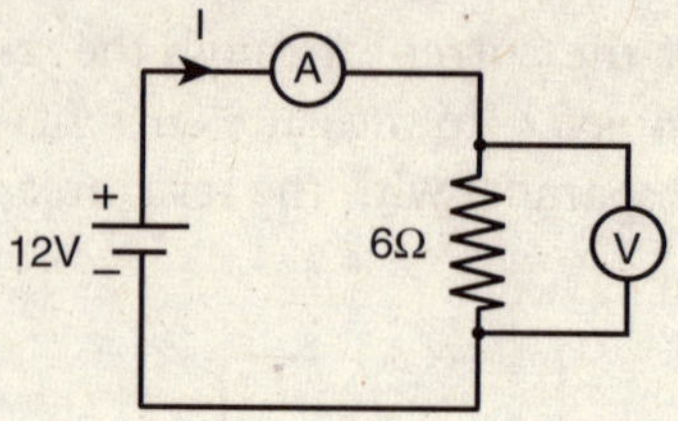

Note that the ammeter is placed in series with the resistor, and the voltmeter is placed in parallel with the resistor.

(B) The ammeter reads the current, which we can calculate using Ohm's law:

$$I = \frac{V}{R} = \frac{12V}{6\Omega} = 2A$$

Power in a circuit is the energy used per unit time, measured in *watts*.

The voltmeter will read 12 V, since the potential difference across the resistor must be equal to the potential difference across the battery. As we will see later, if there were more than one resistor in the circuit, they each would not necessarily get 12 volts.

(C) The power can be found by:

$$P = IV = (2\text{ A})(12\text{ V}) = 24\text{ watts}$$

CIRCUITS WITH MORE THAN ONE RESISTOR

When two or more resistors are placed in a circuit, there are basically three ways to connect them: in series, in parallel, or a combination of series and parallel.

Series Circuits (串連)

Two or more resistors of any value placed in a circuit in such a way that the same current passes through each of them is called a *series circuit*. Consider the series circuit below, which includes a voltage source ε (which stands for *emf*, an older term for voltage) and three resistors R_1, R_2, and R_3.

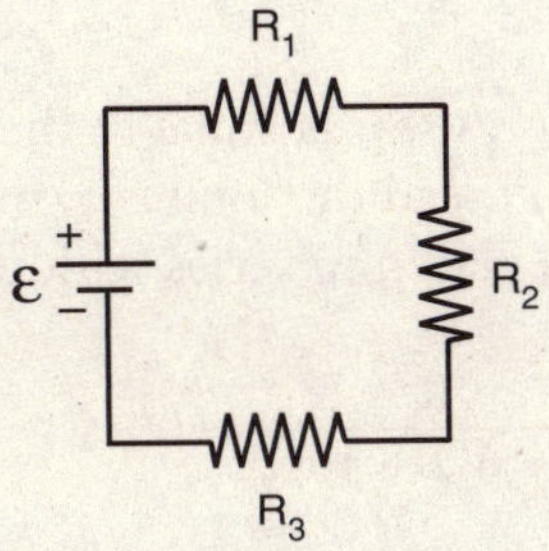

The rules for dealing with series circuits are as follows:

1. The total resistance in a series circuit is the sum of the individual resistances:

$$R_{total} = R_1 + R_2 + R_3$$

2. The total current in the circuit is:

$$I_{total} = \frac{V_{total}}{R_{total}}$$

This current must pass through each of the resistors, so each resistor also gets I_{total}, that is, $I_{total} = I_1 = I_2 = I_3$.

3. The voltage divides proportionally among the resistances according to Ohm's law:

$$V_1 = I_1R_1;\ V_2 = I_2R_2;\ V_3 = I_3R_3$$

Example: Three resistors of 2 Ω, 6 Ω, and 10 Ω are connected in series with a 9 volt battery.

(A) Draw a schematic diagram of the circuit that includes an ammeter to measure the current through the 2 Ω resistor, and a voltmeter to measure the voltage across the 10 Ω resistor.

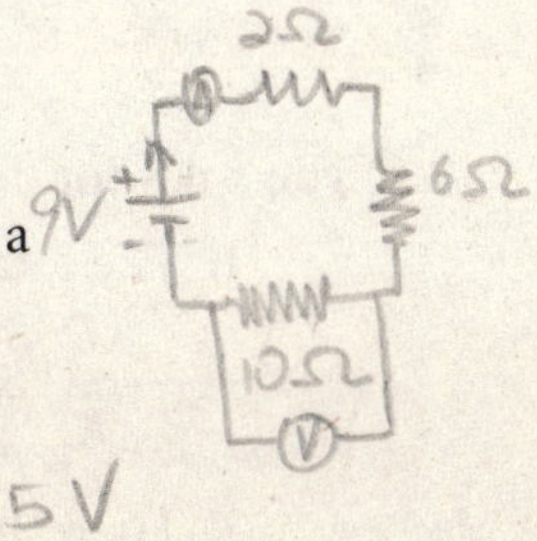

(B) What will the ammeter read? $I = \frac{9}{2+6+10} = 0.5\text{ A}$

(C) What will the voltmeter read? $V = IR_3 = 0.5 \times 10 = 5\text{ V}$

Solution:

(A) The schematic diagram for the circuit looks like this:

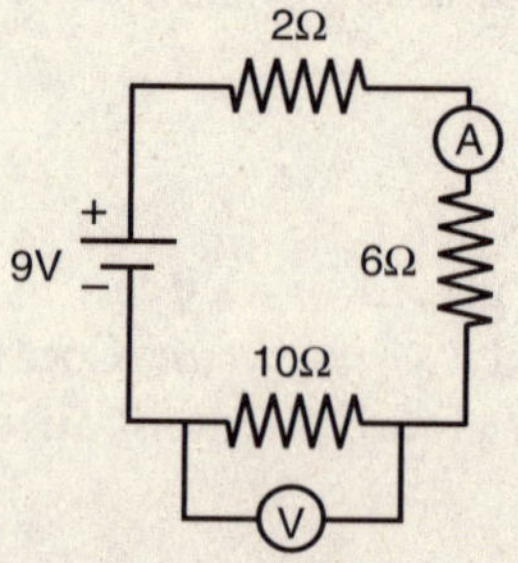

(B) The current through each resistor is equal to the total current in the circuit, so the ammeter will read the total current regardless of where it is placed, as long as it is placed in series with the resistances.

$$I_{total} = \frac{V_{total}}{R_{total}} = \frac{9\text{V}}{2\Omega + 6\Omega + 10\Omega} = 0.5A$$

(C) The voltmeter will read the voltage across the 10 Ω resistor, which is NOT 9 volts. The 9 volts provided by the battery is divided proportionally among the resistances. The voltage across the 10 resistor is:

$$V_{10} = I_{10}R_{10} = (0.5\text{ A})(10\Omega) = 5\text{ V}$$

Parallel Circuits

Two or more resistors of any value placed in a circuit in such a way that each resistor has the same potential difference across it is called a *parallel circuit*. Consider the parallel circuit below, which includes a voltage source ε and three resistors R_1, R_2, and R_3.

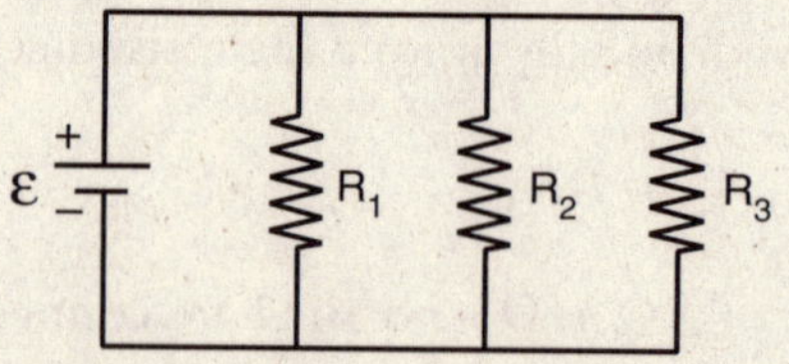

The rules for dealing with parallel circuits are as follows:

1. The total resistance in a parallel circuit is given by the equation:

$$\frac{1}{R_{total}} = \frac{1}{R_1} + \frac{1}{R_2} + \frac{1}{R_3}$$

2. The voltage across each resistance is the same: $V_{total} = V_1 = V_2 = V_3$

3. The current divides in an inverse proportion to the resistance:

$$I_1 = \frac{V_1}{R_1};\ I_2 = \frac{V_2}{R_2};\ I_3 = \frac{V_3}{R_3}$$

where R_1, R_2, and R_3 are equal to each other.

Example: The resistors of resistance 2 Ω, 3 Ω, and 12 Ω are connected in parallel to a battery of voltage 24 V.

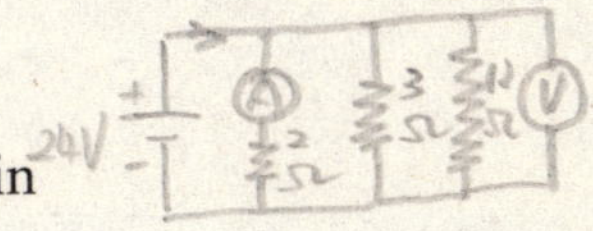

(A) Draw a schematic diagram of the circuit which includes an ammeter to measure only the current through the 2 Ω resistor, and a voltmeter to measure the voltage across the 12 Ω resistor.

(B) What is the total resistance in the circuit?

(C) What will the ammeter read?

(D) What will the voltmeter read?

In a *series* circuit, the resistors get the same current; in a *parallel* circuit, the resistors get the same voltage.

$R_t = \frac{12}{11}\Omega \quad \frac{1}{R_t} = \frac{1}{2} + \frac{1}{3} + \frac{1}{12} = \frac{6+4+1}{12} = \frac{11}{12}(\Omega)$

$I = \frac{24}{2} = 12\ A$

$V = 24V$

Solution:

(A) The schematic diagram for the circuit looks like this:

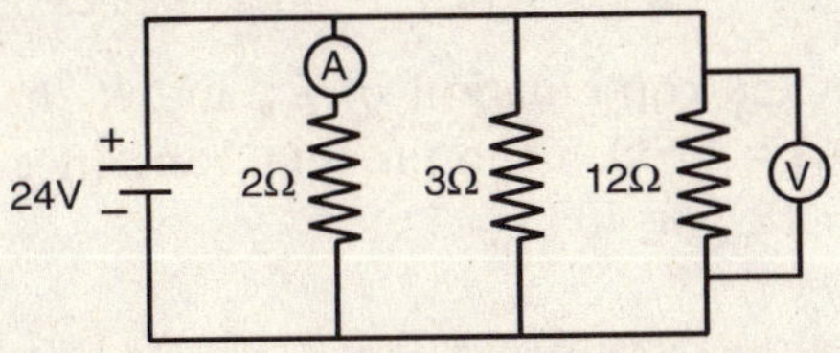

(B) The total resistance in the circuit is found by:

$$\frac{1}{R_{total}} = \frac{1}{R_1} + \frac{1}{R_2} + \frac{1}{R_3} = \frac{1}{2\Omega} + \frac{1}{3\Omega} + \frac{1}{12\Omega} = \frac{6}{12\Omega} + \frac{4}{12\Omega} + \frac{1}{12\Omega}$$

$$= \frac{11}{12\Omega}$$

Notice that this fraction is not the total resistance, but $\frac{1}{R_{total}}$. Thus, the total resistance in this circuit must be:

$$\frac{12}{11}\Omega$$

The equation for the total resistance in parallel comes from the fact that the total current is the sum of the individual currents in the circuit, and each resistor gets the same voltage:

$$\frac{V_{total}}{R_{total}} = \frac{V_1}{R_1} + \frac{V_2}{R_2} + \ldots,$$

where all the *V*'s are equal.

(C) Since the ammeter is placed in series only with the 2 Ω resistance, it will measure the current passing only through the 2 Ω resistance. Recognizing that the voltage is the same (24 V) across each resistance, we calculate that:

$$I_2 = \frac{V_2}{R_2} = \frac{24V}{2\Omega} = 12A$$

(D) Each resistance is connected across the 24 V battery,

Combination Series and Parallel Circuits (串并连)

Consider the combination circuit below:

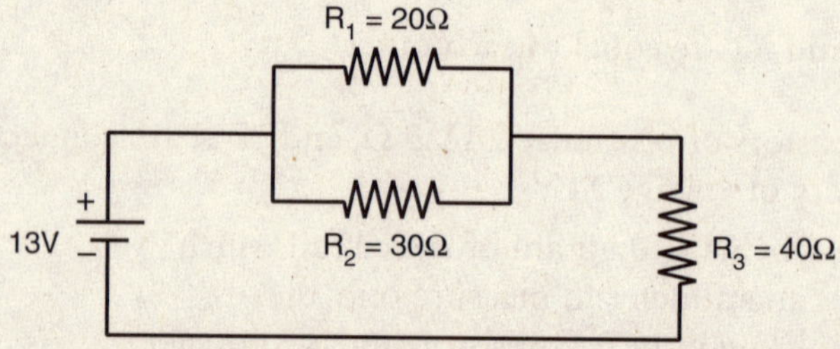

We see that R_1 is in parallel with R_2, and R_3 is in series with the parallel combination of R_1 and R_2. Let's find (A) the total resistance, (B) the total current in the circuit, (C) the voltage across each resistor, and (D) the current through each resistor.

(A) Before we can find the total resistance of the circuit, we need to find the equivalent resistance of the parallel combination of R_1 and R_2:

$$\frac{1}{R_{12}} = \frac{1}{R_1} + \frac{1}{R_2} = \frac{1}{20\Omega} + \frac{1}{30\Omega} = \frac{3}{60\Omega} + \frac{2}{60\Omega} = \frac{5}{60\Omega}$$

which implies that the combination of R_1 and R_2 has an equivalent resistance of 60/5 Ω = 12 Ω. Then the total resistance of the circuit is $R_{total} = R_{12} + R_3 = 12\ \Omega + 40\ \Omega = 52\ \Omega$.

(B) The total current in the circuit is:

$$I_{total} = \frac{V_{total}}{R_{total}} = \frac{13V}{52\Omega} = 0.25A$$

(C) The voltage provided by the battery is divided proportionally among the parallel combination of R_1 and R_2 (with R_1 and R_2 having the same voltage across them), and R_3. Since R_3 has the total current passing through it, we can calculate the voltage across R_3:

$$V_3 = I_3 R_3 = (0.25\ \text{A})(40\ \Omega) = 10\ \text{V}$$

This implies that the voltage across R_1 and R_2 is the remainder of the 13 V provided by the battery. Thus, the voltage across R_1 and R_2 is:

$$13\ \text{V} - 10\ \text{V} = 3\ \text{V}$$

(D) The current through R_3 is the total current in the circuit, 0.25 A. Since we know the voltage and resistance of the other resistances, we can use Ohm's law to find the current through each.

$$I_1 = \frac{V_1}{R_1} = \frac{3V}{20\Omega} = 0.15A \text{ and } I_2 = \frac{V_2}{R_2} = \frac{3V}{30\Omega} = 0.10A$$

Notice that I_1 and I_2 add up to the total current in the circuit, 0.25 A.

CAPACITANCE AND RESISTANCE-CAPACITANCE CIRCUITS

Capacitance

Let's revisit the conducting parallel plates we discussed in chapter 11. Recall that if we connect a battery to the plates, charge flows from the battery onto the plates, with one plate becoming positively charged with a charge $+q$, and the other plate negatively charged with a charge $-q$.

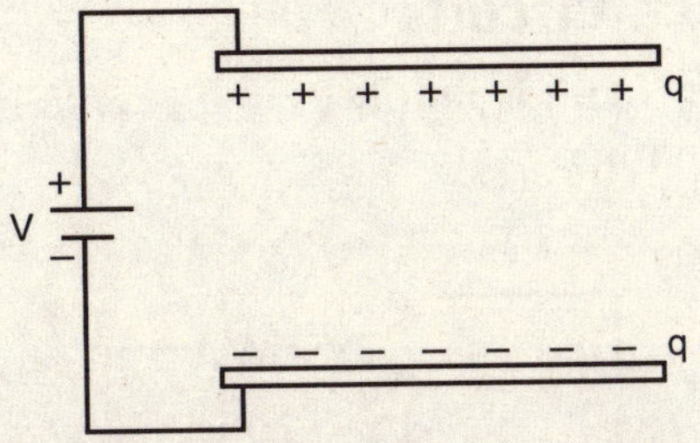

The capacitance of the plates is defined as:

$$C = \frac{q}{V}$$

where q is the charge on one of the plates, and V is the voltage across the plates. The unit for capacitance is the coulomb/volt, or *farad*. The capacitance of a capacitor is proportional to the area of each plate and inversely proportional to the distance between the plates. In symbols:

> ***Capacitance*** **is charge per unit voltage, and depends only on the geometry of the plates.**

$$C \propto \frac{A}{d}$$

The purpose of a capacitor is to store charge and electric field in a circuit that can be used at a later time.

Example: A capacitor has a capacitance C_0. What is the effect on the capacitance if (A) the area of each plate is doubled? (B) the distance between the plates is halved?

(A) doubled (B) doubled

Solution:

(A) If the area of each plate is doubled (assuming the plates have equal area), we can find the new capacitance by:

$$C \propto \frac{(2)A}{d} = 2C_0$$

Twice the area gives twice the capacitance.

(B) If the distance between the plates is halved (i.e., the plates are brought closer together), the new capacitance is:

$$C \propto \frac{A}{\frac{1}{2}d} = 2C_0$$

Bringing the plates closer together increases the capacitance, in this case by a factor of two.

Resistance-Capacitance Circuits

A resistance-capacitance (RC) circuit is simply a circuit containing a battery, a resistor, and a capacitor in series with one another.

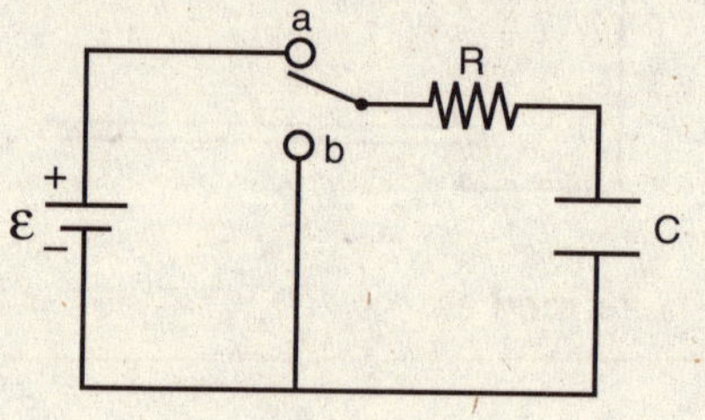

An RC circuit can store charge and release it later.

When the switch is moved to position *a*, current begins to flow from the battery, and the capacitor begins to fill up with charge. Initially, the current is ε/R by Ohm's law, but then it decreases as time goes on until the capacitor is full of charge and will not allow any more charge to flow out of the battery.

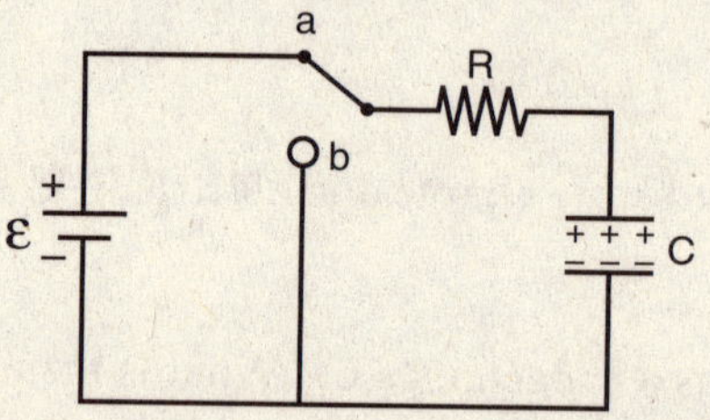

This leads us to a couple of rules we can follow when dealing with capacitors in an RC circuit:

1. An empty capacitor does not resist the flow of current, and thus acts like a wire.
2. A capacitor that is full of charge will not allow current to flow, and thus acts like a broken wire.

If we move the switch to position *b*, the battery is taken out of the circuit, and the capacitor begins to drain its charge through the resistor, creating a current in the opposite direction to the current flowing when the battery was connected.

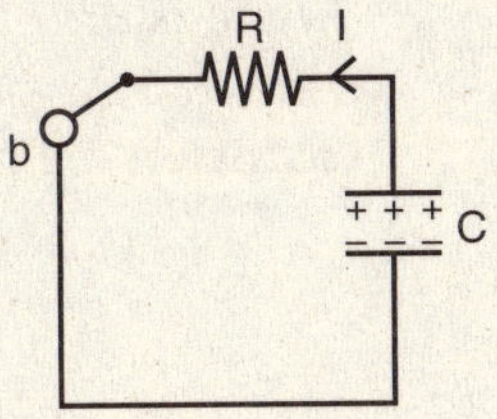

Eventually, the current will die out because of the heat energy lost through the resistor.

Example: A 12 V battery is connected in series to a 3 Ω resistor and an initially uncharged capacitor. Determine the current in the circuit (A) immediately after the battery is connected to the resistor and capacitor, and (B) a long time later.

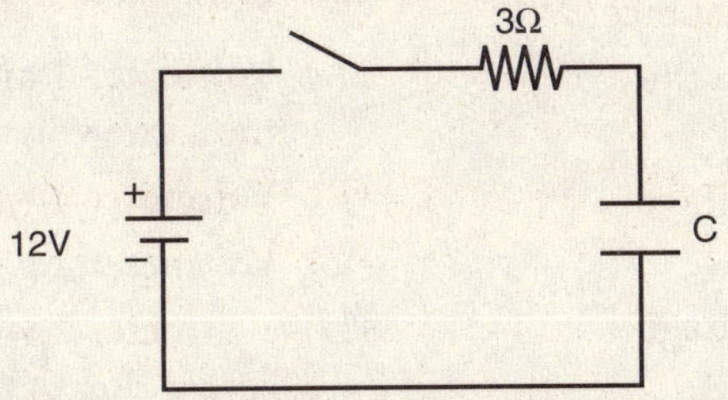

Solution:

(A) Immediately after the circuit is connected, the capacitor is still empty and thus acts like a wire. We can redraw the circuit like this:

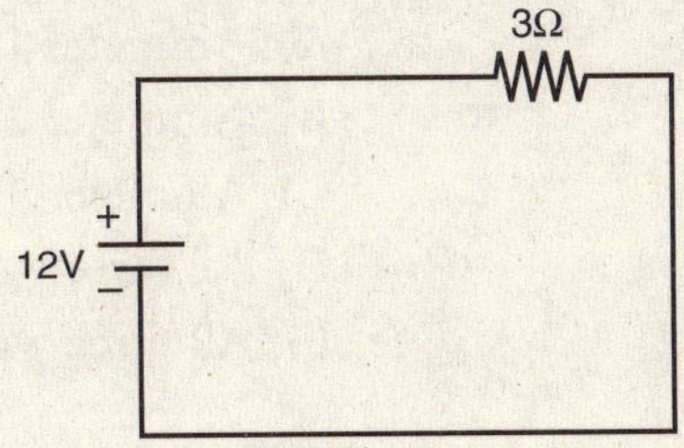

The current, then, is $I = \frac{V}{R} = \frac{12V}{3\Omega} = 4A$.

(B) The current begins to decrease as the capacitor fills up with charge, and after a long time, the capacitor is full of charge, and the current stops flowing completely. Thus, $I = 0$ a long time later.

CIRCUITS REVIEW QUESTIONS

Questions 1–3:
Two resistors of 2 Ω and 4 Ω are placed in series with a 12 V battery.

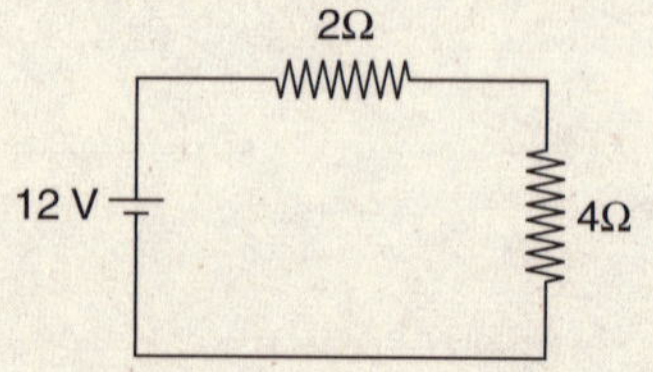

1. Which of the following statements is true?
   (A) The 2 Ω will get more current than the 4 Ω resistor since it has less resistance.
   (B) The 4 Ω will get more current than the 2 Ω resistor since it has more resistance.
   (C) The 2 Ω and 4 Ω resistors will get the same voltage.
   (D) The 2 Ω resistor will get more voltage than the 4 Ω resistor.
   (E) The 4 Ω resistor will get more voltage than the 2 Ω resistor.

2. The current in the 2 Ω resistor is
   (A) 6 A
   (B) 4 A
   (C) 3 A
   (D) 2 A
   (E) 1 A

3. The voltage across the 4 Ω resistor is
   (A) 2 V
   (B) 4 V
   (C) 6 V
   (D) 8 V
   (E) 12 V

Questions 4–6:
Two resistors of 3 Ω and 6 Ω are placed in parallel with a 6 V battery. Three ammeters and two voltmeters are placed in the circuit as shown.

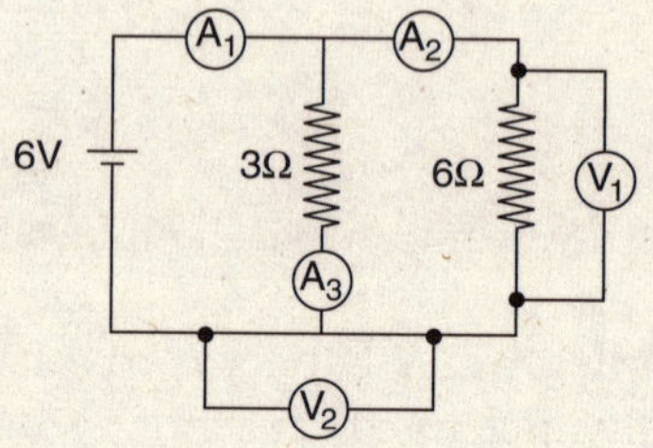

4. Which of the following statements is true of the voltmeters?
   (A) Voltmeter 1 and voltmeter 2 will read the same voltage.
   (B) Voltmeter 1 will read 6 V.
   (C) Voltmeter 2 will read 6 V.
   (D) Voltmeter 2 will read the correct voltage across the 3 Ω resistor.
   (E) Both voltmeters will read the correct voltage across the 6 Ω resistor.

5. Which ammeter will read the highest amount of current?
   (A) Ammeter 1
   (B) Ammeter 2
   (C) Ammeter 3
   (D) All three will read the same current.
   (E) All three will read zero current.

6. What is the correct reading on ammeter 2?
   (A) 0
   (B) 1 A
   (C) 2 A
   (D) 3 A
   (E) 6 A

7. What is the total current flowing in the circuit shown?

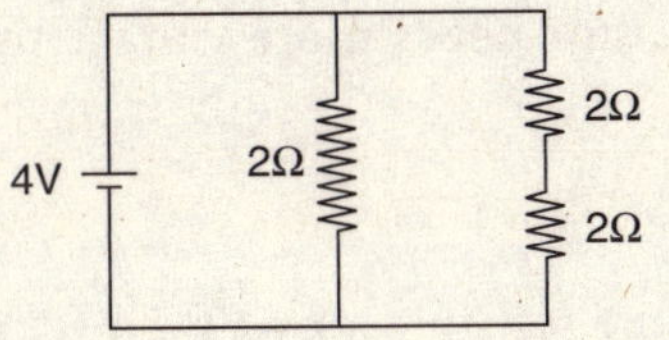

(A) 4 A
(B) 3 A
(C) 2 A
(D) 4/3 A
(E) 2/3 A

8. A resistance-capacitance circuit is connected as shown.

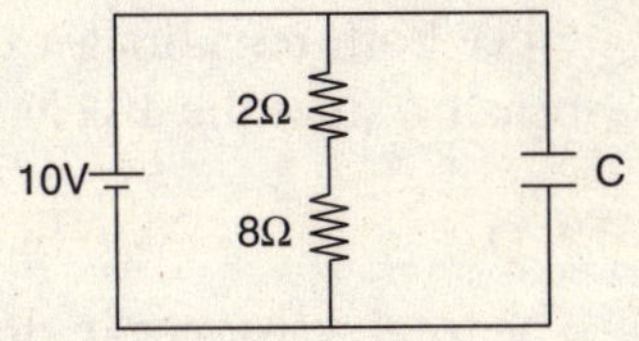

After a long time, the current in the 2 Ω resistor is

(A) 0
(B) 10 A
(C) 8 A
(D) 2 A
(E) 1 A

SOLUTIONS TO CIRCUITS REVIEW QUESTIONS

1. E

Since both resistors get the same current, the 4 Ω resistor needs more voltage than the 2 Ω resistor, since the 4 Ω has more resistance.

2. D

The total resistance in the series circuit is the sum of the resistors, or 6 Ω. The current in the 2 Ω resistor is the same as the total current in the circuit, which is:

$$I = \frac{V}{R} = \frac{12\text{V}}{6\Omega} = 2\text{ A}$$

3. D

Since the 4 Ω resistor represents 2/3 of the total resistance, it will get 2/3 of the total voltage, or 8 V.

4. B

The voltage across the 6 Ω resistor is 6 V since it is in parallel with the battery. Voltmeter 2 is incorrectly placed in the circuit and will read zero volts.

5. A

Ammeter 1 is in a position to measure the total current in the circuit before it splits up to go through the 3 Ω and 6 Ω resistors.

6. B

Ammeter 2 measures only the current through the 6 Ω resistor: $I = \frac{6\text{V}}{6\Omega} = 1\text{ A}$.

7. B

The total resistance can be found by:

$$\frac{1}{R_t} = \frac{1}{2\Omega} + \frac{1}{4\Omega} = \frac{3}{4\Omega}$$

which means that $R_t = 4/3\ \Omega$. So the total current in the circuit is:

$$I = \frac{V}{R} = \frac{4V}{4/3\ \Omega} = 3A$$

8. E

After a long time, the capacitor is full of charge and does not allow any more current to flow through it. Thus, the current from the battery will only flow through the resistors and the current will be:

$$I = \frac{10\text{V}}{10\Omega} = 1\text{ A}$$

CHAPTER THIRTEEN

Magnetic Fields and Forces

MAGNETISM

More than 2,000 years ago in the Greek province of Magnesia were found pieces of iron ore called *lodestones* that would attract or repel each other. When one of these magnetic stones was suspended from a string, one side of the stone would align itself to point in a northerly direction. Eventually, this end of the magnet was labeled *north* and the other end *south*. Later, it was suggested that the earth itself is a magnet, and the stones were simply aligning themselves with the earth's magnetic field. A *magnetic field* ***B*** is defined as the space around a magnet in which another magnet will feel a force, and is measured in *Teslas* (T). Magnetic field lines have no beginning and no end, but generally are drawn from the north pole of a magnet to its south pole. The magnetic fields most often appearing on the SAT II: Physics Test are those of a bar magnet, the earth, and a horseshoe magnet, each shown below.

Magnetic field lines generally point from north to south.

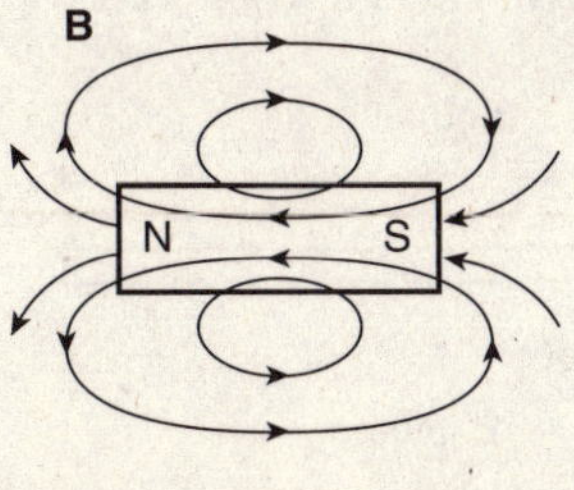

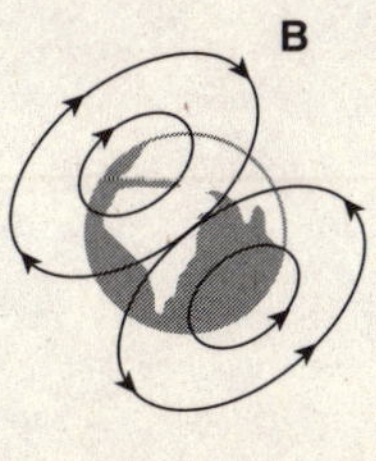

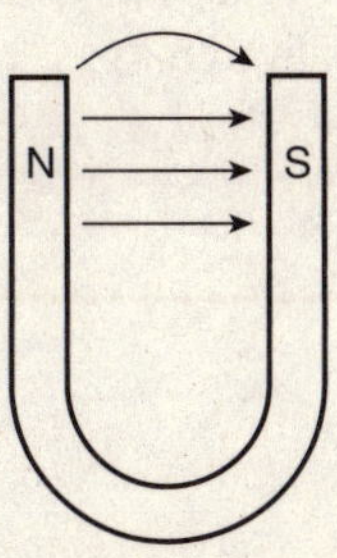

A *domain* is a cluster of magnetically aligned atoms.

A material becomes magnetic when it is placed in a strong external magnetic field, and the clusters of atoms with similar magnetic orientations, called *domains*, become aligned with the external magnetic field. Obviously, some materials are more easily magnetized than others, but the reasons for this are typically not covered on the SAT II: Physics Test.

Magnetic Field Produced by a Current-Carrying Wire

Fundamentally, magnetic fields are produced by moving charges. This is why all atoms are tiny magnets, since the electrons around the nucleus of the atom are moving charges and are therefore magnetic. Prior to 1820, it was generally believed that there was no physical connection between electricity and magnetism. But in that year, Oersted discovered that a current-carrying wire is magnetic, or as we would say in today's language, creates a magnetic field around itself. The magnetic field due to a current-carrying wire circulates around the wire in a direction given by what we will refer to as the *first right-hand rule.*

First Right-Hand Rule: Place your thumb in the direction of the current I, and your fingers will curl around in the direction of the magnetic field produced by that current.

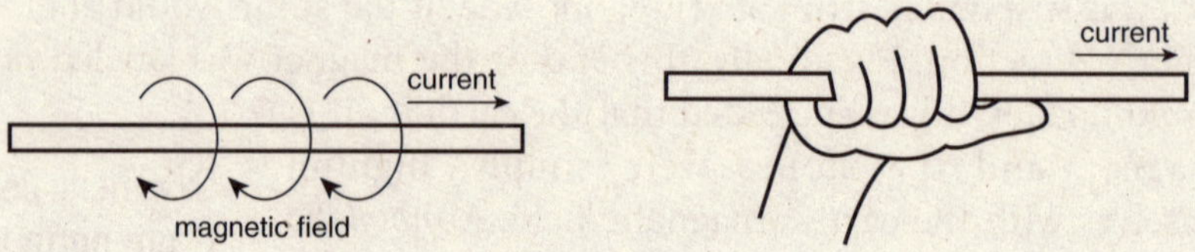

In determining the direction of a magnetic field due to the flow of electrons in a wire, we would use the left hand instead of the right hand.

Use your right hand for conventional current flow or moving positive charges, and your left hand for electron flow or moving negative charges.

Another way of denoting any vector that points into the page or out of the page, in this case the magnetic field vector, is to use an X to represent a vector pointing into the page, and a dot (•) to represent a vector pointing out of the page. Thus, the magnetic field around the current-carrying wire above can be drawn as:

Force on a Current-Carrying Wire in an External Magnetic Field

Since a current-carrying wire creates a magnetic field around itself according to the first right-hand rule, every current-carrying wire is a magnet. Thus, if we place a current-carrying wire in an external magnetic field, it will experience a force. The direction of the force acting on the wire is given by what we will call the *second right-hand rule.*

> **Second Right-Hand Rule:** Place your fingers in the direction of the magnetic field (north to south) with your thumb in the direction of the current in the wire, and the magnetic force on the wire will come out of your palm.

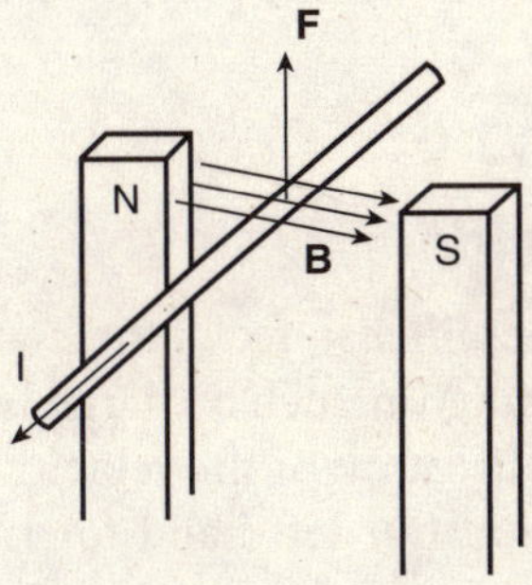

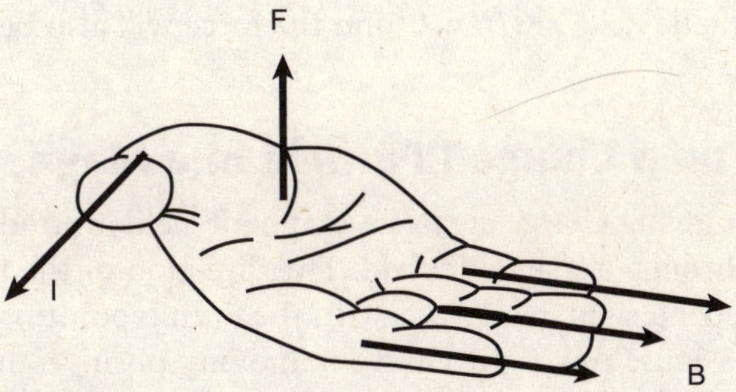

Again, you would use your left hand to find the direction of the magnetic force if you were given electron flow instead of conventional current.

The equation for finding the force on a current-carrying wire in a magnetic field is:

$$F = ILB\sin\theta$$

where I is the current in the wire, L is the length of wire which is in the magnetic field, B is the magnetic field, and θ is the angle between the length of wire and the magnetic field. If the angle is 90°, the equation becomes simply $F = ILB$.

> **Example:** A wire carrying a 20 A current has a length L = 0.10 m between the poles of a magnet at an angle of 45°, as shown. The magnetic field is uniform and has a value of 0.5 T. What is the magnitude and direction of the magnetic force acting on the wire? (sin 45° = cos 45° = 0.7)

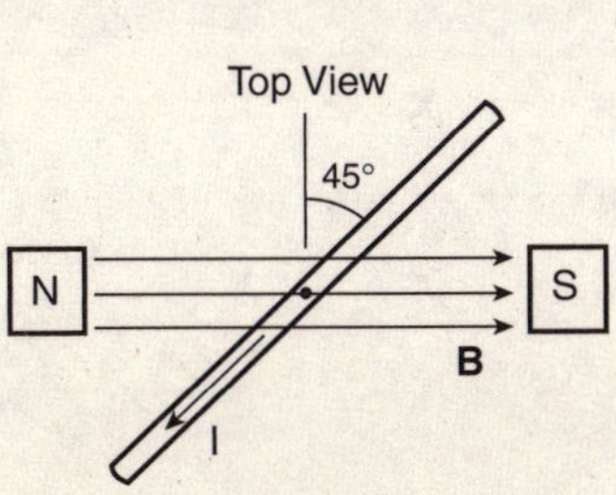

F = ILB sin45°
= 20 × 0.1 × 0.5 × 0.7
= 0.35 × 2
= 0.7 N
out of page

Solution: The magnitude of the force on the wire is found by:

$$F = ILB\sin\theta = (20\ \text{A})(0.10\ \text{m})(0.5\ \text{T})\sin 45°$$

$$F = 0.7\ \text{N}$$

The direction of the force can be found by the second right-hand rule. Place your fingers in the direction of the magnetic field, and your thumb in the direction of the length (and current), which is perpendicular to the magnetic field, and we see that the force is out of the page.

Note that the length must have a component that is perpendicular to the magnetic field, or there will be no magnetic force on the wire. In other words, if the wire is placed parallel to the magnetic field, $\sin 0° = 0$, and the force will also be zero.

Force on a Charged Particle in a Magnetic Field

Since a moving charge creates a magnetic field around itself, it will also feel a force when it moves through a magnetic field. The direction of the force acting on such a charge is given by the second right-hand rule, with the thumb pointing in the direction of the velocity of the charge. We use our right hand for moving positive charges, and our left hand for moving negative charges. The equation for finding the force on a charge moving through a magnetic field is:

$$F = qvB\sin\theta$$

where q is the charge in Coulombs, v is the velocity in m/s, B is the magnetic field in Teslas, and θ is the angle between the velocity and the magnetic field. If the angle is 90°, the equation becomes $F = qvB$.

Use the second right-hand rule to find the force on a charge in a magnetic field.

Example: A proton enters a magnetic field **B** that is directed into the page. The proton has a charge $+q$ and a velocity v which is directed to the right, and enters the magnetic field perpendicularly. Describe the resulting path of the proton.

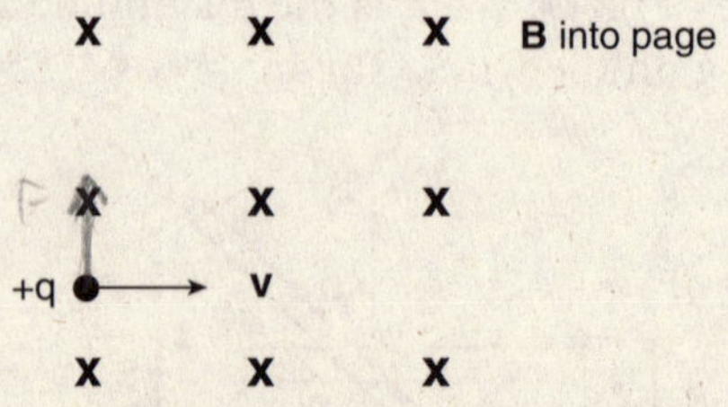

Solution: As the proton enters the magnetic field, it will initially feel a force that is directed upward, as we see from using the second right-hand rule. The path of the proton will curve upward in a circular path, with the magnetic force becoming a centripetal force, changing the direction of the velocity to form the circular path. The radius of this circle can be found by setting the magnetic force equal to the centripetal force:

A charge must be moving across magnetic field lines to feel a force.

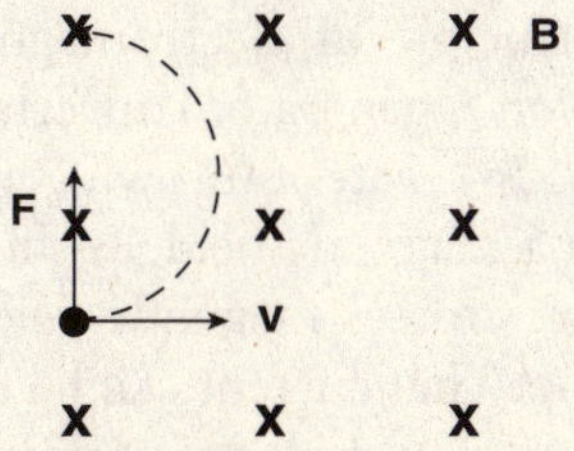

magnetic force = centripetal force

$$qvB = \frac{mv^2}{r}$$

$$r = \frac{mv}{qB}$$

$F = qvB = \frac{mv^2}{r}$

$r = \frac{mv}{qB}$

ELECTROMAGNETIC INDUCTION

For several years after Oersted discovered that a current flowing through a wire produced a magnetic field, several scientists, including Faraday in England and Henry in New York, tried to verify whether or not the opposite were true: Can a magnetic field produce a current? Faraday and Henry independently discovered that an electric current could be produced by moving a magnet through a coil of wire, or, equivalently, by moving a wire through a magnetic field. Generating a current this way is called *electromagnetic induction*.

Moving a magnet through a coil of wire and generating a current is called *electromagnetic induction*.

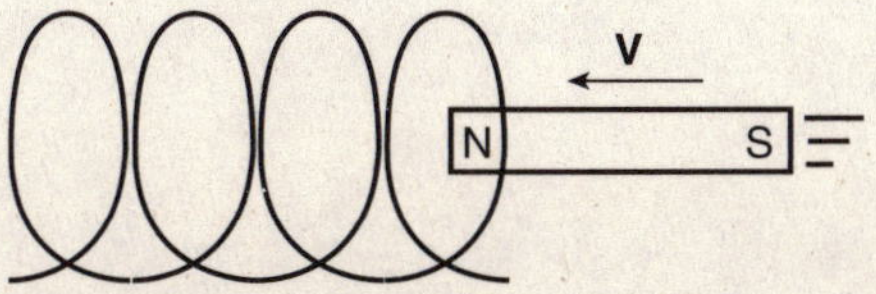

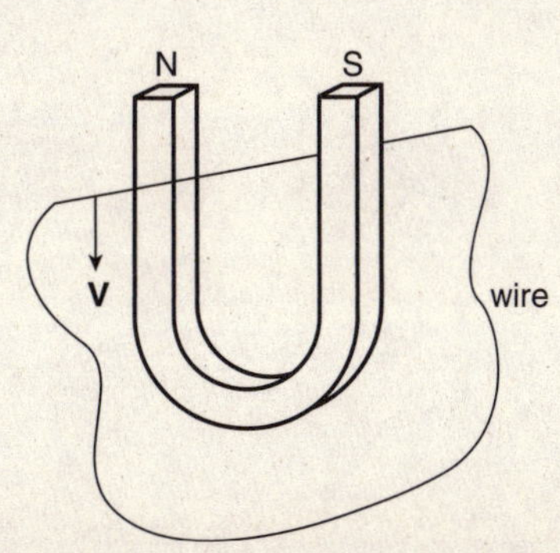

Moving a wire into a magnetic field.

A *generator* converts mechanical energy into electrical energy.

The amount of voltage and current produced in a coil of wire depends on how quickly the magnetic field lines are crossed by the wire. For example, if the magnet is moved through the coil slowly, hardly any current is produced in the coil. If the magnet is moved through the coil quickly, a larger amount of current is produced in the coil. Also, a greater number of coils will produce a greater induced voltage and current.

Quick Quiz:

Which of the following can increase the amount of voltage induced in a coil of wire?

(A) move the magnet faster through the coil

(B) move a stronger magnet through the coils of wire

(C) move the magnet through more coils of wire

(D) move more coils of wire around a magnet

(E) move more magnets simultaneously through a coil of wire

Answer: All of the above can increase the amount of voltage generated in a coil of wire.

The principle of electromagnetic induction is the basis for a *generator*. A generator converts mechanical energy into electrical energy. To create your own generator, place a loop of wire on an axle in a magnetic field, as shown below. As the loop is rotated, the wire crosses magnetic field lines and generates a current in the loop. That current can be used to light a light bulb or power a city. All of our electrical power is generated in a similar way.

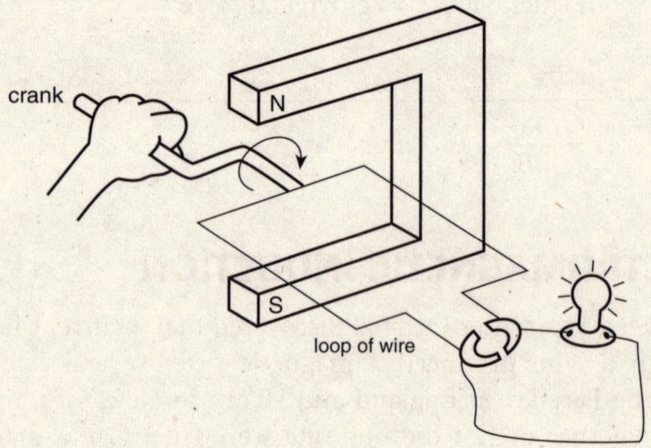

MAGNETIC FIELDS AND FORCES REVIEW QUESTIONS

1. A wire carries a current, creating a magnetic field around itself as shown.

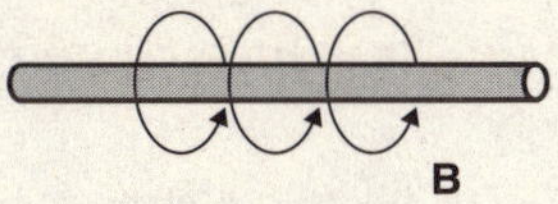

The current in the wire is

(A) directed to the right.
(B) directed to the left.
(C) equal to the magnetic field.
(D) in the same direction as the magnetic field.
(E) zero.

Questions 2–3:
A wire carrying a current of 2 A is placed in a magnetic field of 0.1 T as shown.

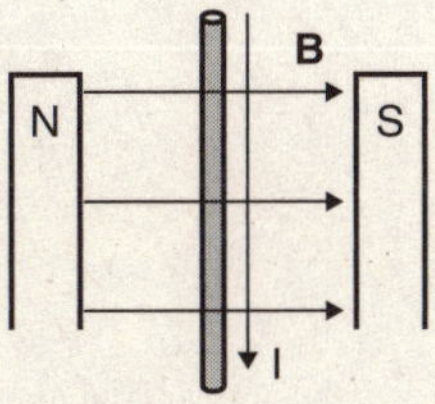

The length of wire in the magnetic field is 0.3 m.

2. The force on the wire is directed

(A) into the page.
(B) out of the page.
(C) toward the top of the page.
(D) toward the bottom of the page.
(E) to the left.

3. The magnitude of the force on the wire is

(A) 0.06 N
(B) 2.0 N
(C) 6.7 N
(D) 0.15 N
(E) 0.015 N

Questions 4–5:
An electron enters a magnetic field as shown.

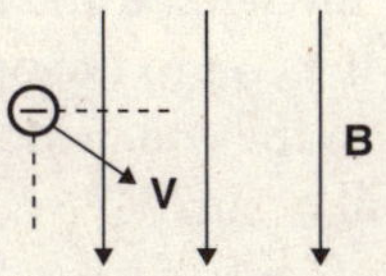

4. The electron will experience a force that is initially

(A) into the page.
(B) out of the page.
(C) toward the top of the page.
(D) toward the bottom of the page.
(E) to the left.

5. The subsequent path of the electron is a

(A) parabola.
(B) straight line.
(C) spiral or helix.
(D) hyperbola.
(E) circle.

6. For electromagnetic induction to occur,
   - (A) a magnet must be at rest within a coil of wire.
   - (B) a coil of wire must be at rest relative to the magnet.
   - (C) a magnet must move through a coil of wire.
   - (D) a magnet and a coil must have the same velocity.
   - (E) a magnet must be pointing north.

7. A magnet moves into a coil of wire inducing a current in the wire. If the magnet pulled back out of the coil in the opposite direction as it went into the coil, which of the following will occur?
   - (A) There will be a current produced in the coil in the same direction as before.
   - (B) There will be a current produced in the coil in the opposite direction as before.
   - (C) There will be no current produced in the coil.
   - (D) The current produced must be stronger than before.
   - (E) The current produced must be weaker than before.

8. A generator
   - (A) converts mechanical energy into electrical energy.
   - (B) converts electrical energy into mechanical energy.
   - (C) converts heat energy into mechanical energy.
   - (D) converts electrical energy into heat energy.
   - (E) converts nuclear energy into heat energy.

SOLUTIONS TO MAGNETIC FIELDS AND FORCES REVIEW QUESTIONS

1. A

By the first right-hand rule, if the fingers curl up and over the wire toward you, the current must be to the right.

2. B

According to the second right-hand rule, the fingers point to the right in the direction of the magnetic field, the thumb points toward the bottom of the page in the direction of the current, and the force comes out of the palm and out of the page.

3. A

$$F = ILB = (2\text{ A})(0.1\text{ T})(0.3\text{ m}) = 0.06\text{ N}$$

4. B

We use the second left-hand rule, since the electron is a negative charge, placing our fingers toward the bottom of the page, and the thumb to the right, in the direction of the component of the velocity which crosses the magnetic field lines. The force, then, comes out of the palm and out of the page.

5. C

The electron will orbit a magnetic field line but will also continue to move toward the bottom of the page, spiraling downward.

6. C

The magnet and coil must be moving relative to each other.

7. B

Pushing the north end of a magnet into a coil will produce a current in the opposite direction than if the north end were pulled out of the coil. Opposite magnet velocities will create opposite currents.

8. A

Mechanical energy is put into a generator, and electrical energy is produced in the form of current.

CHAPTER FOURTEEN

General Wave Properties

A *mechanical wave* is a traveling disturbance in a medium that transfers energy from one place to another. A *medium* is the substance through which a wave moves, such as water for a water wave, or air for a sound wave. An *electromagnetic wave* is a vibration of an electric and magnetic field that travels through space at an extremely high speed, and does not need a medium through which to travel. Visible light, radio waves, and microwaves are all examples of electromagnetic waves. We will return to electromagnetic waves later.

Waves transfer energy from one place to another.

BASIC PROPERTIES OF MECHANICAL WAVES (机械波的基本性质)

There are two types of mechanical waves. *Transverse waves* vibrate in a direction that is perpendicular to the direction of motion of the wave. For example, if you hold the end of a horizontal spring and vibrate your hand up and down, you create a transverse wave in the spring that travels from one end to the other.

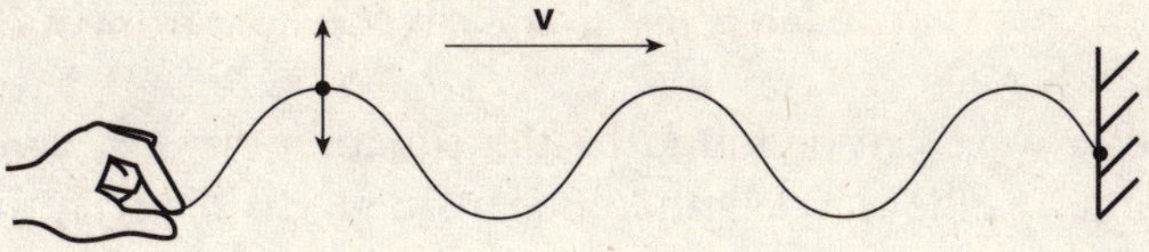

If you gather the spring up into a bunch and let it go, you create a *longitudinal wave*, in which the spring vibrates in a direction which is parallel to the direction of motion of the wave.

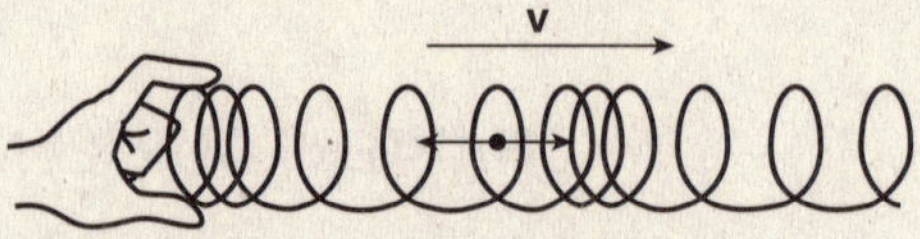

Sound is a common example of a longitudinal wave, since the air through which a sound wave moves is repeatedly compressed and expanded.

The two types of waves are *transverse* and *longitudinal*.

Since an object vibrating with simple harmonic motion can create a wave in a medium, it is not surprising that many of the terms we discussed in chapter 9 can also be applied to waves. A wave has a *period*, the time it takes for a wave to vibrate once; a *frequency*, the number of waves that pass a given point per second; and an *amplitude*, the maximum displacement of a wave, or its height. But a wave also has length. The length of one complete vibration of a wave is called the *wavelength*, and is denoted by the Greek letter *lambda*, λ. The figure below illustrates these quantities.

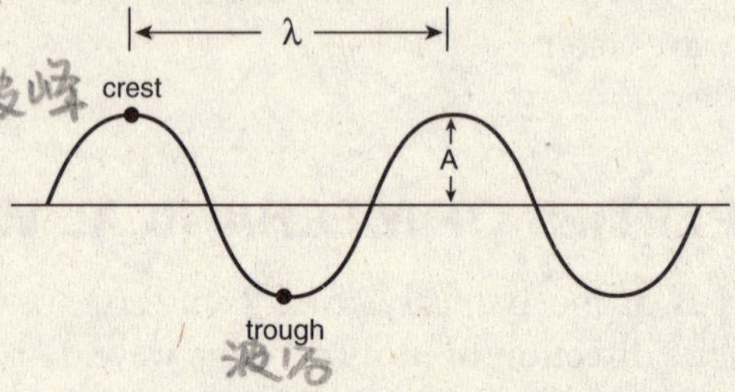

Harmonic motion and wave motion are very closely related.

The *crest* is the highest point on the wave and the *trough* is the lowest point on the wave. The wavelength can be measured from one crest to the next crest, or from one trough to the next trough.

The speed of a wave can be found by the equation $v = f\lambda$, where v is the speed, f is the frequency, and λ is the wavelength. Since frequency is the reciprocal of period, we can also write this equation as:

$$v = \frac{\lambda}{T}$$

The speed of all types of waves can be found using this equation.

Example: Consider the following diagram of a wave:

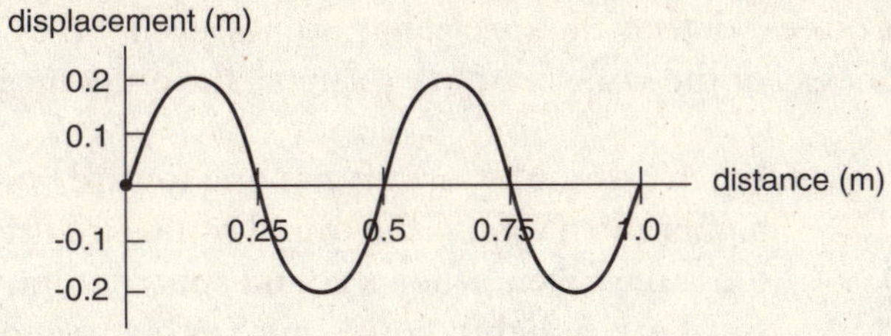

(A) What is the wavelength of the wave?
(B) What is the amplitude of the wave?
(C) If the speed of the wave is 12 m/s, what is the frequency of the wave?
(D) What is the period of the wave?

The wave equation is $v = f\lambda$.

Solution:

(A) The wavelength is the length of one full cycle of the wave, 0.5 m.

(B) The amplitude of the wave is the vertical height from the zero line, which is 0.2 m.

(C) $f = \frac{v}{\lambda} = \frac{12 \text{ m/s}}{0.5 \text{ m}} = 24 \frac{\text{cycles}}{\text{s}} = 24 \text{ Hertz}$

The *period* is the time for one vibration of the wave; the *wavelength* is the length of one wave.

(D) Period $T = \frac{1}{f} = \frac{1}{24 \text{ Hz}} = 0.04 \text{ s}$

Thus, it takes 0.04 seconds for one full vibration of this wave.

SOUND WAVES AND THE DOPPLER EFFECT

Sound is a mechanical longitudinal wave, and therefore must have a medium to travel through. Sound generally travels at about 340 m/s in air, but travels at considerably higher speeds in more dense media such as water or steel. The characteristics of sound which are produced and how we detect and perceive these characteristics are summarized in the table below.

Sound produced as:	Sound detected as:
Frequency	Pitch
Amplitude	Loudness or volume
Harmonics	Quality or tone

Sound is a mechanical longitudinal wave.

The third characteristic, *harmonics*, is the combination of several simultaneous frequencies that give a sound its special tone. For example, we can tell the difference between a trumpet and a clarinet because our ear detects the special harmonics, even if they are playing the same pitch at the same loudness. For the same reason, we can tell the difference between two voices.

The *Doppler effect* is the apparent change in pitch due to relative motion between a sound source and a listener.

When a sound source is moving toward you, you hear a slightly higher pitch than if the sound source is at rest relative to you. By the same token, when a sound source is moving away from you, you hear a slightly lower pitch. This phenomenon is called the *Doppler effect*. For example, if a train is traveling toward you while blowing its horn at a certain pitch, the waves will appear to be arriving at your ear more frequently, increasing the pitch you perceive.

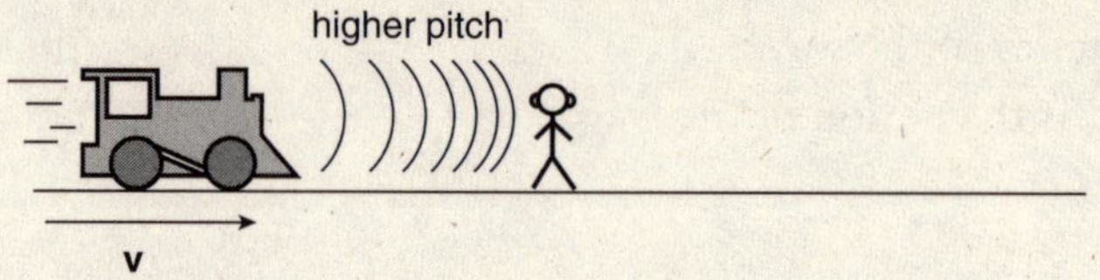

If the train blows its whistle while traveling away from you, the waves will appear to be arriving at your ear less frequently, decreasing the pitch you perceive.

Weather forecasters use radio waves and the Doppler effect to detect the movement of clouds in our atmosphere.

Of course, the frequency of the whistle is not actually changing, but you perceive a change due to the relative motion between you and the train. The Doppler effect also occurs when a light source is moving toward or away from us. The light spectrum of a star, for example, is shifted toward the red (low frequency) end if the star is moving away from us, and toward the blue (high frequency) end of the spectrum if it were to move toward us.

REFLECTION, REFRACTION, DIFFRACTION, AND INTERFERENCE OF MECHANICAL WAVES (机械波的反射，折射，衍射和干涉)

Let's discuss four ways you can affect a wave. *Reflection* is the bouncing of a wave off of a barrier. The *law of reflection* states that the angle of incidence of a wave is equal to the angle of reflection of the wave as measured from a line normal (perpendicular) to the barrier. This simply means that the wave will bounce off at the same angle at which it came in.

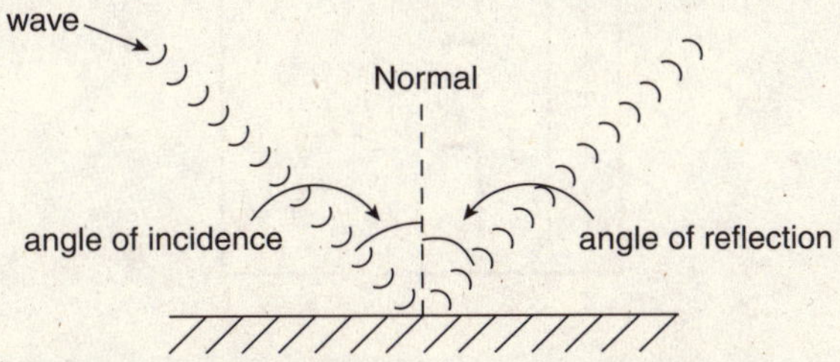

Refraction is the bending of a wave due to a change in medium. A water wave moving from deep water to shallow water will bend its path, slow down, and shorten its wavelength.

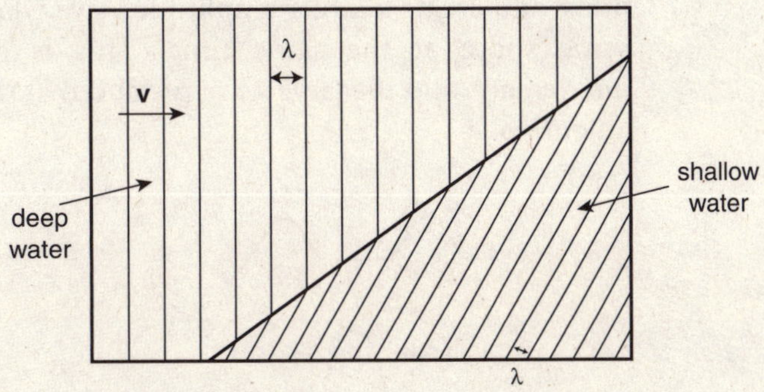

The speed of the wave and its wavelength always change when a wave undergoes refraction, but its frequency does not change. According to the equation $v = f\lambda$, a slower speed means having a shorter wavelength, and a higher speed means having a longer wavelength.

Four things you can do with a wave: reflection, refraction, diffraction, interference.

Diffraction is the bending of a wave around a barrier, such as water waves bending around a rock in a stream, or sound waves bending around the corner of a building. When water waves pass through a narrow opening, the sides of the waves drag on the walls of the opening, causing these parts of the wave to lag behind the center of the wave and creating a semi-circular wave pattern, as shown in the figure below.

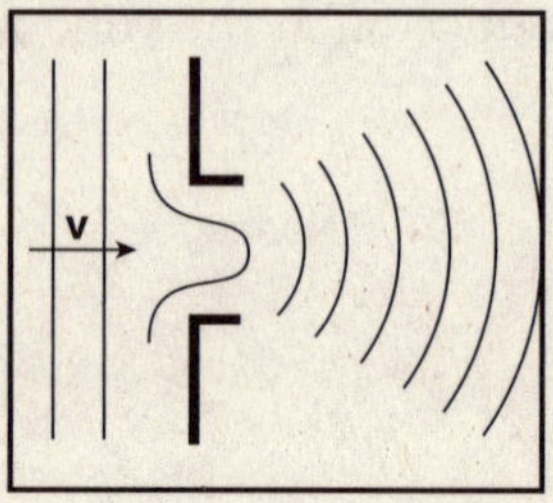

Refraction is the bending of a wave due to a change in medium; *diffraction* is the bending of a wave around a barrier.

When two waves are traveling in the same medium at the same time, they *interfere* with each other. For example, consider two waves moving toward each other in the same rope. If the waves are moving on the same side of the rope, they build on each other and create a larger amplitude wave when they occupy the same space at the same time. This is called *constructive interference,* and the large wave produced at that instant is called an *antinode.*

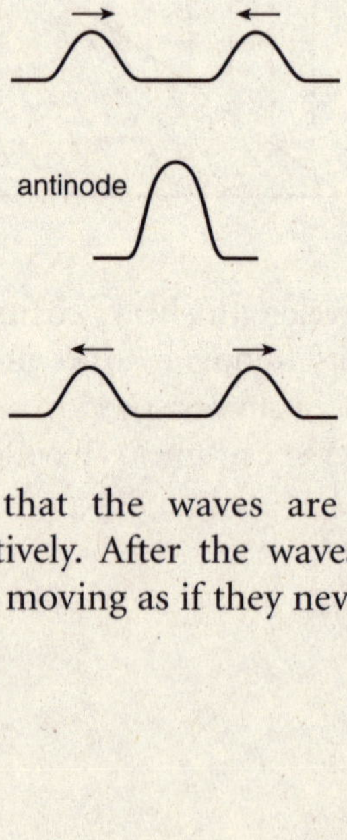

Two types of interference: constructive and destructive.

We say that the waves are *in phase* when they interfere constructively. After the waves pass through each other, they continue moving as if they never interfered.

If two waves of equal amplitude approach each other on opposite sides of the rope, they interfere *destructively*, and the waves destroy each other for the instant they are occupying the same space, and a *node* is created at that point. A node is a point of no displacement, in this case resulting in a flat rope.

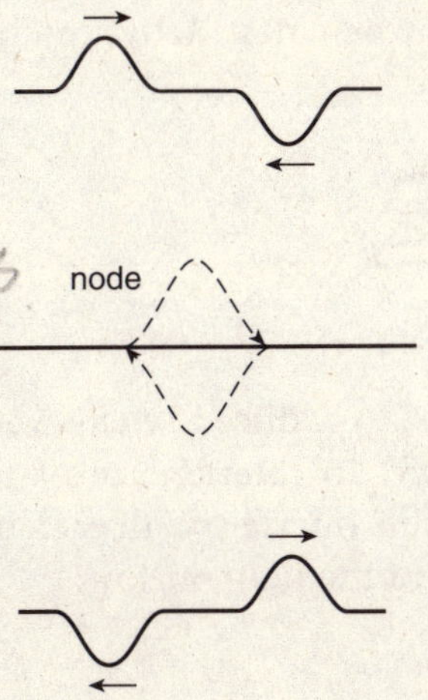

We say that the waves are *out of phase* when they interfere destructively. Once again, after the waves pass through each other, they continue moving as if they never interfered.

Quick Quiz:

Can two waves of different amplitudes interfere destructively?

Answer: Yes, but if they have different amplitudes, they won't cancel each other completely while they are interfering.

STANDING WAVES

The term *standing wave* is an oxymoron, since waves must move and never actually stand still. But waves can appear to stand still when two identical waves traveling in opposite directions in the same medium at the same time create a series of nodes and antinodes. Consider a rope tied to a wall. If we send a wave down the rope toward the wall, the wave reflects off the wall on the opposite side from which it was sent, according to the law of reflection.

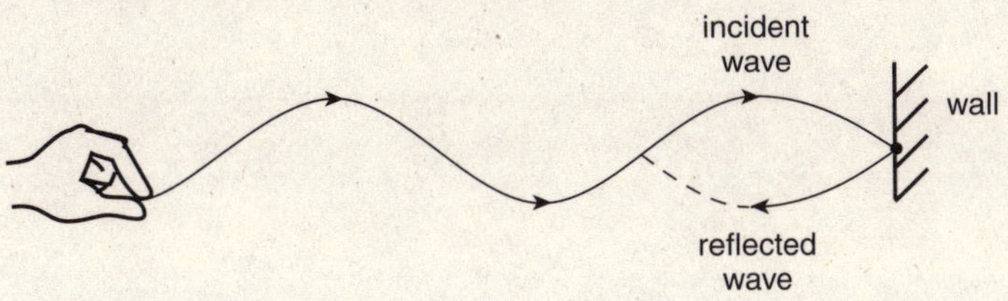

A *standing wave* is a series of nodes and antinodes.

If we continue to send regular waves down the rope and they continue to reflect off the wall, the incident and reflected waves will reinforce each other in some places and cancel each other in other places. The result is a series of antinodes (loops) where constructive interference is occurring, and nodes (points of no displacement between the loops) where destructive interference is occurring, and we call the pattern produced a standing wave.

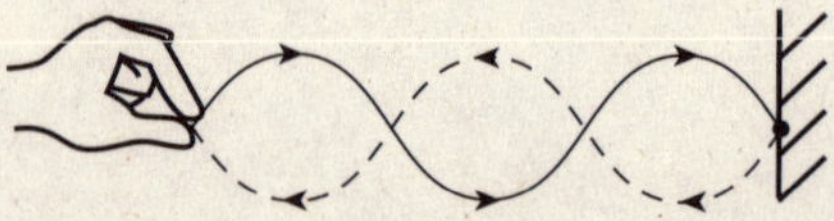

Another type of standing wave is produced when water waves are passed through two openings, called a double-slit, and an interference pattern results. The semicircular wave patterns which emerge from the slits interfere with each other, creating nodes and antinodes, and we see a pattern like the one in the figure below:

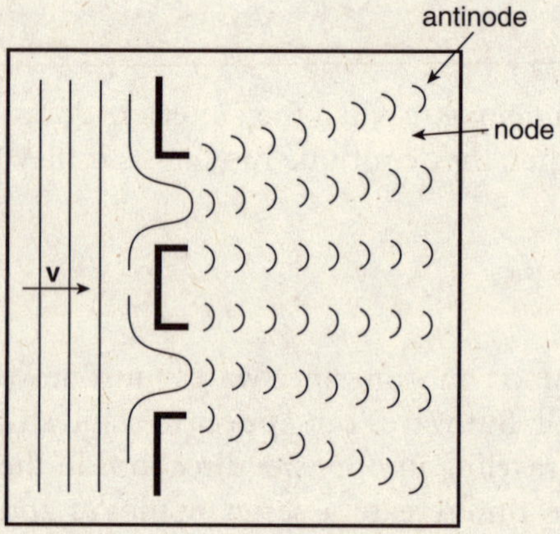

Since light is a wave, we will apply all of the characteristics of waves that we have discussed to the characteristics of light in the next few chapters.

GENERAL WAVE PROPERTIES REVIEW QUESTIONS

Questions 1–3:

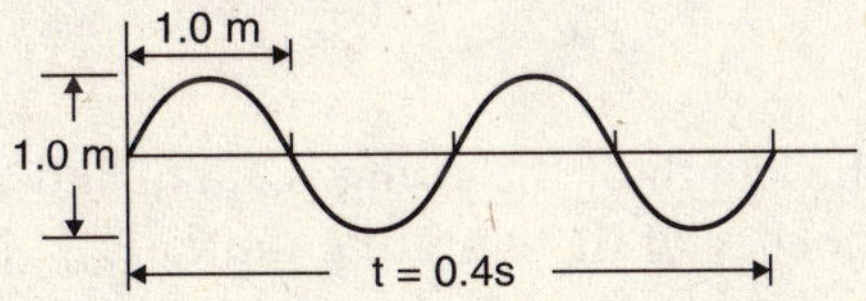

1. The wavelength of the wave shown above is

   (A) 0.5 m
   (B) 1.0 m
   (C) 2.0 m
   (D) 4.0 m
   (E) 6.0 m

2. The amplitude of the wave shown above is

   (A) 0.5 m
   (B) 1.0 m
   (C) 2.0 m
   (D) 4.0 m
   (E) 6.0 m

3. The frequency of the wave shown above is

   (A) 2 Hz
   (B) 4 Hz
   (C) 5 Hz
   (D) 0.4 Hz
   (E) 0.2 Hz

4. A girl on the beach watching water waves sees 4 waves pass by in 2 seconds, each with a wavelength of 0.5 m. The speed of the waves is

   (A) 0.25 m/s
   (B) 0.5 m/s
   (C) 1.0 m/s
   (D) 2.0 m/s
   (E) 4.0 m/s

5. The Doppler effect produces apparent changes in

   (A) loudness.
   (B) pitch.
   (C) amplitude.
   (D) velocity.
   (E) acceleration.

6. As a wave passes from a spring to another spring with a greater tension,

   (A) the speed of the wave decreases.
   (B) the frequency of the wave increases.
   (C) the frequency of the wave decreases.
   (D) the amplitude of the wave increases.
   (E) the speed of the wave increases.

7. The diffraction of a wave through a single opening produces

   (A) refraction.
   (B) an angle of incidence.
   (C) a semicircular wave pattern.
   (D) an increase in speed of the wave.
   (E) a decrease in speed of the wave.

B 8. Two waves approach each other in the same rope at the same time, as shown.

When the two waves are exactly between points P and Q, the shape of the rope will be

(A) 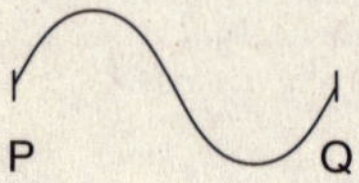

(B) 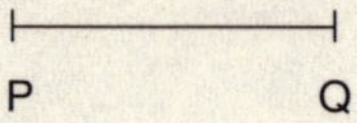

(C) 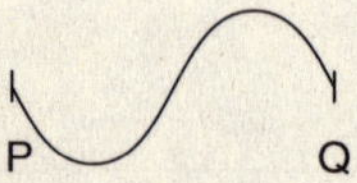

(D) 

(E) 

D 9. A standing wave is produced in a vibrating string as shown.

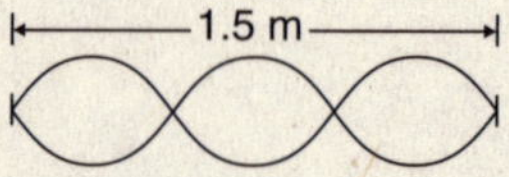

If the length of the string is 1.5 m and the frequency of the vibrating motor is 60 Hz, the speed of the wave is

(A) 15 m/s
(B) 20 m/s
(C) 40 m/s
(D) 60 m/s
(E) 90 m/s

$\lambda = \frac{2}{3} \times 1.5 = 1$ m

$v = f\lambda$
$= 60 \times 1$
$= 60$ m/s

SOLUTIONS TO GENERAL WAVE PROPERTIES REVIEW QUESTIONS

1. C

The wavelength is the length of one full cycle.

2. A

The amplitude is half the distance from trough to crest, or $\frac{1}{2}(1.0 \text{ m}) = 0.5 \text{ m}$.

3. C

The wave vibrates two times per 0.4 s, so the frequency is 2 cycles/0.4 s = 5 Hz.

4. C

The frequency is 4 waves/2 s = 2 Hz, and the speed $v = f\lambda = (2 \text{ Hz})(0.5 \text{ m}) = 1.0 \text{ m/s}$.

5. B

The pitch of a moving sound source sounds higher if it is moving toward you, and lower if it is moving away from you.

6. E

A wave will travel faster in a tight spring than a loose spring. !!!

7. C

The center of the wave passes through the opening while the edges of the wave drag on the sides of the opening.

8. B

When both waves are exactly between P and Q, one wave's trough is superposed on the other wave's crest, causing the waves to interfere destructively, canceling each other out.

9. D

The wavelength is the length of two loops and is 1 m. The speed of the wave is $v = f\lambda = (60 \text{ Hz})(1 \text{ m}) = 60 \text{ m/s}$.

CHAPTER FIFTEEN

Reflection and Refraction of Light

（光的反射与折射）

ELECTROMAGNETIC WAVES

（电磁波）

As we briefly discussed in the last chapter, an electromagnetic wave is a vibration of an electric and magnetic field that moves through space at an extremely high speed. The electromagnetic wave spectrum, listed from lowest frequency to highest frequency, include radio waves, microwaves, infrared, visible light, ultraviolet, x-rays, and gamma rays. They are all a result of the same phenomena, and although they have different wavelengths and frequencies, they all travel through a vacuum at exactly the same speed: 3×10^8 m/s, or about 670 million miles per hour. This speed is often referred to as the speed of light, although light is just one example of an electromagnetic wave. More accurately, this speed is the speed of any electromagnetic wave in a vacuum. In any case, the speed of an electromagnetic wave is given the symbol c, from the Latin word *celeritas*, meaning "swift."

All electromagnetic waves travel at the same speed in a vacuum, 3×10^8 m/s.

Example: Red light has a wavelength of 7×10^{-7} m. What is the frequency of red light?

Solution: We are given the wavelength of the red light, and we know the speed of the light as well, $c = 3 \times 10^8$ m/s. The frequency can be found using the wave equation:

$$f = \frac{c}{\lambda} = \frac{3 \times 10^8 \text{ m/s}}{7 \times 10^{-7} \text{ m}} = 4.3 \times 10^{14} \text{ Hz}$$

If the visible colors of light are listed from long wavelength (low frequency) to short wavelength (high frequency), they would follow this order: red, orange, yellow, green, blue, and violet (ROYGBV).

REFLECTION OF LIGHT (光的反射)

Any wave that bounces off of a barrier follows the *law of reflection*: The angle of incidence is equal to the angle of reflection as measured from a line normal (perpendicular) to the barrier. In the case of light, the barrier is often a mirror.

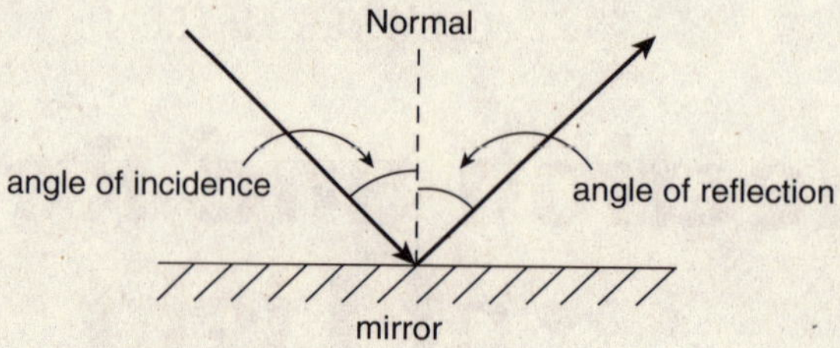

Law of reflection: The angle of incidence equals the angle of reflection as measured from the normal line.

This is why if someone can see you in a mirror, you can see them in the mirror as well.

REFRACTION OF LIGHT (光的折射)

If you put a pencil in a clear glass of water, the image of the pencil in the water appears to be bent and distorted. The light passing from the air into the water is *refracted*, bending due to the fact that it's passing from one medium to another. If we consider a single beam of laser light, we can observe it as it passes from air into a piece of glass.

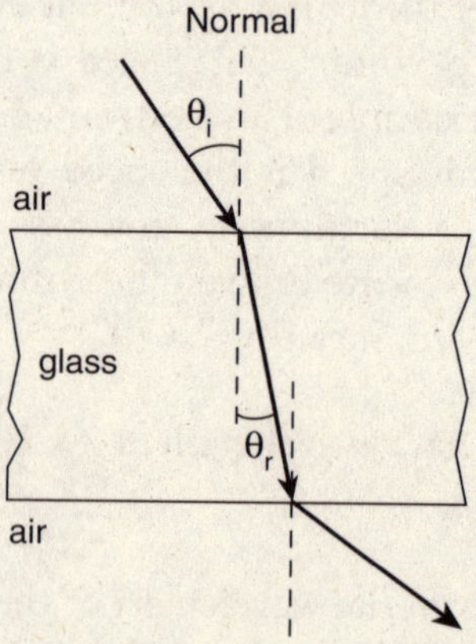

The angle θ_i from the normal line at which the beam approaches the glass from the air is called the *angle of incidence*. The angle θ_r from the normal line in the glass is the *angle of refraction*. As the light passes from the air (a less dense medium) into the glass (a more dense medium), the beam bends toward the normal. When the beam of light exits the glass and passes back into the air, it bends away from the normal at the same angle it entered the glass from the air.

Refraction is the bending of a light ray due to a change in medium.

The light bends toward the normal in the glass because the beam slows down as it enters the glass. Light travels more slowly in a more dense medium. Recall that sound travels faster in a more dense medium; however, sound is a mechanical wave, while light is an electromagnetic wave. The ratio of the speed of light in air (approximately a vacuum) to the speed of light in the glass (or any other medium) is called the *index of refraction n*:

The *angle of reflection* is the angle at which light bounces off a barrier, and the *angle of refraction* is the angle at which light bends through a substance. Both are measured from a line normal to the surface, but they are not equal.

$$n = \frac{c}{v_{glass}}$$

Example: The index of refraction for water is 1.3. What is the speed of light in water?

The *index of refraction* tells us how much light slows down in a more dense medium.

Solution: Since $n = \frac{c}{v_{water}}$, then:

$$v_{water} = \frac{c}{n} = \frac{3 \times 10^8 \text{ m/s}}{1.3} = 2.3 \times 10^8 \text{ m/s}$$

Prism Dispersion (棱镜的色散)

Each color in the spectrum refracts just a little differently than every other color. This is why we can separate white light into its component colors by passing it through a prism. The shorter wavelengths slow down and bend more than the longer wavelengths, so violet bends the most, and red the least:

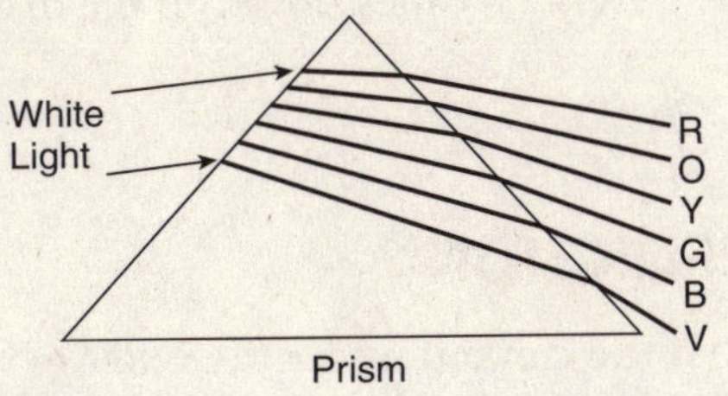

Snell's Law of Refraction

We can relate the index of refraction to the angles of incidence and refraction by using *Snell's law of refraction*:

$$n_1 \sin \theta_1 = n_2 \sin \theta_2$$

where n_1 and n_2 are the indices of refraction of the first and second media, and θ_1 and θ_2 are the angles of incidence and refraction, respectively.

Snell's law of refraction relates the angles of incidence and refraction to the index of refraction.

Example: A beam of light enters the flat surface of a diamond from the air at an angle of 30° from the normal. The angle of refraction in the diamond is measured to be 12° from the normal. What is the index of refraction of this diamond? Interpret your answer physically. (sin 30° = 0.5, sin 12° = 0.2)

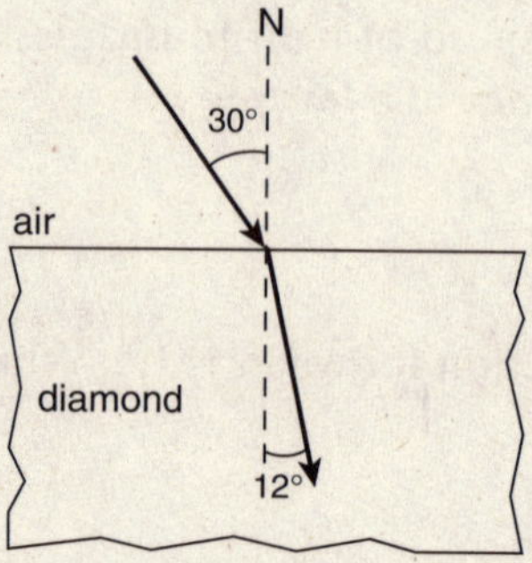

Solution: Let medium 1 be air and medium 2 be diamond. Using Snell's law, we can find n_2, the index of refraction of the diamond.

$$n_1 \sin \theta_1 = n_2 \sin \theta_2$$

where $n_1 = 1$ for air, $\theta_1 = 30°$, and $\theta_2 = 12°$. Then we have:

$$n_2 = \frac{n_1 \sin \theta_1}{\sin \theta_2} = \frac{(1)\sin 30}{\sin 12} = 2.5$$

This means that the light travels 2.5 times slower in the diamond than it does in air.

TOTAL INTERNAL REFLECTION

Consider a waterproof laser that you can put under the water to shine a beam of light up out of the water into the air. If you shine the light at a small angle relative to the normal, the light will emerge from the water and bend away from the normal as it enters the air.

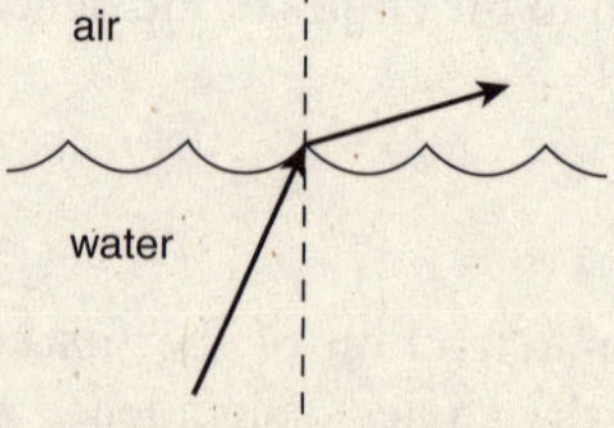

As you increase the angle at which the laser is pointed at the surface of the water, the refracted angle also increases, eventually causing the refracted ray to bend parallel to the surface of the water:

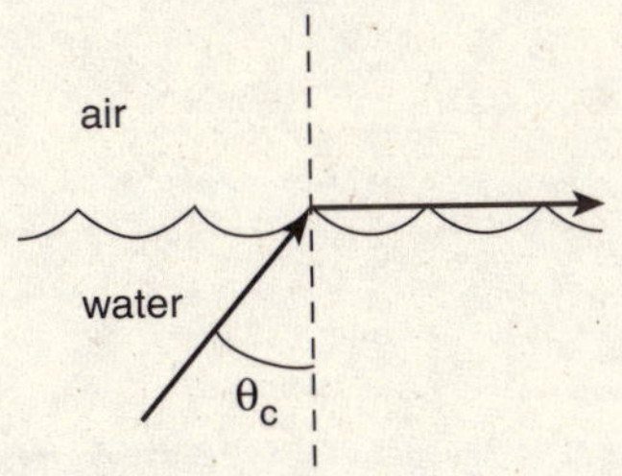

The angle of incidence in this case is called the *critical angle*, θ_c. If the laser is pointed at an angle greater than the critical angle, the beam will not emerge from the water, but will reflect back into the water, or, equivalently, have an index of refraction of 90° from the normal.

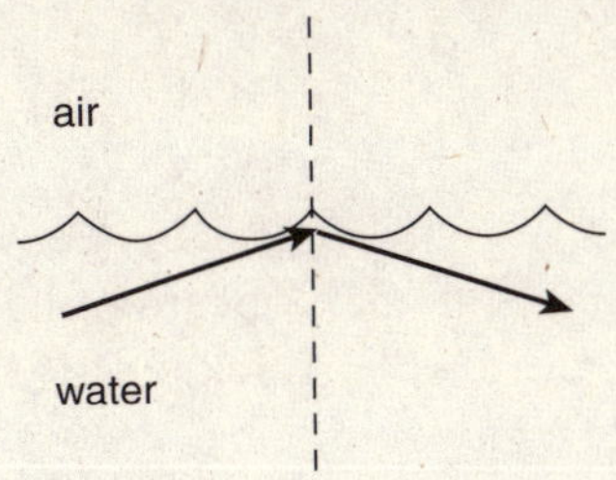

This phenomenon is called *total internal reflection*. The inside surfaces of a glass prism in a pair of binoculars can become like mirrors, reflecting light inside the prism if the light is pointed at the surface at an angle greater than the critical angle. Total internal reflection is the also the principle behind the transmitting of light waves through transparent fiber optic cable for communication purposes.

REFLECTION AND REFRACTION OF LIGHT REVIEW QUESTIONS

1. Which of the following lists the electromagnetic waves from longest wavelength to shortest wavelength?

   (A) x-ray, ultraviolet, visible light
   (B) visible light, ultraviolet, x-ray
   (C) ultraviolet, x-ray, visible light
   (D) ultraviolet, visible light, x-ray
   (E) x-ray, visible light, ultraviolet

2. In a vacuum, all electromagnetic waves have the same

   (A) amplitude.
   (B) frequency.
   (C) wavelength.
   (D) intensity.
   (E) speed.

3. Violet light has a frequency of 7.5×10^{14} Hz. The wavelength of violet light is most nearly

   (A) 2.25×10^{23} m
   (B) 2.5×10^{6} m
   (C) 4×10^{-7} m
   (D) 7.5×10^{-7} m
   (E) 2.25×10^{7} m

4. In the figure shown, the angle of incidence is θ.

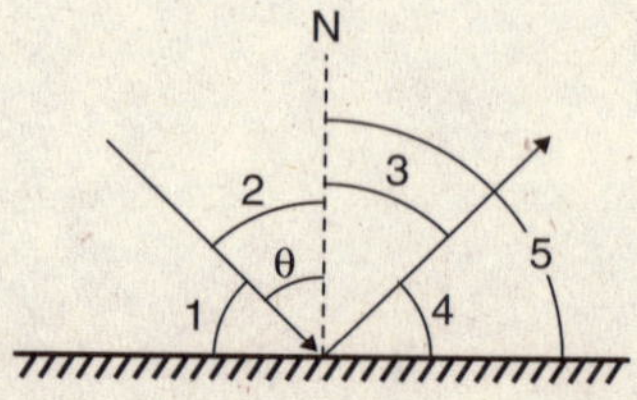

   Which angle is the angle of reflection?

   (A) 1
   (B) 2
   (C) 3
   (D) 4
   (E) 5

5. A beam of light passes from the air through a thick piece of glass as shown.

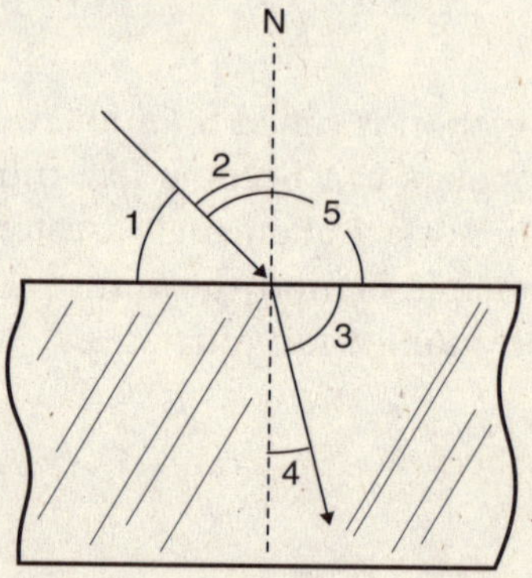

   Which of the following angles is the angle of refraction?

   (A) 1
   (B) 2
   (C) 3
   (D) 4
   (E) 5

B 6. The speed of light in a piece of crown glass is 2×10^8 m/s. What is the index of refraction of crown glass?

(A) 2
(B) 1.5
(C) 0.67
(D) 0.33
(E) 0.2

$n = \frac{3 \times 10^8}{2 \times 10^8} = 1.5$

A 7. A beam of light passes from air into glass. Which of the following statements is true?

(A) The angle of incidence is greater than the angle of refraction in the glass.
(B) The angle of incidence is less than the angle of refraction in the glass.
(C) The angle of incidence is equal to the angle of refraction in the glass.
(D) The frequency of the light decreases.
(E) The frequency of the light increases.

} The frequency remains constant.

D 8. Total internal reflection occurs when

(A) light passes from air into water. ×
(B) light refracts as it exits glass into air.
(C) light reflects off of a mirror. ×
(D) light passing through glass is reflected inside the glass.
(E) the angle of incidence is less than the critical angle. ×

☆: lowest frequency to the highest frequency:

radio waves, microwaves, infrared, visible light, ultraviolet, x-rays, gamma rays

longest wavelength to shortest wavelength

radio waves, microwaves, infrared, visible light, ultraviolet, x-rays, gamma rays.

SOLUTIONS TO REFLECTION AND REFRACTION OF LIGHT REVIEW QUESTIONS

1. B

Visible light has the longest wavelength and lowest frequency of the three, and x-rays have the shortest wavelength and the highest frequency of the three.

2. E

All electromagnetic waves travel at 3×10^8 m/s in a vacuum.

3. C

$$\lambda = \frac{c}{f} = \frac{3 \times 10^8 \text{ m/s}}{7.5 \times 10^{14} \text{ Hz}} = 4 \times 10^{-7} \text{ m}$$

4. C

The angles of incidence and reflection are measured from the normal and are equal.

5. D

The angles of incidence and refraction are measured from the normal, and angle 2 is the angle of incidence and angle 4 is the angle of refraction in the glass.

6. B

$$n = \frac{c}{v} = \frac{3 \times 10^8 \text{ m/s}}{2 \times 10^8 \text{ m/s}} = 1.5$$

7. A

When the light enters the glass, the beam bends toward the normal line, causing the angle of refraction to be less than the angle of incidence. The frequency remains constant.

8. D

Total internal reflection implies that no light exits the glass since it is reflected inside the glass.

CHAPTER SIXTEEN

Polarization, Diffraction, and Interference of Light

POLARIZATION

Light is a vibration of an electric and a magnetic field. If you could watch light waves coming toward you, you would see that they actually vibrate in many directions. We say the light is *polarized* when it is forced to vibrate in only one plane:

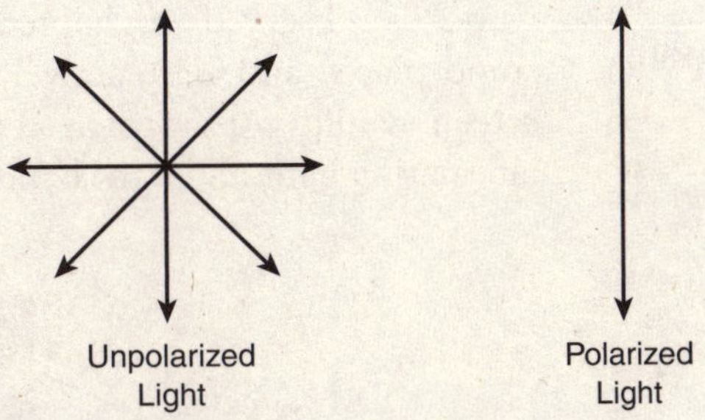

Polarizing sunglasses block out all but one of the vibrations of light so that considerably less light reaches your eyes after passing through the lenses. This is kind of like passing a rope through picket fence and shaking the rope in many directions. The only wave that passes through the fence is vertical.

We say that light is *polarized* when it is made to vibrate in only one plane.

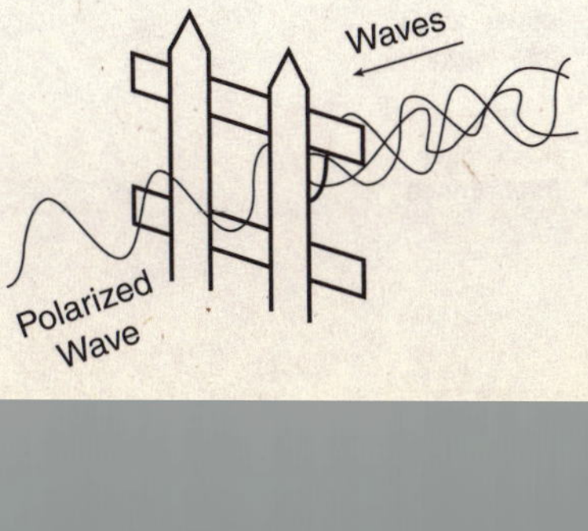

DIFFRACTION AND INTERFERENCE

Diffraction is the bending of a wave around a barrier or through an opening. If we pass a light wave through a narrow single slit, it will behave very similarly to the water waves discussed earlier, with the edges of the light waves lagging behind the center of the waves. If we place a screen opposite to the single-slit opening, we would see a bright light near the center of the screen with narrow lines becoming dimmer toward the edges of the screen. These lines form because of interference from light waves arriving from different locations within the slit.

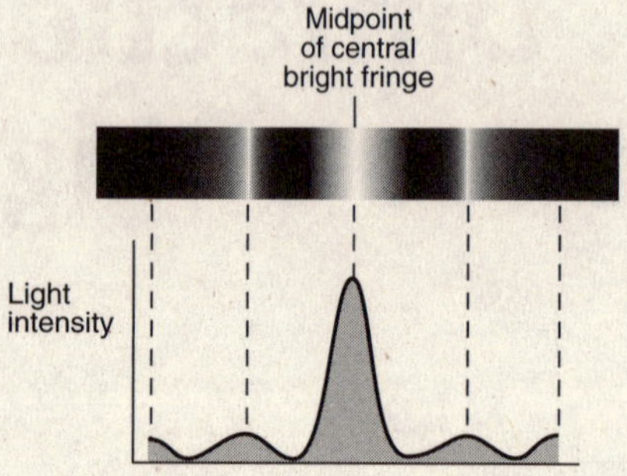

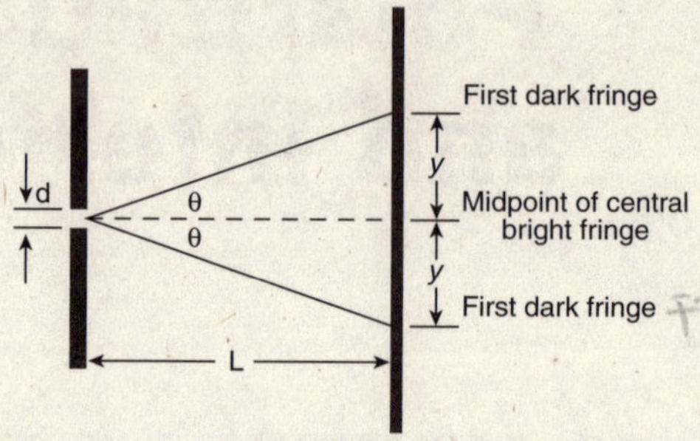

A single-slit will produce bright light in the center of a screen getting dimmer toward the edges, and a double-slit will produce a central bright band with alternating light and dark bands toward the edges of the screen.

If we replace the single-slit opening with a double-slit opening, the pattern on the screen changes. As the light passes through the two openings, it becomes two sources of light waves instead of one. These two light waves behave like the semicircular water waves we observed in chapter 14, interfering constructively in some places and destructively in others. The pattern on the screen would consist of a central bright band of light, with alternating light and dark bands toward the edges of the screen:

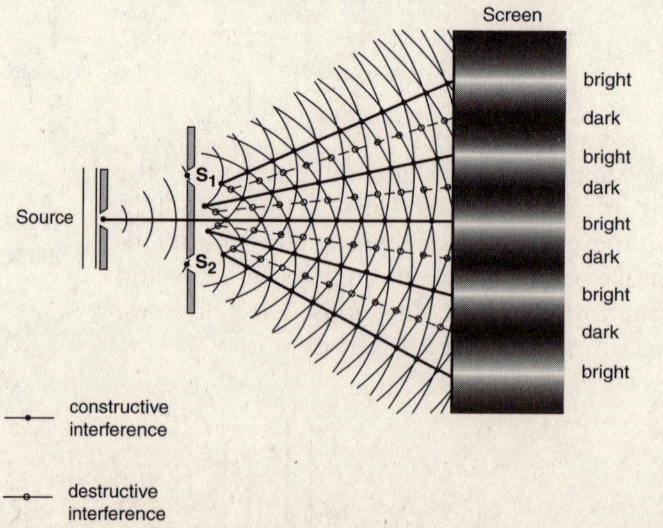

The bright bands on the screen are the places where constructive interference is occurring (antinodes), and the dark bands are a result of destructive interference (nodes). In 1801, Thomas Young was able to measure the wavelength of light waves using this double-slit diffraction pattern. He found that for a given distance between the slits and length from the slits to the screen, the width of the bright central antinode on the screen is proportional to the wavelength of the light. Thus, red light would produce a wider bright central band than violet light.

The width of the central antinode produced in an interference pattern is proportional to the wavelength of the light.

POLARIZATION, DIFFRACTION, AND INTERFERENCE OF LIGHT REVIEW QUESTIONS

1. Making a light wave vibrate in only one plane is called

    (A) refraction.
    (B) reflection.
    (C) interference.
    (D) diffraction.
    (E) polarization.

2. If light is passed through a narrow, single-slit opening onto a screen, the pattern of light produced on the screen is

    (A) alternating bright and dark lines of equal width.
    (B) a bright central band of light with much smaller, dimmer bands toward the edges.
    (C) concentric circles of light.
    (D) one circle of light.
    (E) one band of light.

3. If light is passed through a double-slit opening onto a screen, the pattern produced on the screen is

    (A) a bright central band of light with slightly diminished, alternating bright and dark bands called nodes and antinodes.
    (B) a bright central band of light with tiny lines toward the edges of the screen.
    (C) a large circle of light with tiny circles around it.
    (D) equally sized concentric circles of light.
    (E) one antinode and no nodes.

4. Which of the following colors, when passed through a double-slit opening, will produce the widest central band of light?

    (A) red
    (B) orange
    (C) yellow
    (D) green
    (E) blue

SOLUTIONS TO POLARIZATION, DIFFRACTION, AND INTERFERENCE OF LIGHT REVIEW QUESTIONS

1. E

All planes of vibration of the light except one are blocked when a light wave is polarized.

2. B

The pattern would look like this:

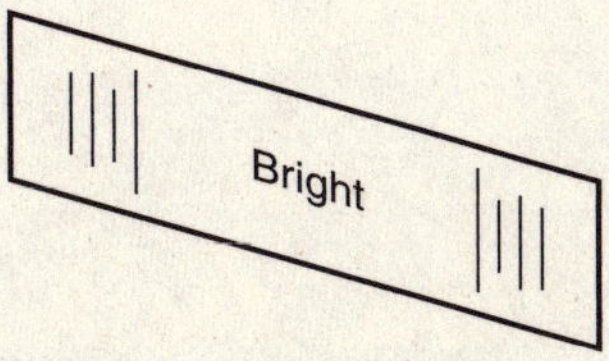

3. A

The pattern would look like this:

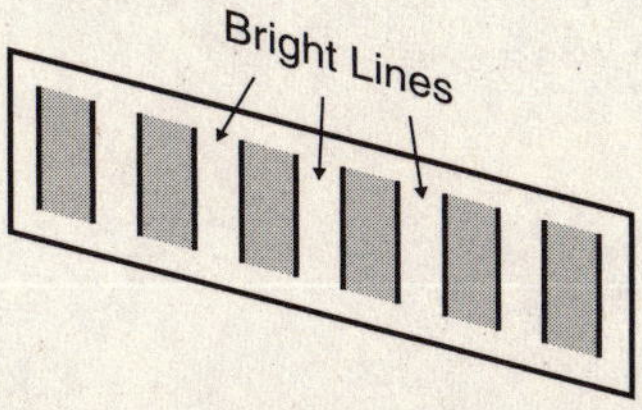

4. A

The width of the central band is proportional to the wavelength of the light, and red has the longest wavelength of the colors listed.

CHAPTER SEVENTEEN

Ray Optics

We can apply the principles of reflection and refraction to the images formed by mirrors and lenses.

MIRRORS

We will discuss three types of mirrors in this chapter: *plane, diverging,* and *converging.* A *plane mirror* is simply a flat mirror. From your everyday experience, you know that a plane mirror always produces an image which is the same size as the object (such as yourself in the morning), left-right reversed, and the same distance behind the surface of the mirror as the object is in front of the mirror. We say that the image formed by a plane mirror is *virtual,* since we cannot place a screen behind the mirror and see the image projected on the screen. A virtual image is one that cannot be projected onto a screen.

A *diverging mirror* is sometimes referred to as a *convex mirror*. You may have seen this type of mirror in the corner of a convenience store. The mirror diverges the rays of light which strike it, allowing the clerk at the store to see practically the entire store in one mirror. Light rays coming into the mirror parallel to its principal axis will diverge, or spread apart:

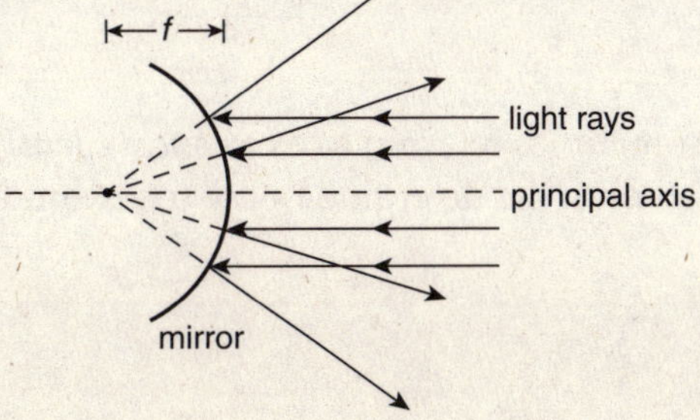

> The *focal length* is the distance from the center of the mirror or lens to the focal point of the rays.

Notice that the rays appear to originate from a point behind the mirror. This point is called the *virtual focus*, and the distance between the surface of the mirror at its center and the focal point is called the *focal length f*.

A *converging mirror* is sometimes referred to as a *concave mirror*. Light rays coming in parallel to the principal axis of the mirror will converge to a focal point.

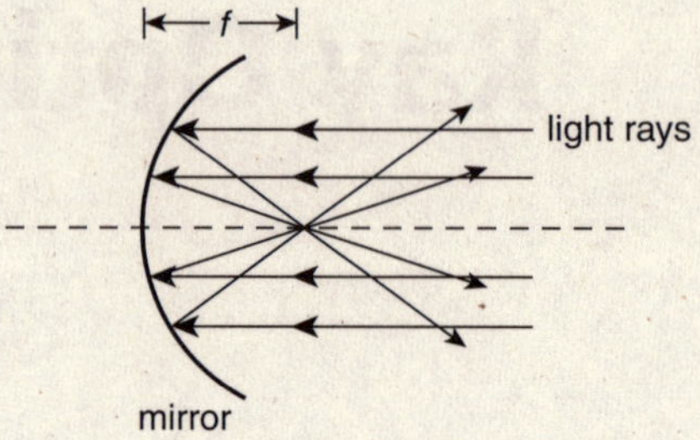

> A diverging mirror is also called a convex mirror; a converging mirror is also called a concave mirror.

If you look into a converging mirror, you will at first see an image of yourself that is inverted (upside-down), but then as you move closer to the mirror, you will see your image turn upright as you pass the focal point of the mirror. Satellite dishes act as converging mirrors for radio and TV waves, gathering them at a detector located at the focal point of the dish. Most research telescopes also use converging mirrors rather than lenses to focus light and study images.

LENSES

Lenses operate on the principle of refraction. A *diverging lens* is a lens that is thicker on the edges than it is in the middle, and it diverges the light rays that pass through it:

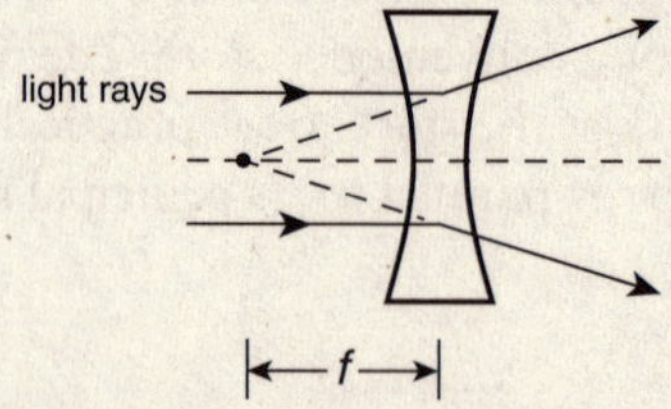

A diverging lens is sometimes referred to as a *concave lens*. The focal point of a diverging lens can be found by extending the diverging rays back behind the lens until they seem to meet.

A *converging lens* is a lens which is thicker in the middle than on the edges, and it converges parallel rays that pass through it:

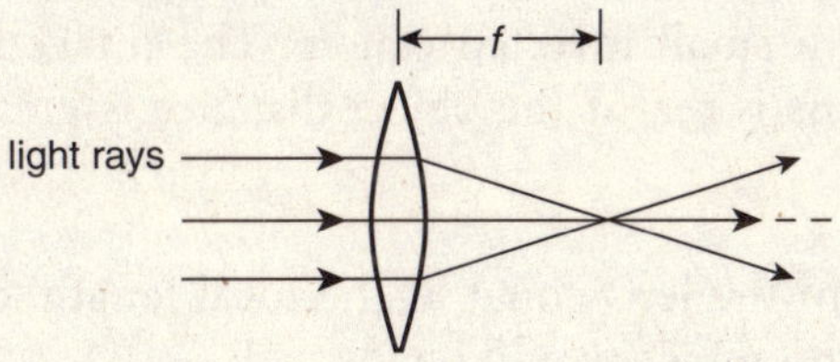

A converging lens is sometimes referred to as a *convex lens.*

A diverging lens is also called a concave lens; a converging lens is also called a convex lens.

Images Formed by a Converging (Convex) Lens

凸透镜（会聚透镜）成的像

Let's take a closer look at the images formed by a converging lens. We said earlier that a *virtual image* is one which cannot be projected onto a screen. On the other hand, a *real image* is one that can be projected onto a screen. If your teacher uses an overhead projector in your classroom, the image projected on the screen is a real image. The image projected on a screen at a movie theater is also real. Under certain circumstances, a converging lens can create an image that can be real or virtual, upright or inverted, enlarged or reduced in size, or the same size as the object. It all depends on where the object is placed relative to the focal length of the lens.

When you read the words *diverging* and *converging*, or *convex* and *concave*, be sure you identify whether the question is asking you about lenses or mirrors. The answers associated with lenses might be quite different from those associated with mirrors!

For our example, let's choose a burning candle as our object, since it produces light from the flame at the top of the candle. We'll first draw a side view of a converging lens, include a horizontal principal axis through the center of the lens, and mark the focal length of the lens on either side of it. We'll place the candle at a distance greater than twice the focal length from the center of the lens:

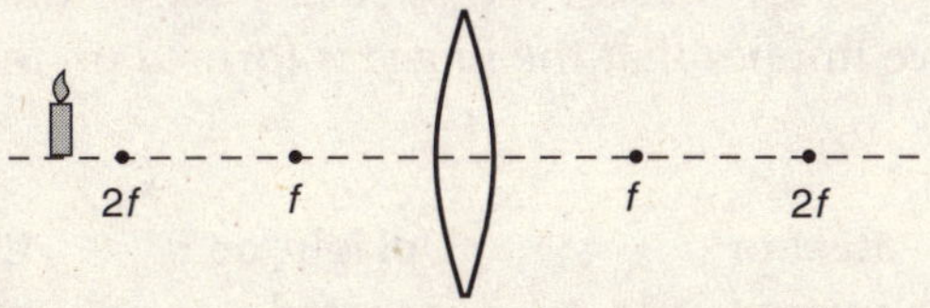

To find out what kind of image will be formed by the lens, we will draw two rays: one ray from the flame entering the lens parallel to the principal axis and bending through the focal point, and another ray from the flame which passes straight through the center of the lens without bending. The image is formed at the location of the intersection of these two rays:

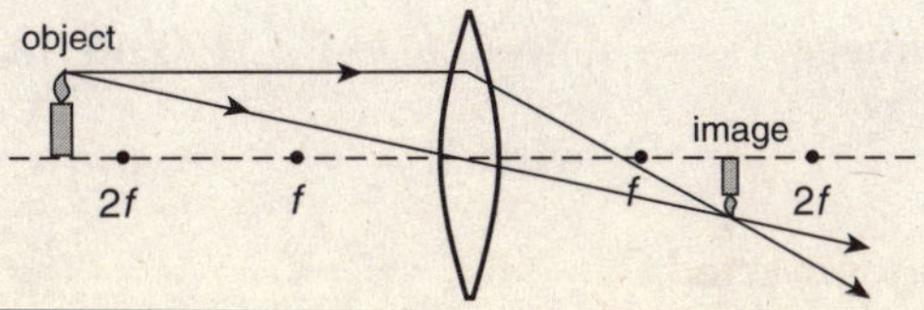

We see that in the case where the object (candle) distance from the lens is greater than twice the focal length, the image is inverted and reduced in size. The image is also real, so if we placed a screen at the location of the image, we would see the projection of a small inverted candle. The image formed by a converging lens is real if the object distance is greater than the focal length.

It takes at least two drawn rays to locate an image formed by a lens.

If we placed the candle at a distance from the lens equal to the focal length, our two rays would emerge parallel to each other, and no image would be formed:

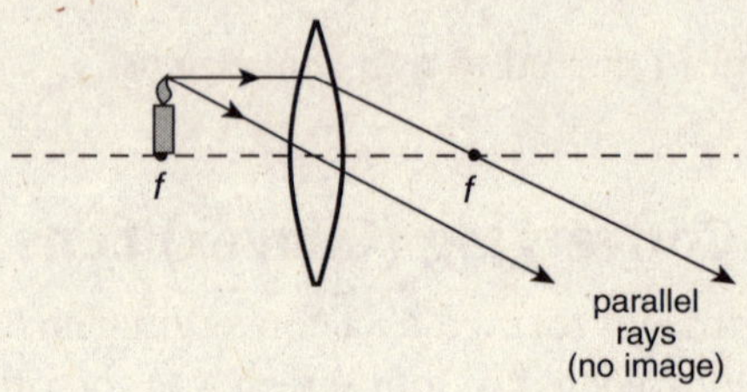

If we placed the candle inside the focal length, our two rays would diverge as they emerged from the lens. No image would be formed on the side opposite to the candle, but if we extended the rays backward, we would find that they seem to originate on the same side as the candle. The point from which they seem to originate is where a virtual image of the candle is formed:

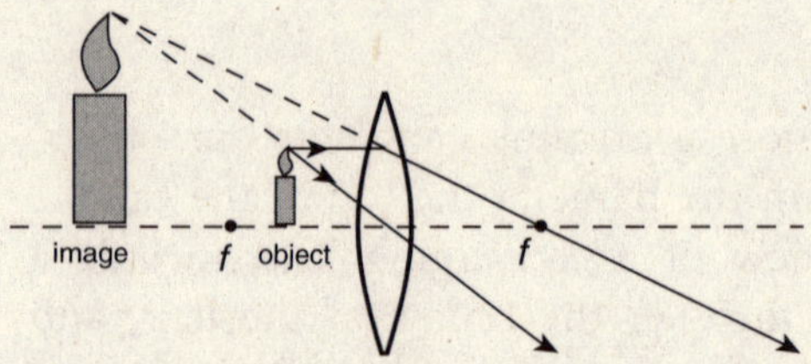

A summary of the images formed by a converging lens is listed in the table below, where d_o is the distance from the candle to the object and f is the focal length of the lens. A positive (+) image distance d_i implies that the image is formed on the opposite side of the lens as the object, and a negative (–) image distance implies that the image is formed on the same side as the object.

Object placement	Image distance d_i	Real or virtual	Upright or inverted	Enlarged or reduced
$d_o > 2f$	+	real	inverted	reduced
$d_o = 2f$	+	real	inverted	same size
$f < d_o < 2f$	+	real	inverted	enlarged
$d_o = f$	no image	no image	no image	no image
$d_o < f$	–	virtual	upright	enlarged

Note that all real images are inverted.

Magnification Using a Converging Lens (会聚透镜的放大率)

The amount by which an image is magnified by a converging lens is simply equal to the ratio of the image distance to the object distance, each measured from the lens:

$$M = \left| \frac{d_i}{d_o} \right|$$

Magnification tells us how many times larger or smaller an image is than the object.

Example: A candle is placed a distance of 30 cm from a lens of focal length 20 cm, forming an image 60 cm from the lens. Describe the image formed by the lens and calculate the magnification of the image.

Solution: In this case, the candle is place at 1.5*f*, which is between *f* and 2*f*. Consulting the table above, we see that the image formed is real, inverted, and larger than the object.

The magnification is:

$$M = \left| \frac{d_i}{d_o} \right| = \frac{60 \text{ cm}}{30 \text{ cm}} = 2$$

Thus, the image is twice as large as the object.

RAY OPTICS REVIEW QUESTIONS

1. A plane mirror will produce a virtual image

   (A) when the object distance is greater than the image distance.
   (B) when the object distance is less than the image distance.
   (C) when the object is on the principal axis of the mirror.
   (D) when the rays converge at the focal point of the mirror.
   (E) at all distances from the mirror.

2. Which of the following mirrors converge parallel light rays to a focal point?

   (A) plane
   (B) convex
   (C) concave
   (D) inverted
   (E) upright

3. Which of the following is true of a diverging lens?

   (A) Incoming parallel rays passing through the lens converge to a focal point.
   (B) The lens is thinner in the center than on the edges.
   (C) The lens is thicker in the center than on the edges.
   (D) Light must enter it parallel to the principal axis.
   (E) The lens must be flat on one side.

4. A candle is placed on the principal axis of a convex lens at a distance of 30 cm from the lens. The focal length of the lens is 10 cm. The image formed will be

   (A) real, upright, and enlarged.
   (B) real, inverted, and enlarged.
   (C) real, inverted, and smaller.
   (D) virtual, upright, and enlarged.
   (E) virtual, upright, and smaller.

5. A candle is placed on the principal axis of a convex lens at a distance of 10 cm from the lens. The focal length of the lens is 20 cm. The image formed will be

   (A) real, upright, and enlarged.
   (B) real, inverted, and enlarged.
   (C) real, inverted, and smaller.
   (D) virtual, upright, and enlarged.
   (E) virtual, upright, and smaller.

6. A candle is placed on the principal axis of a convex lens at a distance of 20 cm from the lens. The image formed is magnified 3 times. The image distance is

   (A) 7 cm
   (B) 20 cm
   (C) 60 cm
   (D) 90 cm
   (E) 120 cm

SOLUTIONS TO RAY OPTICS REVIEW QUESTIONS

1. E

A plane (flat) mirror will reflect light and produce an image at any distance from the object.

2. C

A concave mirror is also called a converging mirror.

3. B

A diverging lens is also called a concave lens, and is thinner in the center than on the edges.

4. C

The candle is placed at a distance greater than twice the focal length, and so the image formed will be real, inverted, and smaller than the candle.

5. D

The candle is placed at a distance less than the focal length, and so the image formed will be virtual, upright, and enlarged. This is the case for a magnifying glass.

6. C

Since $Magnification = \frac{d_i}{d_o}$, then:

$$d_i = (M)(d_o) = (3)(20 \text{ cm}) = 60 \text{ cm}$$

CHAPTER EIGHTEEN

Heat, Temperature, and Thermodynamics

HEAT AND TEMPERATURE (热和温度)

There are three states of matter we will be discussing in these next few chapters: solid, liquid, and gas. The molecules of a solid are fixed in a rigid structure. The molecules of a liquid are loosely bound and may mix with one another freely, but the liquid still has a definite volume, although it takes the shape of its container. The molecules of a gas interact with each other only slightly, and generally move at high speeds compared the molecules of a liquid or solid. But in all three states of matter, the molecules are moving and therefore have energy. They have potential energy because of the bonds between them and kinetic energy because the molecules have mass and speed. Relatively speaking, the potential energy between gas molecules can be ignored, and so later we will focus only on their kinetic energy. The sum of the potential and kinetic energies of the molecules in a substance is called the *internal energy* of the substance. When a warmer substance is brought in contact with a cooler substance, some of the kinetic energy of the molecules in the warmer substance is transferred to the cooler substance. The energy representing the kinetic energy of molecules that is transferred spontaneously from a warmer substance to a cooler substance is called *heat energy.*

Heat is the kinetic energy of molecules transferred from a warmer substance to a cooler one.

Temperature is the measure of how hot or cold a substance is, relative to some standard. It is the measure of the *average kinetic energy* of the molecules in a substance.

Temperature is a measure of the average kinetic energy of the molecules in a substance.

The two temperature scales used most widely in scientific applications is the Celsius scale and the Kelvin scale. The only difference between them is where each starts. On the Celsius scale, the freezing point of water is 0° C, and the boiling point of water (at standard pressure) is 100° C. The Kelvin scale has temperature units which are equal in size to the Celsius degrees, but the temperature of 0 Kelvin is *absolute zero*, defined as the temperature at which all molecular motion in a substance ceases. Zero Kelvin is equal to –273.15° C, so we can convert between the Kelvin scale and the Celsius scale by this equation:

$$K = {}^\circ C + 273$$

(Note that we have rounded 273.15 to 273.)

Quick Quiz:

What is the boiling point of water in Kelvins?

Answer: K = 100° C + 273 = 373 K.

Mechanical Equivalent of Heat

The unit most often used for mechanical energy is the *joule.* Historically, the unit for heat has been the *calorie.* One calorie is defined as the heat needed to raise the temperature of one gram of water by one degree Celsius. In the mid-19th century, the British engineer James Joule showed there is a relationship between energy in the form of work in joules and energy in the form of heat in calories. Joule performed experiments that revealed that doing mechanical work on a substance can make its temperature rise. For example, rubbing your hands together causes them to heat up, and stirring a drink adds heat to it. The conversion between joules and calories is:

$$1 \text{ calorie} = 4.186 \text{ joules}$$

The numbers here are not really important for the SAT II: Physics Test, but only the concept that mechanical work and heat are both forms of energy and can be converted into one another.

Thermal Expansion of a Solid

When a solid is heated, it typically expands. Different substances expand at different rates, which is why it makes sense to heat the lid of a jar when the lid is too tight. The metal lid will expand more than the glass jar when it is heated, making it easier to loosen. Solids undergo two types of expansion when heated: *linear thermal expansion*, which is the increase in any one dimension of the solid, and *volume thermal expansion*, which results in an increase in the volume of the solid. In the case of linear expansion, the change in length Δl is proportional to the original length and the change in temperature of the solid. The same is true for volume expansion. The change in volume ΔV is proportional to the original volume of the solid and its change in temperature.

Heat Transfer (热传递)

There are three ways of transferring heat from one place to another:

Three ways of transferring heat are *conduction*, *convection*, and *radiation*.

(传导) 1. *Conduction.* Conduction is the transfer of heat directly through a material, or by actual contact between two materials. Metals are typically good heat *conductors*. In fact, materials which are good electrical conductors are usually good heat conductors as well. A material which is not a good heat conductor, like wood or air, is called an *insulator.* If you place an iron skillet on a fire, heat is transferred by conduction to the handle of the skillet. If you grasp the iron handle with your bare hand, you will feel it transfer heat to your hand by conduction.

(对流) 2. *Convection.* Convection is the transfer of heat by the bulk movement of a fluid (liquid or gas). If the air near the floor of a cool room is heated, it expands and becomes less dense than the air above it, causing it to rise. As it rises, it cools, becomes more dense again, and falls toward the floor. If the air near the floor is continually heated, the cycle will repeat itself. Water heated in a pan is another example of heat transfer by convection, since water near the bottom of the pan near the fire is heated, rises, cools, then falls again. If the temperature gets high enough, the water begins to boil as it cools itself by transferring heat to the air by convection.

(辐射) 3. *Radiation.* Radiation is the process by which heat is transferred by electromagnetic waves. We receive heat from the sun by radiation principally in the form of visible light, infrared, and ultraviolet waves. Microwave ovens use microwaves to transfer heat to food. If you stand near a roaring campfire, you will feel the heat radiating from the fire in the form of light and infrared rays.

SPECIFIC HEAT AND HEAT CAPACITY

Heat is supplied to (or absorbed by) a system to raise its temperature; conversely, heat is released if it cools. The heat absorbed or released by an object as a result of a change in temperature is calculated from the equation:

$$Q = mc\Delta T$$

where Q is the symbol for heat, m is the mass of the object, ΔT is the change in temperature and is equal to the final temperature minus the initial temperature, and c is a quantity known as the *specific heat* of the substance.

The more massive a substance is, the more heat is required to bring about a particular change in temperature. Recall that the temperature of a substance is a measure of the average kinetic energy of the particles in the substance. This applies to all states of matter. In solids, for example, the atoms vibrate about their equilibrium positions; the stronger these vibrations, the higher the temperature of the solid. A small amount of heat supplied to a large number of particles would not increase their average energy by much; however, if there were only a small number of particles in the system, that same amount of heat would not be spread as widely, and thus each particle would gain a larger amount of energy, resulting in a greater increase in temperature.

Quick Quiz:

Which would take more heat to raise its temperature by 1° C:

1 kg of aluminum or 1 kg of wood?

Answer: 1 kg of wood would take more heat, since its specific heat is higher. It doesn't take much heat to make aluminum feel hot, but it takes more heat to make a piece of wood feel hot.

Yet, not every substance is responsive to heat to the same degree. Even though we expect that raising 2 kg of a substance by 1° C requires more heat than raising 1 kg of the same substance by 1° C (in fact, it requires twice the amount of heat), we would not expect that the same amount of heat is required to raise the temperature of 1 kg of steel versus 1 kg of plastic. The *specific heat c* is a proportionality constant that gives an indication of the ease with which one can raise the temperature of something; the larger it is, the larger the amount of heat required to raise its temperature a certain number of degrees, and also the more heat is released if it cools by a certain number of degrees. Its value is a property of the *nature* of the substance and does not change based on the *amount* of the substance, since that has already been taken into account by the mass. The specific heat is often more formally defined as the heat necessary to raise the temperature of 1 kg or 1 g of a material by 1° C or 1 K. Iron, for example, has a specific heat of about 0.1 kilocalorie/kg°C, while water has a specific heat of 1.0 kcal/kg°C. It is therefore much easier to raise the temperature of 1 kg of iron by 10° C than it is to do the same to 1 kg of water; in fact, you should be able to see that 10 times the amount of heat is needed to heat the water.

The more of a substance there is, the more heat is required to bring about a change in its temperature.

The mass and the specific heat are sometimes lumped together to give a quantity known as the *heat capacity*. This quantity describes the heat needed to raise the temperature of the object as a whole by 1° C or 1 K.

Example: What is the heat necessary to raise the temperature of 2 kg of iron from 40° C to 80° C?

Solution: From our discussion above, we know that the specific heat of iron is 0.1 kcal/kg°C. The heat required is:

$$Q = mc\Delta T = mc(T_f - T_i) = (2\text{ kg})(0.1\text{ kcal/kg °C})(80 - 40)$$
$$= 8\text{ kcal or 8,000 calories}$$

This is approximately equal to 33,000 J.

PHASE CHANGES

While heat is associated with the change of thermal energy, a system does not necessarily increase in temperature. Heat can also increase the potential (rather than kinetic) energy of the particles in a system; this occurs during a *phase change*. Heat is required to *melt* something (change its phase from a solid to a liquid), or to *vaporize* something (change its phase from a liquid to a gas), or to *sublimate* something (change its phase directly from a solid to a gas). In all these cases, the molecules are overcoming the attractive forces that hold them together. This is where the energy supplied by heating is being put to use. Conversely, heat is released as a substance *freezes* or *condenses*. During such phase changes, the temperature remains constant, and the heat involved in these processes can be expressed as:

$$Q = mL$$

where m is the mass of the substance undergoing the phase change and L is the heat of transformation, the value of which depends on both the substance and the particular process we are talking about: vaporization, sublimation, or fusion (melting).

Example: The heat of fusion L_f of ice is 3.3×10^5 joules/kg° C, and the specific heat of water is 4.2×10^3 joules/kg° C. How much heat is required to completely melt 10 kg of ice and then raise the temperature of the water from 0° C to 30° C?

Solution: The heat required to melt the ice and then raise the temperature of the water is:

Adding heat can change the temperature of an object or change its phase.

$$Q = mL_f + mc\Delta T$$

$$Q = (10 \text{ kg})(3.3 \times 10^5 \text{ joules/kg° C}) + (10 \text{ kg})(4.2 \times 10^3 \text{ joules/kg° C})(30 - 0)$$

$$Q = 4.56 \times 10^6 \text{ joules}$$

This is approximately one million calories.

If we were to continue to heat the water until it boiled and turned to steam at 100° C, and then continued to add heat to the steam to raise its temperature beyond 100° C, the temperature would not change during the phase changes from ice to water and water to steam. The heat added would go into changing the phase of the water, not its temperature.

A graph of temperature vs. heat added for this process is shown below:

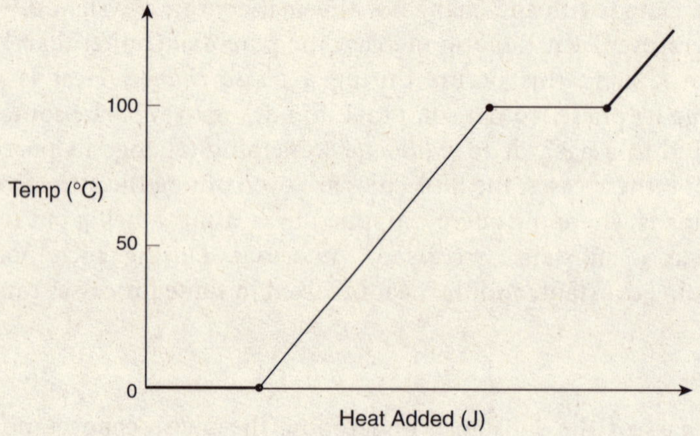

The flat portions of the graph imply that the temperature is not changing, and correspond to the processes of melting and boiling.

THE FIRST AND SECOND LAWS OF THERMODYNAMICS

Thermodynamics is the study of heat transfer. There are actually four laws of thermodynamics, but we will look at the two that are required for the SAT II: Physics Test.

In order to understand the laws of thermodynamics, we first need to know the definitions of the three types of *systems*:

1. *Isolated.* A system is said to be isolated when it cannot exchange energy or matter with the surroundings, like a well-insulated thermos flask.
2. *Closed.* A system is said to be closed when it can exchange energy but not matter with the surroundings, like a test tube with a stopper in it.
3. *Open.* A system is said to be open when it can exchange both matter and energy with the surroundings, like a pot of boiling water allowing water vapor to escape into the air.

The First Law of Thermodynamics

As we've discussed in previous chapters, energy can be transformed into many forms but still be conserved; that is, the total amount of energy will remain constant. This is true of a system only if it is isolated; since energy can neither go in nor go out, it must be conserved. If the system is closed or open, however, the amount of energy in the system can certainly change. A system can exchange energy with its surroundings in two general ways:

Three types of systems: isolated, closed, and open.

as heat or as work. *The first law of thermodynamics states that the change in the internal energy ΔU of a system is equal to the heat Q added to the system minus the work W that a system does:*

$$\Delta U = Q - W$$

If work is done ON a system, W is negative. Note that sometimes W is defined as the done ON, rather than BY the system, in which case the equation is written as $\Delta U = Q + W$, and the work done BY the system is considered negative. Regardless of which convention is used, if work is done on a system, its energy would increase. If work is done by the system, its energy would decrease. Work is generally associated with movement against some force. For ideal gas systems, for example, expansion against some external pressure means that work is done by the system, while compression implies work being done on the system.

The first law of thermodynamics is simply the law of conservation of energy.

Example: A system has 50 J of heat added to it, resulting in 20 J of work being done BY the system. What is the change in internal energy of the system?

Solution: The change in internal energy is $\Delta U = Q - W = 50\text{ J} - 20\text{ J} = 30$ J.

Example: A 50 g sample of metal is heated to 100° C and then dropped into a beaker containing 50 g of water at 25° C. if the specific heat capacity of the metal is 0.25 cal/g° C, what is the final temperature of the water?

Solution: Since heat is transferred and no work is done, the heat lost by the metal must equal the heat gained by the water.

$$m_{metal}c_{metal}(T_f - T_i)_{metal} = m_{water}c_{water}(T_f - T_i)_{water}$$

$$(50\text{ g})(0.25\text{ cal/g° C})(T_f - 100) = (50\text{ g})(1\text{ cal/g° C})(T_f - 25)$$

Since the final equilibrium temperature is the same for the metal and the water, we find that $T_f = 40°$ C.

Heat Engine

A *heat engine* is any device that uses heat to perform work. There are three essential features of a heat engine:

1. Heat is supplied to the engine at a high temperature from a hot reservoir.
2. Part of the input heat is used to perform work.
3. The remainder of the input heat that did not do work is exhausted into a cold reservoir, which is at a lower temperature than the hot reservoir.

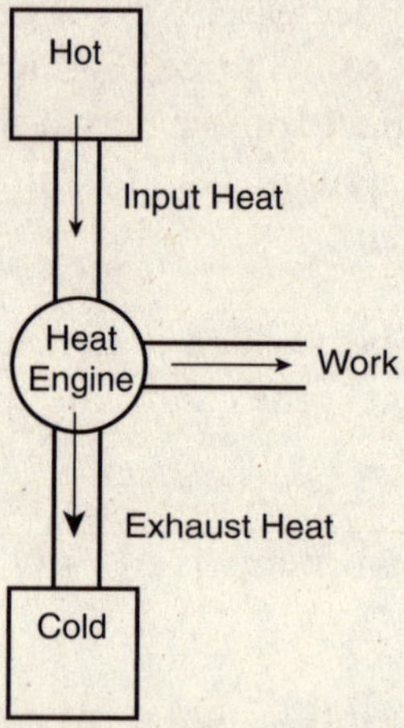

The *percent efficiency* (%ε) of the heat engine is equal to the ratio of the work done to the amount of input heat:

$$\%\varepsilon = \frac{Work}{Q_{hot}} \times 100$$

Example: A heat engine extracts 100 J of energy from a hot reservoir, does work, then exhausts 40 J of energy into a cold reservoir. What is the work done by the engine, and the percent efficiency of the engine?

Solution: The work done is equal to the difference between the input heat and the output heat:

$$W = 100\text{ J} - 40\text{ J} = 60\text{ J}$$

and the percent efficiency is:

$$\frac{60\text{ J}}{100\text{ J}} = 60\%$$

The Second Law of Thermodynamics

Entropy is a measure of the disorder, or randomness, of a system. The greater the disorder of a system, the greater the entropy. If a system is highly ordered, like the particles in a solid, we say that the entropy is low. At any given temperature, a solid will have a lower entropy than a gas, because individual molecules in the gaseous state are moving randomly, while individual molecules in a solid are constrained in place.

Entropy is important because it determines whether a process will occur spontaneously. *The second law of thermodynamics states that all spontaneous processes proceeding in an isolated system lead to an increase in entropy.*

The second law of thermodynamics is the law of entropy.

In other words, an isolated system will naturally pursue a state of higher disorder. If you watch a magician throw a deck of cards into the air, you would expect the cards to fall to the floor around him in a very disorderly manner, since the system of cards would naturally tend toward a state of higher disorder. If you watched a film of a magician, and his randomly placed cards jumped off the floor and landed neatly stacked in his hand, you would believe the film is running backward, since cards do not seek this state of order by themselves. Thus, the second law of thermodynamics gives us a direction for the passage of time.

The second law of thermodynamics also gives us a direction for the flow of heat. A bowl of ice cream will melt if left out on a hot day. The ice cream would not get colder if left in a hot environment, since heat flows spontaneously from hot to cold. As it relates to heat, the second law can be stated like this:

Heat flows spontaneously from a substance at a higher temperature to a substance at a lower temperature and does not flow spontaneously in the reverse direction.

Of course, we can *force* heat to flow from a colder object to a hotter object, which is exactly what your refrigerator in your kitchen does. The flow of heat in the refrigerator is not spontaneous, and so the second law of thermodynamics does not apply.

HEAT, TEMPERATURE, AND THERMODYNAMICS REVIEW QUESTIONS

1. The average kinetic energy of the molecules in a substance is most closely associated with

   (A) heat.
   (B) temperature.
   (C) specific heat.
   (D) absolute zero.
   (E) potential energy.

2. The temperature of absolute zero is equal to

   (A) 0° C
   (B) 100° C
   (C) 273° C
   (D) –273° C
   (E) –100° C

3. The conversion between mechanical energy and heat energy was first developed by

   (A) Celsius.
   (B) Kelvin.
   (C) Joule.
   (D) Newton.
   (E) Watt.

4. In general, when a solid is heated, it

   (A) expands proportionally to the change in temperature.
   (B) contracts proportionally to the change in temperature.
   (C) expands inversely proportionally to the change in temperature.
   (D) contracts inversely proportionally to the change in temperature.
   (E) does not expand nor contract.

5. A liquid is poured into a pan and heated over a flame. The hot liquid begins rising to the top. This type of heat transfer is called

   (A) conduction.
   (B) convection.
   (C) radiation.
   (D) thermal expansion.
   (E) specific heat.

6. You feel heat as you approach a fire by way of

   (A) conduction.
   (B) convection.
   (C) radiation.
   (D) thermal expansion.
   (E) specific heat.

7. A 3 kg block of aluminum is heated so that its temperature increases by 3 degrees. How much heat would be needed to raise the temperature of a 9 kg block of aluminum by 3 degrees?

   (A) 9 times as much heat as the 3 kg block
   (B) 3 times as much heat as the 3 kg block
   (C) the same amount of heat as the 3 kg block
   (D) one-third as much heat as the 3 kg block
   (E) one-ninth as much heat as the 3 kg block

8. What is the specific heat of a substance which requires 10 J of heat to raise the temperature of a 2 kg sample of the substance from 22° C to 24° C?

   (A) 10 J/kg°C
   (B) 5 J/kg°C
   (C) 4 J/kg°C
   (D) 2.5 J/kg°C
   (E) 2 J/kg°C

C 9. Heat is added to a block of ice initially at 0° C until the ice changes completely into water. The same amount of heat continues to be added to the liquid water for 15 minutes. Which of the following statements is true?

(A) The temperature of the ice and water remains constant until the end of the 15 minute time period.
(B) The temperature of the ice rises steadily until all of the ice has melted into water.
(C) The temperature of the ice remains constant until all of the ice has melted into water, then the temperature of the water steadily rises for 15 minutes.
(D) The temperature of the ice rises from 0° C to 32° C, then the temperature of the water rises from 32° C to 100° C.
(E) The temperature of the ice remains at 0° C until the ice has melted, then the temperature of the water remains constant at 32° C.

A 10. The first law of thermodynamics is a form of

(A) the law of conservation of energy.
(B) the law of specific heat.
(C) the ideal gas law.
(D) the law of entropy.
(E) the law of conservation of temperature.

D 11. A system has 60 J of heat added to it, resulting in 15 J of work being done by the system, and exhausting the remaining 45 J of heat. What is the efficiency of this process?

(A) 100%
(B) 60%
(C) 45%
(D) 25%
(E) 15%

$\frac{15}{60} = \frac{1}{4} = 25\%$

A E 12. The law of entropy states that

(A) heat always flows spontaneously from a colder body to a hotter one.
(B) every natural system will tend toward lower entropy. ×
(C) heat lost by one object must be gained by another. ×
(D) the specific heat of a substance cannot exceed a certain value. ×
(E) every natural system will tend toward disorder.

SOLUTIONS TO HEAT, TEMPERATURE, AND THERMODYNAMICS REVIEW QUESTIONS

1. B

The average kinetic energy of the molecules is proportional to the temperature of the substance.

2. D

Kelvin = Celsius + 273, so 0 K = –273° C.

3. C

Joule first measured the mechanical equivalent of heat.

4. A

The expansion of a substance is proportional to the change in temperature.

5. B

Convection is heat transfer by the rising and falling of a fluid as it heats and cools and changes density.

6. C

The fire radiates heat mostly in the form of infrared electromagnetic waves.

7. B

Three times the mass of the same substance will require three times the heat to raise its temperature by a certain amount according to the equation $Q = mc\Delta T$.

8. D

$$Q = mc\Delta T \text{ implies that } c = \frac{Q}{m\Delta T} = \frac{10\text{ J}}{(2\text{ kg})(24°\text{ C} - 22°\text{ C})} = 2.5\text{ J/kg°C}$$

9. C

The heat added to the ice is used only to change the phase of the ice to water, and the heat added to the liquid water is used to change its temperature.

10. A

The first law of thermodynamics states that the heat lost by one object must be gained by another, but the amount of energy in the transfer must remain constant.

11. D

If only 15 J of the 60 J is used to do work, we say that the efficiency of the process is:

$$\frac{15\text{ J}}{60\text{ J}} \times 100 = 25\%$$

12. E

Entropy means disorder, and every natural system will tend toward a state of higher disorder. The fact that heat spontaneously flows from a hotter body to a colder body is a consequence of this law.

CHAPTER NINETEEN

Gases and Kinetic Theory

Among the different phases of matter, the gaseous phase is the simplest to understand and to model, since all gases display similar behavior and follow similar laws regardless of their identity. The atoms or molecules in a gaseous sample move rapidly and are far apart. In addition, intermolecular forces between gas particles tend to be weak; this results in certain characteristic physical properties, such as the ability to expand to fill any volume and to take on the shape of a container. Furthermore, gases are easily, though not infinitely, compressible.

The state of a gas can be defined by pressure, volume, temperature, and amount of gas in moles.

The state of a gaseous sample is generally defined by four variables: pressure (P), volume (V), temperature (T), and number of moles (n), though as we shall see, these are not all independent. The *pressure* of a gas is the force per unit area that the atoms or molecules exert on the walls of the container through collisions. The SI unit for pressure is the *pascal* (Pa), which is equal to one newton per meter squared. Sometimes gas pressures are expressed in *atmospheres* (atm). One atmosphere is equal to 10^5 Pa, and is approximately equal to the pressure the earth's atmosphere exerts on us each day. Volume can be expressed in liters (L) or cubic meters (m^3), and temperature is measured in Kelvins (K) for the purpose of the gas laws. Recall that we can find the temperature in K by adding 273 to the temperature in Celsius. Gases are often discussed in terms of *standard temperature and pressure* (STP), which refers to the conditions of 273 K (0° C) and 1 atm.

IDEAL GASES

When examining the behavior of gases under varying conditions of temperature and pressure, it is most convenient to treat them as ideal gases. An ideal gas represents a hypothetical gas whose molecules have no intermolecular forces; that is, they do not interact with each other and occupy no volume. Although gases in reality deviate from this idealized behavior, at relatively low pressures and high temperatures many gases behave in nearly ideal fashion. Therefore, the assumptions used for ideal gases can be applied to real gases with reasonable accuracy.

Most gases can be treated as ideal.

The Ideal Gas Law

The ideal gas law gives the relationship between the pressure, volume, and temperature of a gas before and after some process. It states that the product of pressure and volume divided by temperature remains constant during any process:

$$\frac{P_1V_1}{T_1} = \frac{P_2V_2}{T_2}$$

If temperature remains constant during a process (*isothermic*), the equation becomes:

$$P_1V_1 = P_2V_2 \text{ (Boyle's Law)}$$

If the volume remains constant during a process (*isochoric*), the equation becomes:

$$\frac{P_1}{T_1} = \frac{P_2}{T_2}$$

If the pressure remains constant during a process (*isobaric*), the equation becomes:

$$\frac{V_1}{T_1} = \frac{V_2}{T_2} \text{ (Charles's law)}$$

Example: Under isothermal conditions, what would be the volume of a 1 liter sample of helium gas after its pressure is changed from 12 atm to 4 atm?

Solution: Isothermal means that the temperature is constant during the process, so the ideal gas law becomes:

$$P_1V_1 = P_2V_2$$

$$(12 \text{ atm})(1 \text{ L}) = (4 \text{ atm})V_2$$

$$V_2 = 3 \text{ L}$$

$P_1V_1 = P_2V_2$

$1 \times 12 = 4 \times V_2$

$V_2 = 3L$

Example: If the temperature of 2 liters of gas is isobarically changed from 10° C to 293° C, what would be the final volume?

Solution: *Isobarically* means that the pressure remains constant during this process. But before we can apply the ideal gas law, or Charles's law in this case, we need to convert Celsius degrees to Kelvins.

The *ideal gas law* states that pressure times volume divided by temperature remains constant.

$$T_1 = 10° \text{ C} + 273 = 283 \text{ K and } T_2 = 293° \text{ C} + 273 = 566 \text{ K}$$

Then

$$\frac{V_1}{T_1} = \frac{V_2}{T_2}$$

$$\frac{2 \text{ L}}{283 \text{ K}} = \frac{V_2}{566 \text{ K}}$$

$$V_2 = 4 \text{ L}$$

Quick Quiz:

True or False: As the temperature of a gas is doubled, the average speed of each molecule is also doubled.

Answer: False. Temperature is proportional to the average kinetic energy of the molecules, and kinetic energy is proportional to the square of the speed of the molecules. Therefore, the temperature is proportional to the *square* of the average speed, not the average speed.

THE KINETIC THEORY OF GASES

As indicated by the gas laws, all gases show similar physical characteristics and behavior. A theoretical model to explain why gases behave the way they do was developed during the second half of the 19th century. The combined efforts of Boltzmann, Maxwell, and others led to the kinetic theory of gases, which gives us an understanding of gaseous behavior on a microscopic, molecular level. Like the gas laws, this theory was developed in reference to ideal gases, although it can be applied with reasonable accuracy to real gases as well.

The assumptions of the kinetic theory of gases are as follows:

1. Gases are made up of particles whose volumes are negligible compared to the container volume.
2. Gas atoms or molecules exhibit no intermolecular attractions or repulsions.
3. Gas particles are in continuous, random motion, undergoing collisions with other particles and the container walls.
4. Collisions between any two gas particles are elastic, meaning that no energy is dissipated and kinetic energy is conserved.
5. The average kinetic energy of gas particles is proportional to the absolute (Kelvin) temperature of the gas, and is the same for all gases at a given temperature.

For gases of the same molecules, the higher the temperature, the higher the kinetic energy and thus the higher the average speed of the molecules.

GASES AND KINETIC THEORY REVIEW QUESTIONS

1. Which of the following is NOT true of an ideal gas?

   (A) Gas molecules have no intermolecular forces.
   (B) Gas particles are in random motion.
   (C) Gas particles have no volume.
   (D) The collisions between any two gas particles are elastic.
   (E) The average kinetic energy of the gas molecules is proportional to the temperature in Celsius degrees.

2. A sample of argon occupies 50 liters at standard temperature. Assuming constant pressure, what volume will argon occupy if the temperature is doubled?

   (A) 25 liters
   (B) 50 liters
   (C) 100 liters
   (D) 200 liters
   (E) 2,500 liters

3. What is the final pressure of a gas that expands from 1 liter at 10° C to 10 liters at 100° C if the original pressure was 3 atmospheres?

   (A) 0.3 atm
   (B) 0.4 atm
   (C) 3 atm
   (D) 4 atm
   (E) 30 atm

4. Which of the following graphs best represents the relationship between the average kinetic energy of the molecules of a gas and its temperature?

   (A) 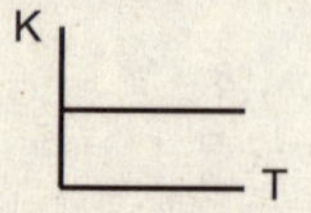

   (B) 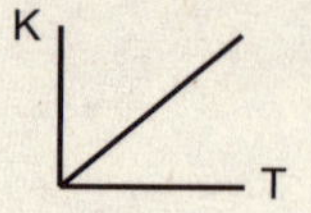

   (C) 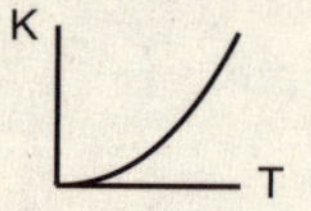

   (D) 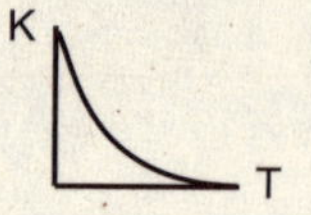

   (E) 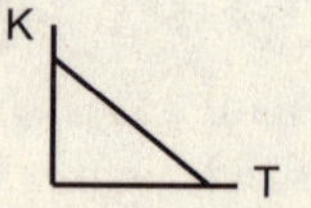

5. Which of the following pressure vs. volume graphs best represents how pressure and volume change when temperature remains constant (isothermal)?

(A) 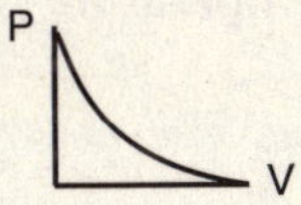

(B) 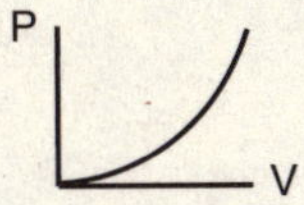

(C) 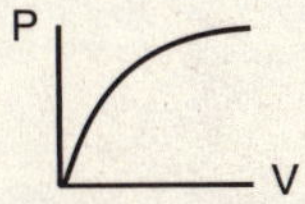

(D) 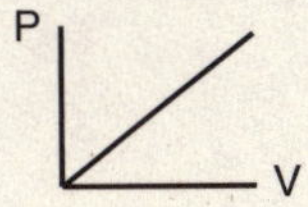

(E) 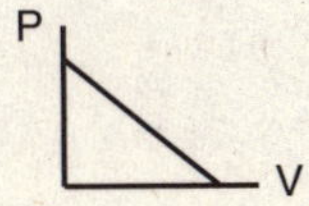

6. Which of the following volume vs. temperature graphs best represents how volume changes with Kelvin temperature if the pressure remains constant (isobaric)?

(A) 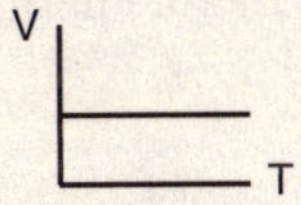

(B) 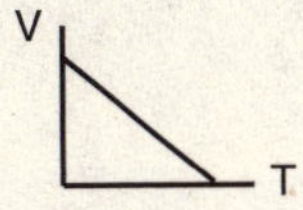

(C) 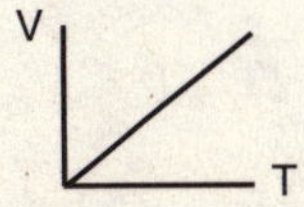

(D) 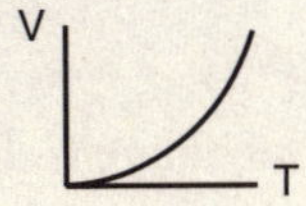

(E) 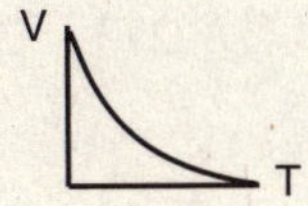

SOLUTIONS TO GASES AND KINETIC THEORY REVIEW QUESTIONS

1. E

The average kinetic energy of the molecules is proportional to the temperature in Kelvins, not Celsius degrees.

2. C

$$\frac{V_1}{T_1} = \frac{V_2}{T_2}$$

$$\frac{50\text{ L}}{T} = \frac{V_2}{2T}$$

$$V_2 = 100\text{ L}$$

3. B

Don't forget to change Celsius temperature to Kelvin temperature:

$$T_1 = 10°\text{ C} + 273 = 283\text{ K and } T_2 = 100°\text{ C} + 273 = 373\text{ K}$$

$$\frac{P_1V_1}{T_1} = \frac{P_2V_2}{T_2}$$

$$\frac{(3\text{ atm})(1\text{ L})}{283\text{ K}} = \frac{P_2(10\text{ L})}{373\text{ K}}$$

Solving for P_2, we get a final pressure of 0.4 atm.

4. B

The average kinetic energy and temperature are proportional to each other, yielding a line with a constant positive slope.

5. A

According to Boyle's law, the product of pressure and volume must remain constant, yielding an inverse relationship.

6. C

According to Charles's law, volume is proportional to Kelvin temperature if pressure is constant, yielding a line with a constant positive slope.

CHAPTER TWENTY

Quantum Phenomena

The word *quantum* simply means *the smallest piece of something*. The quantum of American money is one cent; the quantum of negative charge is the electron, since as far as we know there is no smaller negative charge that exists by itself. There are several quantities in physics which are *quantized*, that is, that occur in multiples of some smallest value. Light is one of these quantities.

The *quantum* is the smallest particle of a quantity.

PHOTONS AND THE PHOTOELECTRIC EFFECT

Back in chapters 15 through 17, we treated light as a wave. But there are circumstances when light behaves more like it is made up of individual bundles of energy, separate from each other, but sharing a wavelength, frequency, and speed. The quantum of light is called the *photon*. In the late 19th century, an effect was discovered by Heinrich Hertz that could not be explained by the wave model of light. He shined ultraviolet light on a piece of zinc metal, and the metal became positively charged. Although he did not know it at the time, the light was causing the metal to emit electrons. This effect of using light to cause electrons to be emitted from a metal is called the *photoelectric effect*. According to the theory of light at the time, light was considered a wave, and should not be able to "knock" electrons off of a metal surface. At the turn of the 20th century, Max Planck showed that light could be treated as tiny

The *photon* is the quantum of light.

bundles of energy called photons, and the energy of a photon was proportional to its frequency. Thus, a graph of photon energy E vs. frequency f looks like this:

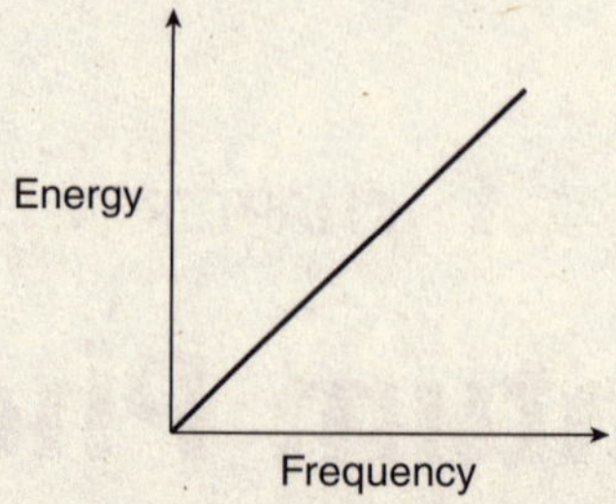

The slope of this line is a constant that occurs many times in the study of quantum phenomena called *Planck's constant.* Its symbol is h, and its value happens to be 6.63×10^{-34} J s. The equation for the energy of a photon is:

$$E = hf$$

or, since $f = \frac{c}{\lambda}$,

$$E = \frac{hc}{\lambda}$$

So the energy of a photon is proportional to its frequency, but inversely proportional to its wavelength.

Example: List the following photons of light from lowest energy to highest energy:

(A) green
(B) violet
(C) yellow

The energy of a photon is proportional to its frequency.

Solution: Listing these photons from lowest to highest energy is the same as listing them from lowest to highest frequency, or, equivalently, from longest to shortest wavelength: yellow, green, violet.

In 1905, Albert Einstein used Planck's idea of the photon to explain the photoelectric effect: One photon is absorbed by one electron in the metal surface, giving the electron enough energy be released from the metal. But not just any photon will knock an electron off a metal surface. The photon must first have enough energy to "dig" the electron out of the metal, and then have some energy left over to give the electron some kinetic energy to escape completely.

The *photoelectric effect* results when light shined on a metal releases electrons from the surface of the metal.

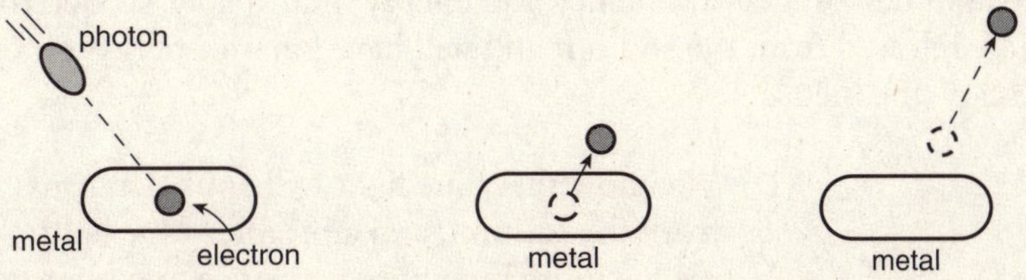

Each metal that can exhibit the photoelectric effect has a minimum energy and frequency called the *threshold frequency* (f_0) that the incoming photon must meet to dig the electron out of the metal, and exceed if the electron is to have kinetic energy to escape. For example, the metal sodium has a threshold frequency that corresponds to yellow light. If yellow light is shined on a sodium surface, the yellow photons will be absorbed by electrons in the metal, causing them to be released, but there will be no energy left over for the electrons to have any kinetic energy. If we shine green light on the sodium metal, the electrons will be released and have some energy left over to use as kinetic energy, and the electrons will jump off the metal completely, since green light has a higher frequency and energy than yellow light.

The *threshold frequency* of a metal is the minimum frequency of incoming light necessary to release an electron from a metal surface.

Example: In the case of the yellow light and sodium metal described above, describe what would happen in the following instances:

(A) We shine orange light on the sodium metal.

(B) We shine very bright red light on the sodium metal.

(C) We shine blue light on the sodium metal.

(D) We shine very bright blue light on the metal.

(E) Sketch a graph of maximum kinetic energy of a photoelectron (an electron emitted in the photoelectric effect) vs. frequency of incident light for this metal.

Solution:

(A) Since orange light is below the threshold frequency for sodium, no emission of electrons will take place.

(B) The brightness of the light doesn't matter if the threshold frequency (minimum color) is not met, and red photons have less frequency than the necessary yellow photons, so no emission of electrons will occur.

(C) Blue light has a higher frequency and energy than yellow, so electrons will be released from the sodium and will have kinetic energy left over to escape the metal.

For light above the threshold frequency, a brighter light means more photons, and thus more electrons released from the metal.

(D) We know that blue light will emit electrons from the sodium, and a brighter blue light means more blue photons. Since one photon can be absorbed by one electron, more electrons will be emitted by a brighter light which exceeds the threshold frequency. If these electrons are funneled into a circuit, we can use them as current in an electrical device.

(E) The graph of maximum kinetic energy of a photoelectron vs. frequency of incident light would look like this:

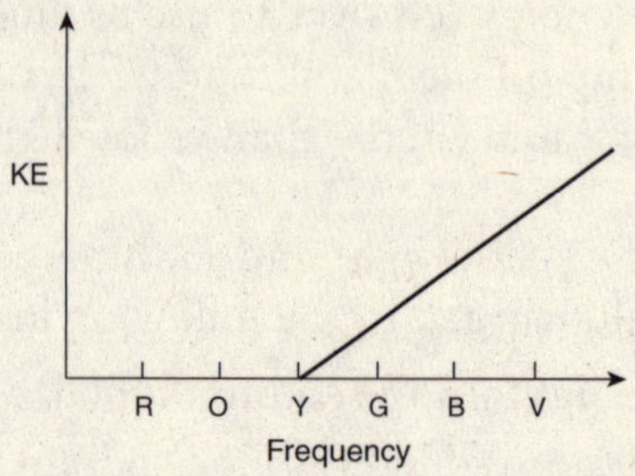

Auto-focus cameras use the photoelectric effect to measure light and adjust the lens to focus on the subject being photographed.

Note that the electrons have no kinetic energy up to the threshold frequency (color), and then their kinetic energy is proportional to the frequency of the incoming light.

THE MOMENTUM OF A PHOTON AND THE HEISENBERG UNCERTAINTY PRINCIPLE

Since a photon has energy, does it follow that it has momentum? Recall that in chapter 6 we defined momentum as the product of mass and velocity. But a photon has no mass. It turns out that in quantum physics, photons do have momentum by virtue of their wavelength. The equation for the momentum of a photon is:

$$p = \frac{h}{\lambda}$$

So the momentum of a photon is inversely proportional to its wavelength. Photons can and do impart momentum to subatomic particles in collisions that follow the law of conservation of momentum. It is difficult to understand how a photon, having only energy and no mass, can collide with a particle such as an electron and change its momentum. But this has been verified experimentally, and we realize that our understanding of quantum phenomena is still catching up with experimental evidence in some cases.

Photons have momentum that is inversely proportional to their wavelength.

When a policeman wants to measure the speed of your car, he aims his radar gun and shoots a radio wave which bounces off your car and returns to his detector, giving an indication of your speed. One assumption here is that the radio wave reflecting off your car does not significantly add to or take away from the speed of your car. Your car is much too massive to be affected by a radio wave bouncing off of it. But what if you wanted to measure the speed or position of a subatomic particle, like an electron? Even if you shoot a lightweight photon, bounce it off the electron, and measure how quickly the photon returns to you, you have imparted a significant amount of the photon's momentum to the electron in the process, and thus changed the momentum and position of the electron just by trying to measure it. Here we have reached a limit of measurement. Since a photon is the smallest and most unobtrusive measuring device we have available to us, and even a photon is too large to make accurate measurements of the speed and position of subatomic particles, we must admit to an uncertainty that will always exist in quantum measurements. This limit to accuracy at this level was formulated by Werner Heisenberg in 1928 and is called the *Heisenberg uncertainty principle*. It can be stated as follows:

The *Heisenberg uncertainty principle* states that we cannot simultaneously measure the position and speed (or momentum) of a subatomic particle with complete accuracy.

> *There is a limit to the accuracy of the measurement of the speed (or momentum) and position of any subatomic particle. The more accurately we measure the speed of a particular particle, the less accurately we can measure its position, and vice versa.*

MATTER WAVES AND THE DE BROGLIE WAVELENGTH OF A PARTICLE

In 1924, Louis de Broglie reasoned that if a wave such as light can behave like a particle, having momentum, then why couldn't particles behave like waves? If the momentum of a photon can be found by the equation $p = h/\lambda$, then the wavelength can be found by $\lambda = h/p$. De Broglie suggested that for a particle with mass m and speed v, we could write the equation as $\lambda = h/mv$, and the wavelength of a moving particle could be calculated. This hypothesis was initially met with a considerable amount of skepticism until it was shown by Davisson and Germer in 1927 that electrons passing through a nickel crystal were diffracted through the crystal, producing a diffraction pattern on a photographic plate. Thus, de Broglie's hypothesis that particles could behave like waves was experimentally verified. Today, nuclear and particle physicists must take into account the wave behavior of subatomic particles in their experiments. We typically don't notice the wave properties of objects moving around us because the masses are large in comparison to subatomic particles and the value for Planck's constant h is extremely small. But the wavelength of any moving mass is inversely proportional to the momentum of the object.

> **Example:** If an electron, proton, and an alpha particle are all moving at the same speed, which will have the longest de Broglie wavelength? Which will have the shortest?
>
> **Solution:** At equal speeds, the particle with the least mass will have the longest wavelength, and the particle with the largest mass will have the shortest wavelength. Listed from longest to shortest wavelength, we have the electron, the proton, and the alpha particle.

QUANTUM PHENOMENA REVIEW QUESTIONS

1. The smallest, discrete value of any quantity in physics is called the

   (A) atom.
   (B) molecule.
   (C) proton.
   (D) electron.
   (E) quantum.

2. The smallest discrete value of electromagnetic energy is called the

   (A) photon.
   (B) proton.
   (C) electron.
   (D) neutron.
   (E) quark.

3. Which of the following photons has the highest energy?

   (A) x-ray
   (B) ultraviolet light
   (C) green light
   (D) microwave
   (E) radio wave

4. The threshold frequency of zinc for the photoelectric effect is in the ultraviolet range. Which of the following will occur if x-rays are shined on a zinc metal surface?

   (A) No electrons will be emitted from the metal.
   (B) Electrons will be released from the metal but have no kinetic energy.
   (C) Electrons will be released from the metal and have kinetic energy.
   (D) Electrons will be released from the metal but will immediately be recaptured by the zinc atoms.
   (E) Electrons will simply move from one zinc atom in the metal to another zinc atom in the metal.

5. A metal surface has a threshold frequency for the photoelectric effect which corresponds to green light. If blue light is shined on this metal,

   (A) no electrons will be emitted from the metal.
   (B) the number of emitted electrons is proportional to the brightness (intensity) of the blue light.
   (C) the electrons will have no kinetic energy.
   (D) more electrons will be emitted than if green light were shined on the metal.
   (E) electrons will be emitted from the metal, but since the light is not green, only a few electrons will be released.

6. Light is shined on a metal surface which exhibits the photoelectric effect according to the graph shown.

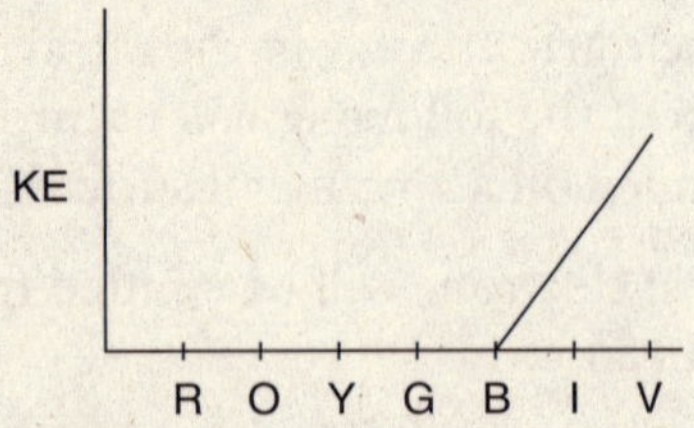

What color(s) correspond to the threshold frequency of the metal?

(A) red only
(B) red and orange
(C) red, orange, yellow, and green
(D) blue only
(E) blue, indigo, and violet

7. Which of the following is true of the momentum of a photon?

(A) It is proportional to the wavelength of the photon.
(B) It is inversely proportional to the wavelength of the photon.
(C) It is inversely proportional to the square of the wavelength of the photon.
(D) It is proportional to the mass of the photon.
(E) It is equal to the energy of the photon.

8. The Heisenberg uncertainty principle implies that

(A) electrons are too small to be studied.
(B) every photon is exactly the same size.
(C) the more you know about the momentum of an electron, the less you can know about its position.
(D) the more you know about the energy of a photon, the less you can know about its frequency.
(E) you cannot state with accuracy the number of electrons in an atom.

9. Which of the following statements is true for the de Broglie wavelength of a moving particle?

(A) It is never large enough to measure.
(B) It is proportional to the speed of the particle.
(C) It is inversely proportional to the momentum of the particle.
(D) It is equal to Planck's constant.
(E) It has no effect on the behavior of electrons.

SOLUTIONS TO QUANTUM PHENOMENA REVIEW QUESTIONS

1. E

The quantum is the smallest discrete value of any quantity, such as the electron for charge and the photon for light.

2. A

A photon is the smallest bundle of light energy.

3. A

The x-ray has the highest frequency of the choices, and since energy is proportional to frequency, has the highest energy as well.

4. C

Since the frequency of x-rays is higher than the ultraviolet threshold frequency, electrons will be emitted from the metal and have kinetic energy left over.

5. B

After the threshold frequency is met, the number of photons (brightness) dictate how many electrons are emitted, since one photon can release one electron. Thus, a brighter light will release more electrons.

6. D

The electrons begin being released when blue light is shined on the metal, so blue has the threshold (minimum) frequency for this metal.

7. B

Since the equation for the momentum of a photon is $p = h/\lambda$, the momentum is inversely proportional to the wavelength of the photon, implying that a photon with a shorter wavelength has a higher momentum than one with a longer wavelength.

8. C

Heisenberg's uncertainty principle states that you have to sacrifice your knowledge of the position of any subatomic particle to know its momentum accurately, and vice versa.

9. C

According to the equations for the de Broglie wavelength, the higher the momentum of the particle, the shorter its wavelength.

CHAPTER TWENTY-ONE

Atomic Physics

RUTHERFORD MODEL OF THE ATOM (卢瑟福原子的模型)

The ancient Greeks were the first to document the concept of the atom. They believed that all matter is made up of tiny indivisible particles. In fact, the word *atom* comes from the Greek word *atomos*, meaning "uncuttable." But a working model of the atom didn't begin to take shape until J. J. Thomson's discovery of the electron in 1897. He found that electrons are tiny negatively charged particles and that all atoms contain electrons. He also recognized that atoms are naturally neutral, containing equal amounts of positive and negative charge, although he was not correct in his theory of how the charge was arranged.

All atoms are naturally neutral, and contain electrons.

You may remember studying Thomson's "plum-pudding" model of the atom, with electrons floating around in positive fluid. A significant improvement on this model of the atom was made by Ernest Rutherford around 1911, when he decided to shoot alpha particles at very thin gold foil to probe the inner structure of the atom in an experiment we now refer to as the *Rutherford alpha-scattering experiment*.

Alpha particles are relatively large positively charged particles which are emitted by certain radioactive elements. Rutherford didn't know at the time that they were actually helium nuclei, consisting of two protons and two neutrons. He aimed a radioactive alpha source at a very thin sheet of gold foil, and fully expected the alpha particles to pass straight through the foil, since he believed Thomson's "plum-pudding" model of the atom should not offer any resistance to the alpha particles. As expected, most of the alpha particles passed straight through the foil undeflected; however some of the particles glanced off at a slight angle. This was a bit surprising to Rutherford, since he didn't know what

The *Rutherford alpha-scattering experiment* established that the atom has a nucleus.

might cause the alpha particles to change their path as they passed through the gold atoms. But the real surprise came when a few of the alpha particles actually bounced back off of the gold foil, and completely reversed their direction.

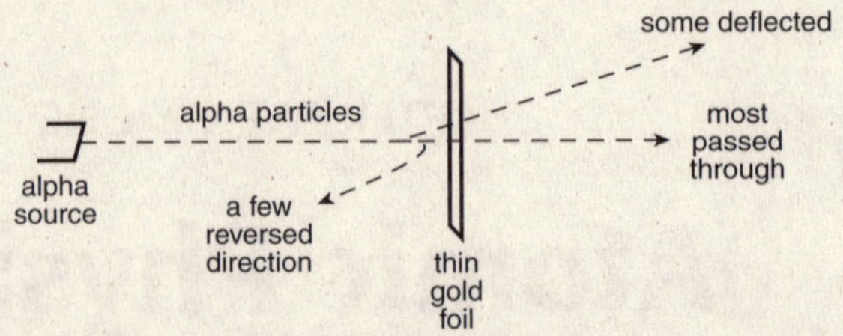

Rutherford drew the following conclusions from the results of this experiment:

1. Since most alpha particles passed through undeflected, the atom must be mostly empty space.
2. Since some of the particles deflected at an angle, and a few of them reversed direction, there must be a dense, positively charged central core of the atom which is repelling the positively charged alpha particles as they come near the center of the gold atoms. Rutherford named this core the *nucleus* of the atom.
3. The atom must consist of a dense, positively charged nucleus which contains most of the mass of the atom. The electrons are scattered at a great distance around the nucleus, and orbit the nucleus like the planets orbit the sun. The empty space in the atom is the space between the nucleus and the electrons.

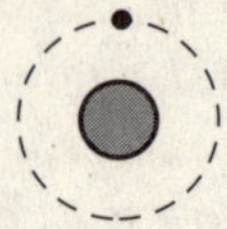

The Rutherford model of the atom places electrons in orbit around the nucleus.

This model of the atom is often called the *Rutherford planetary model of the atom.*

BOHR MODEL OF THE ATOM

In 1913, Niels Bohr made an important improvement to the Rutherford model of the atom. He observed that excited hydrogen gas gave off a spectrum of colors when viewed through a spectrosope. But the spectrum was not continuous; that is, the colors were bright, sharp lines which were separate from each other. It had long been known that every low pressure, excited gas emitted its own special spectrum in this way, but Bohr was the first to associate the bright-line spectra of these gases, particularly hydrogen, with a model of the atom.

Excited, low pressure gases each give off their own bright-line spectrum.

Bohr proposed that the electrons orbiting the nucleus of an atom do not radiate energy in the form of light while they are *in* a particular orbit, but only when they *change* orbits. Furthermore, an electron cannot orbit at just any radius around the nucleus, but only certain selected (quantized) orbits.

The two postulates of the Bohr model of the atom are summarized below:

Bohr's model of the atom places electrons in selected, quantized orbits around the nucleus.

1. Electrons orbiting the nucleus of an atom can only orbit in certain quantized orbits, and no others. These orbits from the nucleus outward are designated n = 1, 2, 3..., and the electron has energy in each of these orbits E_1, E_2, E_3, and so on. The energies of electrons are typically measured in *electron-volts* (*eV*), which is a very small unit of energy. The lowest energy (in the orbit nearest the nucleus) is called the *ground state energy.*

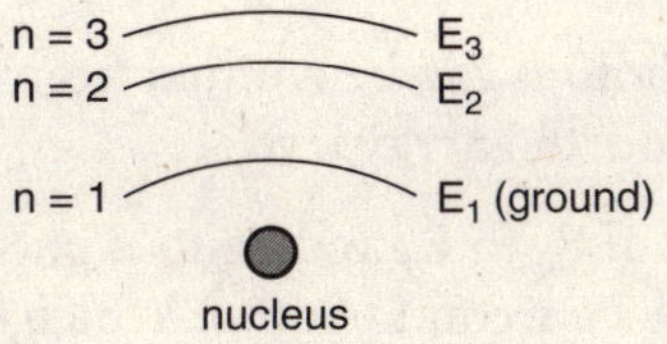

Electron energy levels are measured in electron-volts (eV), a small unit of energy.

2. Electrons can change orbits when they absorb or emit energy.
   (a) When an electron absorbs *exactly* enough energy to reach a higher energy level, it jumps up to that level. If the energy offered to the electron is not exactly enough to raise it to a higher level, the electron will ignore the energy and let it pass.

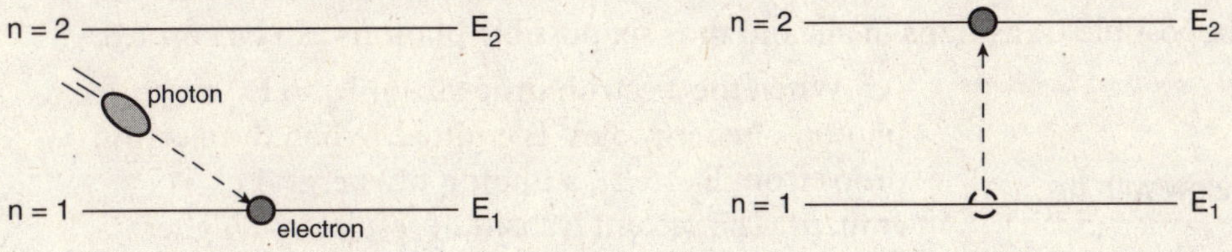

   (b) When an electron is in a higher energy level, it can jump down to a lower energy level by releasing energy in the form of a photon of light. The energy of the emitted photon is exactly equal to the difference between the energy levels the electron moves between.

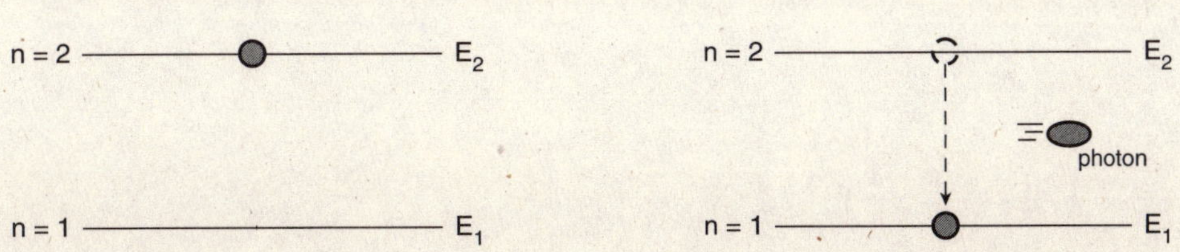

Example: Consider the energy level diagram for a particular atom shown below:

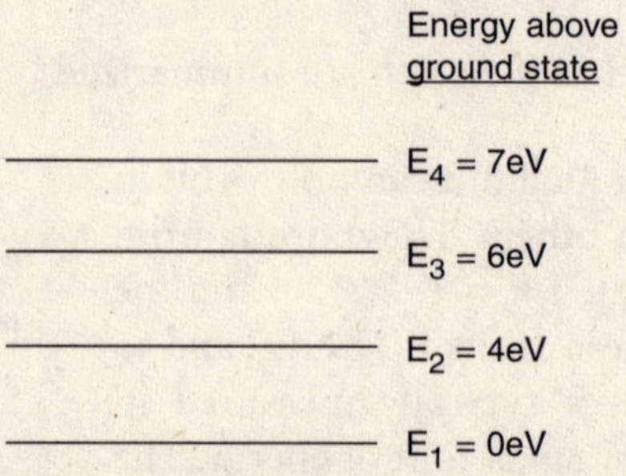

(A) An electron begins in the ground state of this atom. How much energy must be absorbed by the electron to reach the fourth energy level?

(B) How many possible photons can be emitted from this atom if the electron starts in the fourth energy level?

(C) The electron drops from E_4 to E_2 and emits a photon, then drops from E_2 to E_1 and emits a second photon. Which of these emitted photons has the higher frequency?

Solution: (A) Since the energy levels are labeled with the energy of the electron above ground state in each case, the electron would need 7 eV to jump from the ground state energy to E_4.

(B) If the electron starts at E_4, the possible downward transitions are: 4 to 3, 4 to 2, 4 to 1, then if the electron hesitates at a lower level and then continues to drop, we must also include 3 to 2, 3 to 1, and 2 to 1. There are six possible transitions in all, and thus six possible photons can be emitted.

(C) When the electron drops from E_4 to E_2, a photon of energy 3 eV is emitted. When the electron drops from E_2 to E_1, a photon of energy 4 eV is emitted. The second transition results in a higher energy photon than the first, and since energy is proportional to frequency, the second photon must also have a higher frequency.

Quick Quiz:

Match each scientist below with his model of the atom:

1. Thomson	a. Planetary model
2. Rutherford	b. Quantized energy level model
3. Bohr	c. Plum-pudding model

Answer: 1c, 2a, 3b.

ATOMIC PHYSICS REVIEW QUESTIONS

1. Which of the following was a result of Rutherford's alpha-scattering experiment?
   (A) The electrons in the gold foil absorbed the alpha particles and changed energy levels.
   (B) Most of the alpha particles passed through the foil undeflected.
   (C) A few of the alpha particles knocked the nucleus completely out of the atom.
   (D) A large percentage of alpha particles were turned back when striking the foil.
   (E) The gold foil began to emit a bright-line spectrum.

2. Which of the following is NOT one of the conclusions drawn by Rutherford after his alpha-scattering experiment?
   (A) There is a large space between the nucleus of the atom and the electrons.
   (B) The electrons orbit the nucleus of the atom.
   (C) The nucleus is positively charged.
   (D) The atom is neutral.
   (E) Electrons can change orbits when they absorb or emit energy.

3. An emission spectrum is produced when
   (A) electrons in an excited gas jump up to a higher energy level and release photons.
   (B) electrons in an excited gas jump down to a lower energy level and release photons.
   (C) electrons are released from the outer orbitals of an excited gas.
   (D) an unstable nucleus releases energy.
   (E) light is shined on a metal surface and electrons are released.

4. Consider the electron energy level diagram for a particular atom shown.

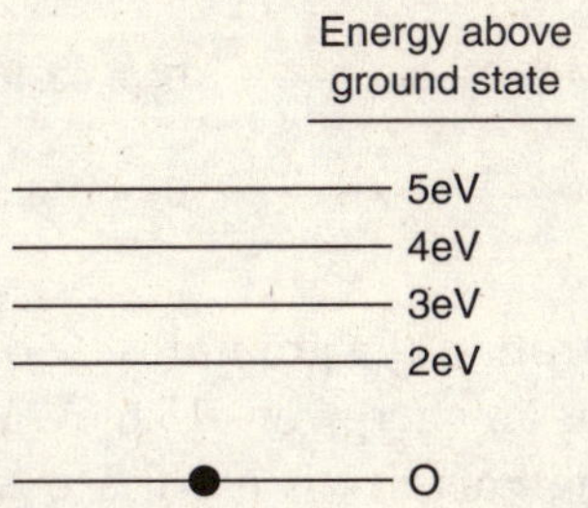

An electron is in the ground state energy level. If a photon of energy 6 eV is given to the electron, which of the following will occur?
   (A) The electron will ignore the photon since the photon's energy does not match the energy levels.
   (B) The electron will absorb the photon, jump up to the 5 eV level shown, and convert the remainder of the photon's energy into kinetic energy, but will stay in the 5 eV energy level.
   (C) The electron will absorb the photon, jump out of the atom completely, and convert the remainder of the photon's energy into kinetic energy.
   (D) The electron will absorb the photon, jump up to the 5 eV level, then go back down to the 4 eV level.
   (E) The electron will jump up to the 3 eV level, then immediately go back down the ground state.

5. Consider the electron energy level diagram for hydrogen shown.

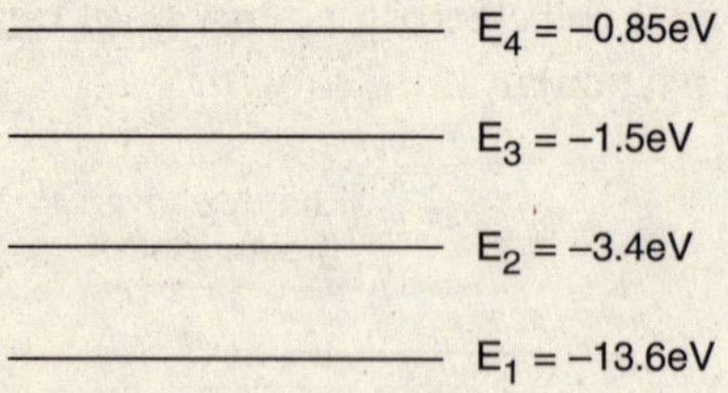

An electron in the ground state of a hydrogen atom has an energy of –13.6 eV. Which of the following energies is NOT a possible energy for a photon emitted from hydrogen?

(A) 1.9 eV
(B) 13.6 eV
(C) 0.65 eV
(D) 11.1 eV
(E) 10.2 eV

SOLUTIONS TO ATOMIC PHYSICS REVIEW QUESTIONS

1. B

Since there is such a large space between the nuclei of adjacent atoms, most of the alpha particles passed through the foil undeflected.

2. E

Rutherford did not yet know that electrons were in energy levels. Bohr's model of the atom placed the electrons in quantized energy levels.

3. B

When electrons jump back to lower energy levels, they emit energy as photons.

4. C

When an electron absorbs enough energy to completely escape the atom, we say that the atom is *ionized*, and the energy remaining, in this case 1 eV, is converted to kinetic energy.

5. D

An electron emits a photon of energy which corresponds exactly to the difference in two energy levels, and 11.1 eV does not correspond to any energy differences in the hydrogen atom.

CHAPTER TWENTY-TWO

Nuclear and Particle Physics

We've established that the atom consists of a positively charged nucleus surrounded by electrons which are in quantized energy levels. Let's take a look at the nucleus of the atom.

NUCLEAR STRUCTURE (原子核结构)

The nucleus is made up of positively charged protons, as well as neutrons, which have no charge. The proton has exactly the same charge as an electron, but is positive. The neutron is actually made up of a proton and an electron bound together to create the neutral particle. A proton is about 1,800 times more massive than an electron, which makes a neutron only very slightly more massive than a proton. We say that a proton has a mass of approximately one *atomic mass unit*, (u). The *atomic number* (Z) of an element is equal to the number of protons found in an atom of that element, and fundamentally is an indication of the charge on the nucleus of that element. All atoms of a given element have the same atomic number. In other words, the number of protons an atom has defines what kind of element it is. The total number of neutrons and protons in an atom is called the *mass number* (A) of that element. The symbol ^A_ZX is used to show both the atomic number and the mass number of an X atom, where Z is the atomic number and A is the mass number.

The nucleus contains protons and neutrons.

Even though the number of protons must be the same for all atoms of an element, the number of neutrons, and thus the mass number, can be different. Atoms of the same element with different masses are known as isotopes of one another. For example, carbon-12 is a carbon atom with 6 protons and 6 neutrons, while carbon-14 is a carbon atom with 6 protons and 8 neutrons. We would write these two isotopes of carbon as $^{12}_{6}C$ and $^{14}_{6}C$.

***Isotopes* are atoms containing the same number of protons but different numbers of neutrons.**

The table below summarizes the basic features of protons, neutrons and electrons. Notice that we use an *H* to symbolize the proton, since the proton is a hydrogen nucleus.

Particle	Symbol	Relative mass	Charge	Location
proton	${}^{1}_{1}H$	1	+1	nucleus
neutron	${}^{1}_{0}n$	1	0	nucleus
electron	${}^{0}_{-1}e$ or e^-	0	–1	electron orbitals around the nucleus

The atomic number *Z* is the number of protons; the mass number *A* is the number of protons and neutrons.

Example: Determine the number of protons, neutrons, and electrons in a neutral atom of iron ${}^{56}_{26}Fe$.

Solution: This isotope of iron has an atomic number of 26 and a mass number of 56. Therefore, it will have 26 protons, 26 electrons, and 56 – 26 = 30 neutrons.

NUCLEAR BINDING ENERGY

Since positive charges repel each other, you might wonder why protons stay together in the nucleus of the atom. There must be a force holding the protons together which is greater than the electrostatic repulsion between them. This force is called the *strong nuclear force*, and is a result of the binding energy of the nucleus. But where does this energy that holds the nucleus together come from?

According to Einstein's famous equation $E = mc^2$, mass and energy can be converted into one another. When a nucleus is assembled, each proton and neutron give up a little of their mass to be converted into binding energy. For example, if you start with two protons and two neutrons, you have a total of 4 atomic mass units. But if these particles are combined into a helium ${}^{4}_{2}He$ nucleus, the resulting mass of the helium nucleus is less than 4 atomic mass units, since some of the mass of the protons and neutrons has been converted into binding energy to hold the nucleus together. Likewise, when a nucleus is split, it doesn't need all of its original binding energy anymore, and some of that energy is released as heat.

Mass is converted into nuclear binding energy to hold the nucleus together.

NUCLEAR REACTIONS

Radioactive Decay

At the end of the 19th century, scientists discovered elements that continuously emitted mysterious rays. These elements were identified as being *radioactive*. A radioactive element spontaneously emits particles from its nucleus because the energy of the nucleus is unstable. Examples of naturally occurring radioactive elements are uranium ${}^{238}_{92}U$, radium ${}^{226}_{88}Ra$, and carbon ${}^{14}_{6}C$.

There are four types of particles that can be emitted when an element undergoes radioactive decay:

Four types of radioactive decay: alpha, beta, gamma, and positron.

1. *Alpha decay*. Uranium, for example, undergoes alpha decay, meaning that it emits an *alpha particle* from its nucleus. An alpha particle is a helium nucleus, consisting of 2 protons and 2 neutrons. When an element emits an alpha particle, its nucleus loses 2 atomic numbers and 4 mass numbers, and thus changes into another element, called the *daughter element*. But what would this element be? We can write the nuclear equation for the radioactive decay of uranium as:

$$ {}^{238}_{92}U \rightarrow {}^{A}_{Z}X + {}^{4}_{2}He $$

where X is the daughter element and ${}^{4}_{2}He$ is the alpha particle. The atomic number on the left must equal the sum of the atomic numbers on the right, since charge and mass are conserved in this process. The same is true for the mass numbers on the left and right. So, the daughter element has an atomic number $Z = 92 - 2 = 90$ and a mass number $A = 238 - 4 = 234$. Uranium decays into the daughter element ${}^{234}_{90}Th$, thorium.

2. *Beta decay*. A *beta particle* is the name given to an electron emitted from the nucleus of a radioactive element. But what is an electron doing in the nucleus of an atom? Remember that we described the neutron in the nucleus of an atom as being a proton and an electron bound together. Beta decay is really just a neutron emitting an electron and becoming a proton. Thus, the daughter element resulting from beta decay is one atomic number higher than the parent nucleus, but the mass number essentially does not change. For example, carbon ${}^{14}_{6}C$ is a radioactive element that undergoes beta decay. The decay equation is:

$$ {}^{14}_{6}C \rightarrow {}^{A}_{Z}X + {}^{0}_{-1}e $$

We use the same symbol for a beta particle as we do for an electron. The daughter element must have an atomic number of $6 - (-1) = 7$ and a mass number of $14 - 0 = 14$. The daughter element is ${}^{14}_{7}N$, nitrogen.

3. *Gamma decay*. Some radioactive elements emit a gamma ray, a very high energy electromagnetic wave which has no charge or mass, so only the energy of the nucleus changes, and neither Z nor A change.

4. *Positron decay*. A positron is exactly like an electron except for the fact that it is positively charged. A positron is not a proton, as their masses and other features are very different. Positron decay equations are typically not included on the SAT II: Physics Test.

Radioactive Decay Half-Life

The *half-life* of a radioactive sample is the time it takes for half the atoms in the sample to decay. For example, let's say we begin with one kilogram of the radioactive element radium in a sealed container. The half-life of radium is 1,600 years. After 1,600 years, our descendants could check the container and find only half a kilogram of radium. There would still be one kilogram of substance in the container, but half of it would be radium and the other half would be the daughter of radium, which happens to be radon gas. After 1,600 more years, the container would contain only one-fourth of a kilogram of radium, and so on. We can plot a graph of the amount of radium remaining vs. time on the axes below:

Half-life **is the time it takes for half the atoms in a radioactive sample to decay.**

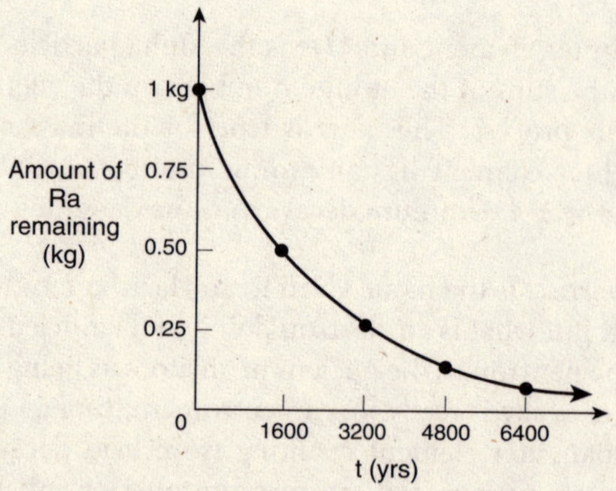

This type of curve is called an *exponential decay curve*.

Example: Cobalt $^{60}_{27}Co$ has a half-life of 5 years. If we start with a 100-gram sample of cobalt, how much cobalt remains after 20 years?

Solution: Let's organize our data in a table:

Amount of *Co* remaining	Half-Lives	Time
100 g	0	0 (today)
50 g	1	5 yrs
25 g	2	10 yrs
12.5 g	3	15 yrs
6.25 g	4	20 yrs

Thus, after 20 years, 4 half-lives have passed, and 6.25 grams of cobalt remain.

Fusion

Fusion occurs when small nuclei combine into larger nuclei and energy is released. Many stars, including our sun, power themselves by fusing four hydrogen nuclei to make one helium nucleus. The fusion process is clean and a large amount of energy is released.

Fusion is the combining of small nuclei into larger ones, releasing energy.

Example: The element tritium $^{3}_{1}H$ is combined with another element to form helium $^{4}_{2}He$ and a neutron, along with the release of energy. The equation for this fusion reaction is:

$$^{3}_{1}H + ^{A}_{Z}X \rightarrow ^{4}_{2}He + ^{1}_{0}n$$

What is the unknown element *X*?

Solution: The numbers on the left must equal the numbers on the right, so the atomic number of element X is $Z = 2 + 0 - 1 = 1$, and the mass number for X is $A = 4 + 1 - 3 = 2$. Element X is $^{2}_{1}H$, a hydrogen isotope called *deuterium*.

Fission (裂变)

Fission is a process in which a large nucleus splits in to smaller nuclei. Fission is usually caused artificially by shooting a slow neutron at a large atom, such as uranium, which absorbs the neutron and splits into two smaller atoms, along with the release of more neutrons and some energy.

Example: A fission reaction occurs when uranium $^{235}_{92}U$ absorbs a slow neutron and then splits into xenon and strontium, releasing three neutrons and some energy. The equation for this fission reaction is:

$$^{235}_{92}U + ^{1}_{0}n \rightarrow ^{140}_{54}Xe + ^{A}_{Z}Sr + 3^{1}_{0}n + energy$$

What are Z and A for strontium?

Fission is the splitting of a large nucleus into smaller ones, with the release of neutrons and energy.

Solution: Once again, the numbers on the left equal the numbers on the right. Then $Z = 92 - 54 = 38$, and $A = 235 + 1 - 140 - 3(1) = 94$.

3

$A = 235 + 1 - 140 - 3$
$= 236 - 143$
$= 93$

$Z = 92 - 54$
$= 38$

NUCLEAR AND PARTICLE PHYSICS REVIEW QUESTIONS

1. The neutral element magnesium $^{24}_{12}Mg$ has

(A) 12 protons, 12 electrons, and 24 neutrons.
(B) 12 protons, 12 electrons, and 12 neutrons.
(C) 24 protons, 24 electrons, and 12 neutrons.
(D) 24 protons, 12 electrons, and 12 neutrons.
(E) 12 protons, 24 electrons, and 24 neutrons.

2. All isotopes of uranium have

(A) the same atomic number and the same mass number.
(B) different atomic numbers but the same mass number.
(C) different atomic numbers and different mass numbers.
(D) the same atomic number but different mass numbers.
(E) no electrons.

3. Six protons and six neutrons are brought together to form a carbon nucleus, but the mass of the carbon nucleus is less than the sum of the masses of the individual particles that make up the nucleus. This missing mass, called the mass defect, has been

(A) converted into the binding energy of the nucleus.
(B) given off in a radioactive decay process.
(C) converted into electrons.
(D) converted into energy to hold the electrons in orbit.
(E) emitted as light.

4. The isotope of thorium $^{234}_{90}Th$ undergoes alpha decay according to the equation:

$$^{234}_{90}Th \rightarrow {}^{A}_{Z}X + {}^{4}_{2}He$$

The element X is

(A) $^{238}_{92}U$
(B) $^{230}_{88}Ra$
(C) $^{236}_{94}Pu$
(D) $^{238}_{88}Ra$
(E) $^{230}_{92}U$

5. The isotope of cobalt $^{60}_{27}Co$ undergoes beta decay according to the equation:

$$^{60}_{27}Co \rightarrow {}^{A}_{Z}X + {}^{0}_{-1}e$$

The element X is

(A) $^{60}_{26}Fe$
(B) $^{56}_{25}Mn$
(C) $^{60}_{28}Ni$
(D) $^{60}_{29}Cu$
(E) $^{61}_{27}Co$

D 6. The half-life of a certain element is 4 years. What fraction of a sample of that isotope will remain after 12 years?

(A) $\frac{1}{2}$
(B) $\frac{1}{3}$
(C) $\frac{1}{4}$
(D) $\frac{1}{8}$
(E) $\frac{1}{12}$

$\frac{1}{8}$

B 7. Consider the following nuclear equation:

$$^{235}_{92}U + {}^{1}_{0}n \rightarrow {}^{139}_{56}Ba + {}^{94}_{36}Kr + 3{}^{1}_{0}n$$

This equation describes the process of

(A) radioactivity.
(B) fission.
(C) fusion.
(D) electron energy level transitions.
(E) reduction and oxidation.

SOLUTIONS TO NUCLEAR AND PARTICLE PHYSICS REVIEW QUESTIONS

1. B

The atomic number 12 implies both 12 protons and 12 electrons, and the mass number 24 is the sum of protons and neutrons, giving 12 neutrons.

2. D

All isotopes of a particular element must have the same atomic number (number of protons), since this number identifies the element, but can have a different mass number (number of neutrons).

3. A

Each particle that makes up the nucleus gives up a little mass to be converted into energy by $E = mc^2$ to bind the nucleus together.

4. B

The atomic number Z of element X is found by $90 = Z + 2$, so $Z = 88$, and the mass number A is found by $234 = A + 4$, so $A = 230$. The element X is radium.

5. C

The atomic number Z of element X is found by $27 = Z + (-1)$, so $Z = 28$, and the mass number A is found by $60 = A + 0$, so $A = 60$. The element X is nickel.

6. D

If the half-life is 4 years, then 12 years represents 3 half-lives. If we cut the sample in half three times, we get 1/8 of the sample remaining.

7. B

A neutron is absorbed by a uranium atom, and the atom splits into barium and krypton and three neutrons.

CHAPTER TWENTY-THREE

Special Relativity

(相对论)

In the summer of 1905, Albert Einstein submitted to a journal of physics a paper titled "On the Electrodynamics of Moving Bodies" that contained a new theory we now refer to as the *theory of special relativity*. This theory completely challenged our notion of space and time, two concepts we previously believed to be absolute and constant throughout the universe. Yet Einstein clearly showed that it is neither space nor time that is absolute in our universe, but another quantity.

THE TWO POSTULATES IN THE THEORY OF SPECIAL RELATIVITY

(相对论的两个假设)

To understand special relativity, we must accept two postulates:

1. The laws of physics are the same in all inertial reference frames. (所有惯性系中，一切物理规律都相同)
2. The speed of light is constant in all reference frames, regardless of any relative motion between an observer and the light source. (光速不变原理)

> **There are two postulates in the theory of special relativity.**

The first postulate is not difficult to accept. An inertial reference frame is one which is at rest or moving with a constant velocity in a straight line. If you throw a ball straight up in the air while you are in a car which is at rest, it comes back down and lands in your hand again. If your car is moving at a constant velocity in a straight line and you throw the ball straight up, it still comes back down in your hand, since the car, you, and the ball all have the same horizontal velocity. This can be extended to any physics experiment that you can think of. In other words, there is not one special inertial reference frame that is better than or different from any other when it comes to making physical measurements. This leads us to the second postulate, which is a little more difficult for us to accept.

The second postulate of special relativity states that everyone in any reference frame will measure the same value for the speed of light regardless of how fast he or she is moving relative to the light source. If I am at rest relative to your flashlight and you turn it on, I measure the speed of the light as being $c = 3 \times 10^8$ m/s, and so do you. But if you remain at rest, and I begin moving at 100 miles per hour, Newtonian physics says that we would not agree on the speed of the light beam since I am moving and you are not. However, Einstein's second postulate states that we will both still measure the speed of the light beam as $c = 3 \times 10^8$ m/s. Even if I move at half the speed of light ($0.5c$), the second postulate still holds true; that is, you and I will still agree on the speed of the light beam, $c = 3 \times 10^8$ m/s. This can be difficult for us to accept, since we are used to taking into account the relative motion between reference frames when calculating velocities. This leads us to some interesting effects on length and time.

An inertial reference frame is one which has a constant velocity, including a velocity of zero.

The basic equation for speed is:

$$v = \frac{length}{time}$$

Substituting the speed of light into the equation, we have:

$$c = \frac{length}{time}$$

We've seen that all observers must agree on the value of the speed of light c, which means that all observers do not have to agree on the value of length or the value of time, only on the ratio of the two. In other words, so that you and I will agree on the speed of light, we will not necessarily have to agree on the length of moving objects or how much time has passed. This gives rise to the *relativistic effects* on length and time:

The speed of light is constant for all observers.

An observer who measures the length of an object moving relative to the observer will measure the length of the object as being shorter (contracted) in the direction of motion compared to the measurement of its length when it is at rest relative to the observer.

In other words, a moving object is shorter than when it is at rest. In order to keep the ratio of length to time constant, time must also change:

Length contracts in the direction of motion for a moving object.

A moving clock will run more slowly than a clock that is at rest.

This is called *time dilation*, and has been verified experimentally many times. This means that the ticks of a moving clock are farther apart than the ticks of a clock that is at rest.

Time is simply the duration between two events, and since any two observers must agree on the speed of light, they may not agree on the length of a moving object or how much time has passed between two events. Relativistic effects occur at all speeds, but they only become measurable at speeds above about ten percent the speed of light.

There is one other relativistic effect we should discuss. In his 1905 paper, Einstein suggested that energy and mass are actually different aspects of the same phenomenon. The famous equation that links energy to mass is:

Moving clocks run more slowly than clocks at rest.

$$E = mc^2$$

This equation tells us that energy and mass can be converted into one another, and that their value is connected by a constant, the speed of light squared. Since the speed of light is a huge number, a little mass can be converted into a lot of energy, as history has witnessed with the release of nuclear binding energy when the nuclei of atoms are split and that energy is released. But not only can mass be converted into energy, energy can be converted into mass.

Mass and energy can be converted into one another.

A moving object has kinetic energy, and experiments have shown that at high velocities, the kinetic energy of a moving object such as a proton begins turning into mass. As a proton approaches a speed near the speed of light, its mass becomes larger and larger, as verified by momentum measurements. This puts an ultimate speed limit on moving objects, the speed of light. The equations of relativity tell us that if it were possible for an object to achieve a speed equal to the speed of light, it would have zero length, its clock would stop, and it would have infinite mass. Since none of these are possible in our universe as far as we know, achieving the speed of light is impossible.

Mass increases as speed increases.

Our discussion of special relativity can be summarized as follows:

- *All inertial reference frames are equivalent.*
- *The speed of light is constant for all observers.*
- *An observer watching a moving object will see its length contract in the direction of motion, its clock slow down, and its mass increase by the equation $E = mc^2$.*

Example: Two spaceships pass you at a speed near the speed of light, each containing a meter stick, a clock, and a 1 kg block. For each of the following diagrams, describe the changes in the length of the meter stick, the ticks of the clock, and the mass of the 1 kg block.

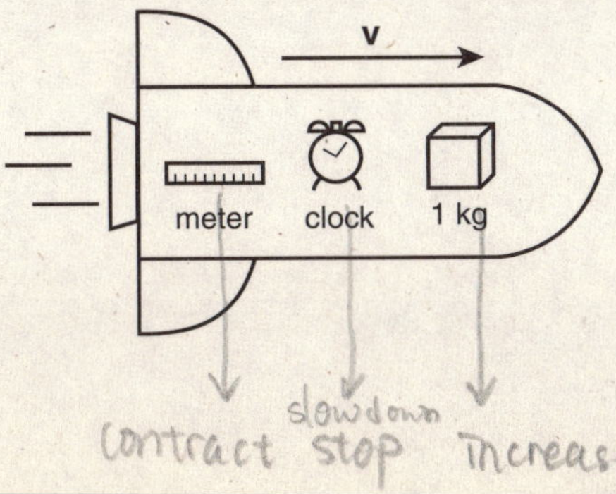

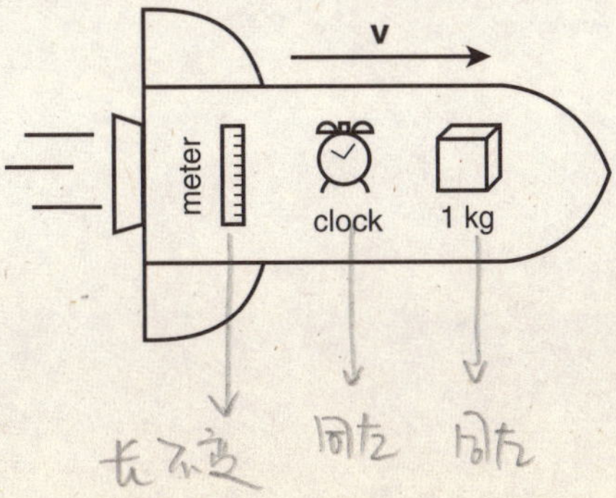

Solution: In Figure A, the meter stick is aligned in the direction of motion of the ship. Thus, you would measure the meter stick as being shorter than one meter. You would also measure the clock as running slower, that is, more time between ticks, and the mass as larger than one kilogram.

In Figure B, you would still measure the clock as running slower and the mass as larger than one kilogram, but since the meter stick is aligned perpendicular to the motion of the ship, you would still measure it as being one meter long, although a little thinner, since length contraction occurs only in the direction of motion.

If an astronaut in the ship Figure A looked at your reference frame as he was passing you, he would see your meter stick as being shorter than one meter, your clock running slower, and your mass as larger. Since his reference frame is as good as yours in which to make measurements, he should measure the same effects in your reference frame as you measured in his.

SPECIAL RELATIVITY REVIEW QUESTIONS

1. A pilot of a spaceship traveling at 90% the speed of light ($0.9c$) turns on its laser headlights just as it passes a stationary observer. Which of the following statements is true?

   (A) The pilot will measure the speed of the light coming out of the headlights as c, and the observer will measure the speed of the light as $0.9c$.
   (B) The pilot will measure the speed of the light coming out of the headlights as c, and the observer will measure the speed of the light as $1.9c$.
   (C) The pilot will measure the speed of the light coming out of the headlights as $0.9c$, and the observer will measure the speed of the light as $1.9c$.
   (D) The pilot will measure the speed of the light coming out of the headlights as $1.9c$, and the observer will measure the speed of the light as $0.9c$.
   (E) The pilot will measure the speed of the light coming out of the headlights as c, and the observer will measure the speed of the light as c.

2. Which of the following graphs best represents the length L of a horizontal meter stick traveling horizontally with speed v as it approaches the speed of light?

   (A) 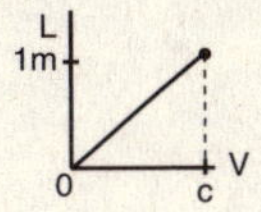

   (B) 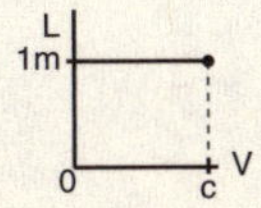

   (C) 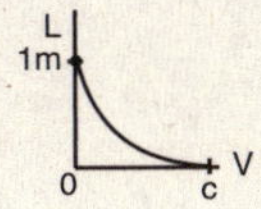

   (D) 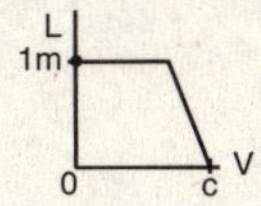

   (E) 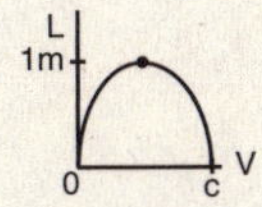

3. Two identical, very precise clocks are started at the same time. One clock is taken on a trip at a very high speed, and the other is left at rest on earth. When the traveling clock is returned to earth, it shows that one hour has passed. Which of the following could be the time that has passed on the earth-bound clock?

(A) 30 minutes
(B) 45 minutes
(C) 59 minutes
(D) 1 hour
(E) 2 hours

4. The mass of an object increases as its speed increases. This increase in mass comes from

(A) nuclear binding energy.
(B) electron energy in the ground state.
(C) potential energy being converted into mass by $E = mc^2$.
(D) kinetic energy being converted into mass by $E = mc^2$.
(E) the lower pressure on the mass.

SOLUTIONS TO SPECIAL RELATIVITY REVIEW QUESTIONS

1. E

The speed of light is constant to all observers, and thus all observers would measure the speed of the light as c.

2. C

The length gets shorter (contracts) as the speed gets higher, so that it is practically zero at $v = c$.

3. E

The traveling clock will have run more slowly than the earth-bound clock, and thus more time will have passed on the earth bound clock than on the traveling clock.

4. D

Some of the kinetic energy converts into mass so that the mass will not exceed the speed of light.

section three

READY, SET, GO!

CHAPTER TWENTY-FOUR

Stress Management

The countdown has begun. Your date with THE TEST is looming on the horizon. Anxiety is on the rise. The butterflies in your stomach have gone ballistic. Perhaps you feel as if the last thing you ate has turned into a lead ball. Your thinking is getting cloudy. Maybe you think you won't be ready. Maybe you already know your stuff, but you're going into panic mode anyway. Worst of all, you're not sure of what to do about it.

Let go of "must-y" thoughts, those notions that you must do something in a certain way–for example, "I must get a great score, or else!" or "I must meet Mom and Dad's expectations!"

Don't freak! It is possible to tame that anxiety and stress—before and during the test. We'll show you how. You won't believe how quickly and easily you can deal with that killer anxiety.

Lack of control is one of the prime causes of stress. A ton of research shows that if you don't have a sense of control over what's happening in your life you can easily end up feeling helpless and hopeless. So, just having concrete things to do and to think about—taking control—will help reduce your stress. This section shows you how to take control during the days leading up to taking the test.

Don't study on your bed, especially if you have problems with insomnia. Your mind might start to associate the bed with work and make it even harder for you to fall asleep.

IDENTIFY THE SOURCES OF STRESS

In the space provided, jot down (in pencil) anything you identify as a source of your test-related stress. The idea is to pin down that free-floating anxiety so that you can take control of it. Here are some common examples to get you started:

- I always freeze up on tests.
- I'm nervous about the math (or the vocabulary or reading, etcetera).
- I need a good/great score to go to Acme College.
- My older brother/sister/best friend/girlfriend or boyfriend did really well. I must match their scores or do better.
- My parents, who are paying for school, will be really disappointed if I don't test well.
- I'm afraid of losing my focus and concentration.
- I'm afraid I'm not spending enough time preparing.
- I study like crazy but nothing seems to stick in my mind.
- I always run out of time and get panicky.
- I feel as though thinking is becoming like wading through thick mud.

Sources of Stress

____________________	____________________
____________________	____________________
____________________	____________________
____________________	____________________

Stress expert Stephen Sideroff, Ph.D., tells of a client who always stressed out before, during, and even after taking tests. Yet, she always got outstanding scores. It became obvious that she was thinking superstitiously—subconsciously believing that the great scores were a result of her worrying. She also didn't trust herself, and believed that if she didn't worry she wouldn't study hard enough. Sideroff convinced her to take a risk and work on relaxing before her next test. She did, and her test results were still as good as ever—which broke her cycle of superstitious thinking.

Take a few minutes to think about the things you've just written down. Then rewrite them in some sort of order. List the statements you most associate with your stress and anxiety first, and put the least disturbing items last. Chances are, the top of the list is a fairly accurate description of exactly how you react to test anxiety, both physically and mentally. The later items usually describe your fears (disappointing Mom and Dad, looking bad, etcetera). As you write the list, you're forming a hierarchy of items so you can deal first with the anxiety provokers that bug you most. Very often, taking care of the major items from the top of the list goes a long way towards relieving overall testing anxiety. You probably won't have to bother with the stuff you placed last.

TAKE STOCK OF YOUR STRENGTHS AND WEAKNESSES

Take one minute to list the areas of the test that you are good at. They can be general ("electricity") or specific ("using Ohm's law to calculate voltage"). Put down as many as you can think of, and if possible, time yourself. Write for the entire time; don't stop writing until you've reached the one-minute stopping point.

Strong Test Subjects

Next, take one minute to list areas of the test you're not so good at, just plain bad at, have failed at, or keep failing at. Again, keep it to one minute, and continue writing until you reach the cutoff. Don't be afraid to identify and write down your weak spots! In all probability, as you do both lists you'll find you are strong in some areas and not so strong in others. Taking stock of your assets and liabilities lets you know which areas you don't have to worry about, and which ones will demand extra attention and effort.

Weak Test Subjects

Facing your weak spots gives you some distinct advantages. It helps a lot to find out where you need to spend extra effort. Increased exposure to tough material makes it more familiar and less intimidating. (After all, we mostly fear what we don't know and are probably afraid to face.) You'll feel better about yourself because you're dealing directly with areas of the test that bring on your anxiety. You can't help feeling more confident when you know you're actively strengthening your chances of earning a higher overall test score.

Don't work in a messy or cramped area. Before you sit down to study, clear yourself a nice, open space. And make sure you have books, paper, pencils—whatever tools you will need—within easy reach before you sit down to study.

Now, go back to the "good" list, and expand it for two minutes. Take the general items on that first list and make them more specific; take the specific items and expand them into more general conclusions. Naturally, if anything new comes to mind, jot it down. Focus all of your attention and effort on your strengths. Don't underestimate yourself or your abilities. Give yourself full credit. At the same time, don't list strengths you don't really have; you'll only be fooling yourself.

Whatever you know comfortably (that is, almost as well as you know the back of your hand) goes on your "good" list. Okay. You've got the picture. Now, get ready, check your starting time, and start writing down items on your expanded "good" list.

Strong Test Subjects: An Expanded List

______________________	______________________
______________________	______________________
______________________	______________________
______________________	______________________

After you've stopped, check your time. Did you find yourself going beyond the two minutes allotted? Did you write down more things than you thought you knew? Is it possible you know more than you've given yourself credit for? Could that mean you've found a number of areas in which you feel strong?

You just took an active step towards helping yourself. Notice any increased feelings of confidence? Enjoy them.

Here's another way to think about your writing exercise. Every area of strength and confidence you can identify is much like having a reserve of solid gold at Fort Knox. You'll be able to draw on your reserves as you need them. You can use your reserves to solve difficult questions, maintain confidence, and keep test stress and anxiety at a distance. The encouraging thing is that every time you recognize another area of strength, succeed at coming up with a solution, or get a good score on a test, you increase your reserves. And, there is absolutely no limit to how much self-confidence you can have or how good you can feel about yourself.

If you want to play music, keep it low and in the background. Music with a regular, mathematical rhythm—reggae, for example—aids the learning process. A recording of ocean waves is also soothing.

IMAGINE YOURSELF SUCCEEDING

This next little group of exercises is both physical and mental. It's a natural follow-up to what you've just accomplished with your lists.

First, get yourself into a comfortable sitting position in a quiet setting. Wear loose clothes. If you wear glasses, take them off. Then, close your eyes and breathe in a deep, satisfying breath of air. Really fill your lungs until your rib cage is fully expanded and you can't take in any more. Then, exhale the air completely. Imagine you're blowing out a candle with your last little puff of air. Do this two or three more times, filling your lungs to their maximum and emptying them totally. Keep your eyes closed, comfortably but not tightly. Let your body sink deeper into the chair as you become even more comfortable.

With your eyes shut, you can notice something very interesting: You're no longer dealing with the worrisome stuff going on in the world outside of you. Now you can concentrate on what happens inside you. The more you recognize your own physical reactions to stress and anxiety, the more you can do about them. You may not realize it, but you've begun to regain a sense of being in control.

When you're in the middle of studying and hit a wall, take a short, brisk walk. Breathe deeply and swing your arms as you walk. Clear your mind. (And don't forget to look for flowers that grow in the cracks of the sidewalk!)

Let images begin to form on the "viewing screens" on the back of your eyelids. You're experiencing visualizations from the place in your mind that makes pictures. Allow the images to come easily and naturally; don't force them. Imagine yourself in a relaxing situation. It might be in a special place you've visited before or one you've read about. It can be a fictional location that you create in your imagination, but a real-life memory of a place or situation you know is usually better. Make it as detailed as possible and notice as much as you can.

Stay focused on the images as you sink farther back into your chair. Breathe easily and naturally. You might have the sensation of any stress or tension draining from your muscles and flowing downward, out your feet and away from you.

Take a moment to check how you're feeling. Notice how comfortable you've become. Imagine how much easier it would be if you could take the test feeling this relaxed and in this state of ease. You've coupled the images of your special place with sensations of comfort and relaxation. You've also found a way to become relaxed simply by visualizing your own safe, special place.

Now, close your eyes and start remembering a real-life situation in which you did well on a test. If you can't come up with one, remember a situation in which you did something (academic or otherwise) that you were really proud of—a genuine accomplishment. Make the memory as detailed as possible. Think about the sights, the sounds, the smells, even the tastes associated with this remembered experience. Remember how confident you felt as you accomplished your goal. Now start thinking about the upcoming test. Keep your thoughts

Don't forget that your school probably has counseling available. If you can't conquer test stress on your own, make an appointment at the counseling center. That's what counselors are there for.

and feelings in line with that successful experience. Don't make comparisons between them. Just imagine taking the upcoming test with the same feelings of confidence and relaxed control.

This exercise is a great way to bring the test down to earth. You should practice this exercise often, especially when the prospect of taking the exam starts to bum you out. The more you practice it, the more effective the exercise will be for you.

EXERCISE YOUR FRUSTRATIONS AWAY

Whether it is jogging, walking, biking, mild aerobics, pushups, or a pickup basketball game, physical exercise is a very effective way to stimulate both your mind and body and to improve your ability to think and concentrate. A surprising number of students get out of the habit of regular exercise, ironically because they're spending so much time prepping for exams. Also, sedentary people—this is a medical fact—get less oxygen to the blood and hence to the head than active people. You can live fine with a little less oxygen; you just can't think as well.

Any big test is a bit like a race. Thinking clearly at the end is just as important as having a quick mind early on. If you can't sustain your energy level in the last sections of the exam, there's too good a chance you could blow it. You need a fit body that can weather the demands any big exam puts on you. Along with a good diet and adequate sleep, exercise is an important part of keeping yourself in fighting shape and thinking clearly for the long haul.

Do eat:

- Fruits and vegetables (raw is best, or just lightly steamed or nuked)
- Low-fat protein such as fish, skinless poultry, beans, and legumes
- Whole grains such as brown rice, whole wheat bread, and pastas

Don't eat:

- Refined sugar; sweet, high-fat snacks (simple carbohydrates like sugar make stress worse and fatty foods lower your immunity)
- Salty foods (they deplete potassium, which you need for nerve functions)

There's another thing that happens when students don't make exercise an integral part of their test preparation. Like any organism in nature, you operate best if all your "energy systems" are in balance. Studying uses a lot of energy, but it's all mental. When you take a study break, do something active instead of raiding the fridge or vegging out in front of the TV. Take a five- to ten-minute activity break for every 50 or 60 minutes that you study. The physical exertion gets your body into the act, which helps to keep your mind and body in sync. Then, when you finish studying for the night and hit the sack, you won't lie there, tense and unable to sleep because your head is overtired and your body wants to pump iron or run a marathon.

One warning about exercise, however: It's not a good idea to exercise vigorously right before you go to bed. This could easily cause sleep onset problems. For the same reason, it's also not a good idea to study right up to bedtime. Make time for a "buffer period" before you go to bed: For 30 to 60 minutes, just take a hot shower, meditate, or simply veg out.

THE DANGERS OF DRUGS

Using drugs (prescription or recreational) specifically to prepare for and take a big test is definitely self-defeating. (And if they're illegal drugs, you may end up with a bigger problem than the SAT II: Physics Test on your hands.) Every drug has major drawbacks—and a false sense of security is only one of them. You may have heard that popping uppers helps you study by keeping you alert. If they're illegal, definitely forget about it. They wouldn't really work anyway, since amphetamines make it hard to retain information. Mild stimulants, such as coffee, cola, or over-the-counter caffeine pills can sometimes help as you study, since they keep you alert. On the down side, they can also lead to agitation, restlessness, and insomnia. Some people can drink a pot of high-octane coffee and sleep like a baby. Others have one cup and start to vibrate. It all depends on your tolerance for caffeine. Remember, a little anxiety is a good thing. The adrenaline that gets pumped into your bloodstream helps you stay alert and think more clearly. But, too much anxiety and you can't think straight at all.

Instead, go for endorphins—the "natural morphine." Endorphins have no side effects and they're free—you've already got them in your brain. It just takes some exercise to release them. Running around on the basketball court, bicycling, swimming, aerobics, power walking—these activities cause endorphins to occupy certain spots in your brain's neural synapses. In addition, exercise develops staying power and increases the oxygen transfer to your brain. Go into the test naturally.

Forcing relaxation is like asking yourself to flap your arms and fly. You can't do it, and every push and prod only gets you more frustrated. Relaxation is something you don't work at. You simply let it happen. Think about it. When was the last time you tried to force yourself to go to sleep, and it worked?

TAKE A DEEP BREATH ...

Here's another natural route to relaxation and invigoration. It's a classic isometric exercise that you can do whenever you get stressed out—just before the test begins, even during the test. It's very simple and takes just a few minutes.

Close your eyes. Starting with your eyes—without holding your breath—gradually tighten every muscle in your body (but not to the point of pain) in the following sequence:

1. Close your eyes tightly.
2. Squeeze your nose and mouth together so that your whole face is scrunched up. (If it makes you self-conscious to do this in the test room, skip the face-scrunching part.)
3. Pull your chin into your chest, and pull your shoulders together.
4. Tighten your arms to your body, then clench your hands into tight fists.
5. Pull in your stomach.
6. Squeeze your thighs and buttocks together, and tighten your calves.
7. Stretch your feet, then curl your toes (watch out for cramping in this part).

At this point, every muscle should be tightened. Now, relax your body, one part at a time, in reverse order, starting with your toes. Let the tension drop out of each muscle. The entire process might take five minutes from start to finish (maybe a couple of minutes during the test). This clenching and unclenching exercise should help you to feel very relaxed.

AND KEEP BREATHING

Conscious attention to breathing is an excellent way of managing test stress (or any stress, for that matter). The majority of people who get into trouble during tests take shallow breaths. They breathe using only their upper chests and shoulder muscles, and may even hold their breath for long periods of time. Conversely, the test taker who by accident or design keeps breathing normally and rhythmically is likely to be more relaxed and in better control during the entire test experience.

A lamp with a 75-watt bulb is optimal for studying. But don't put it so close to your study material that you create a glare.

So, now is the time to get into the habit of relaxed breathing. Do the next exercise to learn to breathe in a natural, easy rhythm. By the way, this is another technique you can use during the test to collect your thoughts and ward off excess stress. The entire exercise should take no more than three to five minutes.

With your eyes still closed, breathe in slowly and deeply through your nose. Hold the breath for a bit, and then release it through your mouth. The key is to breathe slowly and deeply by using your diaphragm (the big band of muscle that spans your body just above your waist) to draw air in and out naturally and effortlessly. Breathing with your diaphragm encourages relaxation and helps minimize tension. Try it and notice how relaxed and comfortable you feel.

QUICK TIPS FOR THE DAYS JUST BEFORE THE EXAM

- The best test takers do less and less as the test approaches. Taper off your study schedule and take it easy on yourself. You want to be relaxed and ready on the day of the test. Give yourself time off, especially the evening before the exam. By that time, if you've studied well, everything you need to know is firmly stored in your memory banks.
- Positive self-talk can be extremely liberating and invigorating, especially as the test looms closer. Tell yourself things such as, "I choose to take this test" rather than "I have to"; "I will do well" rather than "I hope things go well"; "I can" rather than "I cannot." Be aware of negative, self-defeating thoughts and images and immediately counter any you become aware of. Replace them with affirming statements that encourage your self-esteem and confidence. Create and practice doing visualizations that build on your positive statements.

- Get your act together sooner rather than later. Have everything (including choice of clothing) laid out days in advance. Most important, know where the test will be held and the easiest, quickest way to get there. You will gain great peace of mind if you know that all the little details—gas in the car, directions, etcetera—are firmly in your control before the day of the test.
- Experience the test site a few days in advance. This is very helpful if you are especially anxious. If at all possible, find out what room you will be assigned to, and try to sit there (by yourself) for a while. Better yet, bring some practice material and do at least a section or two, if not an entire practice test, in that room. In this case, familiarity doesn't breed contempt, it generates comfort and confidence.

On the day of the test, wear loose layers. That way, you'll be prepared no matter what the temperature of the room is. (An uncomfortable temperature will just distract you from the job at hand.) And if you have an item of clothing that you tend to feel "lucky" or confident in—a favorite shirt, a pair of jeans, whatever—wear it. A little totem couldn't hurt.

- Forego any practice on the day before the test. It's in your best interest to marshal your physical and psychological resources for 24 hours or so. Even race horses are kept in the paddock and treated like princes the day before a race. Keep the upcoming test out of your consciousness; go to a movie, take a pleasant hike, or just relax. Don't eat junk food or tons of sugar. And—of course—get plenty of rest the night before. Just don't go to bed too early. It's hard to fall asleep earlier than you're used to, and you don't want to lie there thinking about the test.

HANDLING STRESS DURING THE TEST

The biggest stress monster will be the test itself. Fear not; there are methods of quelling your stress during the test.

- Keep moving forward instead of getting bogged down in a difficult question or passage. You don't have to get everything right to achieve a fine score. So, don't linger out of desperation on a question that is going nowhere even after you've spent considerable time on it. The best test takers skip difficult material temporarily in search of the easier stuff. They mark the ones that require extra time and thought. This strategy buys time and builds confidence so you can handle the tough stuff later.

- Don't be thrown if other test takers seem to be working more busily and furiously than you are. Continue to spend your time patiently but doggedly thinking through your answers; it's going to lead to higher-quality test taking and better results. Don't mistake the other people's sheer activity as signs of progress and higher scores.

Here's some advice from a Kaplan instructor who won big on *Jeopardy!*™ In the green room before the show, he noticed that the contestants who were quiet and "within themselves" were the ones who did great on the show. The contestants who did not perform as well were the ones who were fact-cramming, talking a lot, and generally being manic before the show. Lesson: Spend the final hours leading up to the test getting sleep, meditating, and generally relaxing.

- Keep breathing! Weak test takers tend to share one major trait: They forget to breathe properly as the test proceeds. They start holding their breath without realizing it, or they breathe erratically or arrhythmically. Improper breathing hurts confidence and accuracy. Just as important, it interferes with clear thinking.
- Some quick isometrics during the test—especially if concentration is wandering or energy is waning—can help. Try this: Put your palms together and press intensely for a few seconds. Concentrate on the tension you feel through your palms, wrists, forearms, and up into your biceps and shoulders. Then, quickly release the pressure. Feel the difference as you let go. Focus on the warm relaxation that floods through the muscles. Now you're ready to return to the task.
- Here's another isometric that will relieve tension in both your neck and eye muscles. Slowly rotate your head from side to side, turning your head and eyes to look as far back over each shoulder as you can. Feel the muscles stretch on one side of your neck as they contract on the other. Repeat five times in each direction.

With what you've just learned here, you're armed and ready to do battle with the test. This book and your studies will give you the information you'll need to answer the questions. It's all firmly planted in your mind. You also know how to deal with any excess tension that might come along, both when you're studying for and taking the exam. You've experienced everything you need to tame your test anxiety and stress. You're going to get a great score.

CHAPTER TWENTY-FIVE

The Final Countdown

Is it starting to feel like your whole life is a build-up to the SAT II test? As the test gets closer, you may find your anxiety is on the rise. You shouldn't worry. After the preparation you've received from this book, you're in good shape for the test.

ITINERARY

To calm any pretest jitters you may have, this chapter leads you through a sane itinerary for the last week.

The Week Before the Test

Review the chapters in this book and in your textbook. Are there major holes in your preparation? If there are, choose a few of these areas to work on—but don't overload.

Take a full-length practice test. If you haven't done so already, take one or more of the practice tests in this book. These are good practice for the real thing.

Two Days Before the Test

Do your last studying—a few more practice problems, a look through the glossary in the back of this book—and then call it quits.

The Night Before the Test

Don't study. Get together the following items:

- A watch
- A few sharpened No. 2 pencils
- Erasers
- Photo ID card
- Your admission ticket

Know exactly where you're going and exactly how you're getting there.

Relax the night before the test. Read a good book, take a bubble bath, watch TV. Get a good night's sleep. Go to bed at a reasonable hour and leave yourself extra time in the morning.

The Morning of the Test

Eat breakfast. Make it something substantial, but not anything too heavy or greasy. Don't drink a lot of coffee if you're not used to it; bathroom breaks cut into your time.

Dress in layers so that you can adjust to the temperature of the test room.

Read something. Warm up your brain with a newspaper or a magazine. You shouldn't let the test questions be the first thing you read that day.

Be sure to get there early.

During the Test

Don't be shaken. If you find your confidence slipping, remind yourself how well you've prepared. You know the structure of the test; you know the instructions; you've studied for every question type.

Even if something goes really wrong, don't panic. If the test booklet is defective—two pages are stuck together or the ink has run—try to stay calm. Raise your hand, tell the proctor you need a new book. If you accidentally misgrid your answer page, again don't panic. Raise your hand, tell the proctor. She might be able to arrange for you to regrid your test after it's over, when it won't cost you any time.

After the Test

Once the test is over, put it out of your mind. Feel free to start thinking about more interesting things than physics!

Call 1-800-KAP-TEST if you have any questions about the SAT II or want more information about our other books or courses. You can also check out our website at www.kaptest.com.

You might walk out of the test room thinking that you blew it. You probably didn't. You tend to remember the questions that stumped you, not the many that you knew.

If you want more help, or just want to know more about the SAT II, college admissions, or Kaplan test prep courses, give us a call at 1-800-KAP-TEST. We're here to answer your questions and to help you in any way that we can.

Good luck!

section four

FULL-LENGTH PRACTICE TESTS

Practice Tests

HOW TO TAKE THESE PRACTICE TESTS

Before taking each of these practice tests, find a quiet room where you can work uninterrupted for one hour. Make sure you have several No. 2 pencils with erasers.

Use the answer sheets provided to record your answers.

Time yourself. Spend no more than one hour per test. Once you start a practice test, don't stop until you've reached the one-hour time limit.

Do not use a calculator. Calculators are not allowed on the SAT II: Physics Test.

You'll find an answer key, score conversion table, and complete explanations following each test.

Good luck!

Answer Sheet for Practice Test 2

1. Ⓐ Ⓑ Ⓒ Ⓓ Ⓔ
2. Ⓐ Ⓑ Ⓒ Ⓓ Ⓔ
3. Ⓐ Ⓑ Ⓒ Ⓓ Ⓔ
D 4. Ⓐ Ⓑ Ⓒ Ⓓ Ⓔ
5. Ⓐ Ⓑ Ⓒ Ⓓ Ⓔ
6. Ⓐ Ⓑ Ⓒ Ⓓ Ⓔ
A 7. Ⓐ Ⓑ Ⓒ Ⓓ Ⓔ
8. Ⓐ Ⓑ Ⓒ Ⓓ Ⓔ
9. Ⓐ Ⓑ Ⓒ Ⓓ Ⓔ
10. Ⓐ Ⓑ Ⓒ Ⓓ Ⓔ
11. Ⓐ Ⓑ Ⓒ Ⓓ Ⓔ
12. Ⓐ Ⓑ Ⓒ Ⓓ Ⓔ
13. Ⓐ Ⓑ Ⓒ Ⓓ Ⓔ
14. Ⓐ Ⓑ Ⓒ Ⓓ Ⓔ
15. Ⓐ Ⓑ Ⓒ Ⓓ Ⓔ
16. Ⓐ Ⓑ Ⓒ Ⓓ Ⓔ
17. Ⓐ Ⓑ Ⓒ Ⓓ Ⓔ
18. Ⓐ Ⓑ Ⓒ Ⓓ Ⓔ
B 19. Ⓐ Ⓑ Ⓒ Ⓓ Ⓔ
20. Ⓐ Ⓑ Ⓒ Ⓓ Ⓔ
21. Ⓐ Ⓑ Ⓒ Ⓓ Ⓔ
22. Ⓐ Ⓑ Ⓒ Ⓓ Ⓔ
23. Ⓐ Ⓑ Ⓒ Ⓓ Ⓔ
24. Ⓐ Ⓑ Ⓒ Ⓓ Ⓔ
25. Ⓐ Ⓑ Ⓒ Ⓓ Ⓔ
26. Ⓐ Ⓑ Ⓒ Ⓓ Ⓔ
27. Ⓐ Ⓑ Ⓒ Ⓓ Ⓔ
B 28. Ⓐ Ⓑ Ⓒ Ⓓ Ⓔ
29. Ⓐ Ⓑ Ⓒ Ⓓ Ⓔ
30. Ⓐ Ⓑ Ⓒ Ⓓ Ⓔ
31. Ⓐ Ⓑ Ⓒ Ⓓ Ⓔ
32. Ⓐ Ⓑ Ⓒ Ⓓ Ⓔ
33. Ⓐ Ⓑ Ⓒ Ⓓ Ⓔ
34. Ⓐ Ⓑ Ⓒ Ⓓ Ⓔ
35. Ⓐ Ⓑ Ⓒ Ⓓ Ⓔ
36. Ⓐ Ⓑ Ⓒ Ⓓ Ⓔ
37. Ⓐ Ⓑ Ⓒ Ⓓ Ⓔ
38. Ⓐ Ⓑ Ⓒ Ⓓ Ⓔ
39. Ⓐ Ⓑ Ⓒ Ⓓ Ⓔ
40. Ⓐ Ⓑ Ⓒ Ⓓ Ⓔ
41. Ⓐ Ⓑ Ⓒ Ⓓ Ⓔ
42. Ⓐ Ⓑ Ⓒ Ⓓ Ⓔ
43. Ⓐ Ⓑ Ⓒ Ⓓ Ⓔ
44. Ⓐ Ⓑ Ⓒ Ⓓ Ⓔ
45. Ⓐ Ⓑ Ⓒ Ⓓ Ⓔ
46. Ⓐ Ⓑ Ⓒ Ⓓ Ⓔ
47. Ⓐ Ⓑ Ⓒ Ⓓ Ⓔ
48. Ⓐ Ⓑ Ⓒ Ⓓ Ⓔ
49. Ⓐ Ⓑ Ⓒ Ⓓ Ⓔ
50. Ⓐ Ⓑ Ⓒ Ⓓ Ⓔ
51. Ⓐ Ⓑ Ⓒ Ⓓ Ⓔ
52. Ⓐ Ⓑ Ⓒ Ⓓ Ⓔ
53. Ⓐ Ⓑ Ⓒ Ⓓ Ⓔ
54. Ⓐ Ⓑ Ⓒ Ⓓ Ⓔ
55. Ⓐ Ⓑ Ⓒ Ⓓ Ⓔ
56. Ⓐ Ⓑ Ⓒ Ⓓ Ⓔ
A 57. Ⓐ Ⓑ Ⓒ Ⓓ Ⓔ
58. Ⓐ Ⓑ Ⓒ Ⓓ Ⓔ
59. Ⓐ Ⓑ Ⓒ Ⓓ Ⓔ
60. Ⓐ Ⓑ Ⓒ Ⓓ Ⓔ
61. Ⓐ Ⓑ Ⓒ Ⓓ Ⓔ
62. Ⓐ Ⓑ Ⓒ Ⓓ Ⓔ
63. Ⓐ Ⓑ Ⓒ Ⓓ Ⓔ
C 64. Ⓐ Ⓑ Ⓒ Ⓓ Ⓔ
65. Ⓐ Ⓑ Ⓒ Ⓓ Ⓔ
C 66. Ⓐ Ⓑ Ⓒ Ⓓ Ⓔ
67. Ⓐ Ⓑ Ⓒ Ⓓ Ⓔ
68. Ⓐ Ⓑ Ⓒ Ⓓ Ⓔ
D 69. Ⓐ Ⓑ Ⓒ Ⓓ Ⓔ
70. Ⓐ Ⓑ Ⓒ Ⓓ Ⓔ
71. Ⓐ Ⓑ Ⓒ Ⓓ Ⓔ
72. Ⓐ Ⓑ Ⓒ Ⓓ Ⓔ
73. Ⓐ Ⓑ Ⓒ Ⓓ Ⓔ
74. Ⓐ Ⓑ Ⓒ Ⓓ Ⓔ
75. Ⓐ Ⓑ Ⓒ Ⓓ Ⓔ

Raw: 67 − 2 = 65

Practice Test 2

Part A

Directions: Each set of lettered choices below relates to the numbered questions immediately following it. Select the one lettered choice that best answers each question. A choice may be used once, more than once, or not at all in each set.

Questions 1–3 relate to the following graphs of displacement *s* vs. time *t* and velocity *v* vs. time *t*.

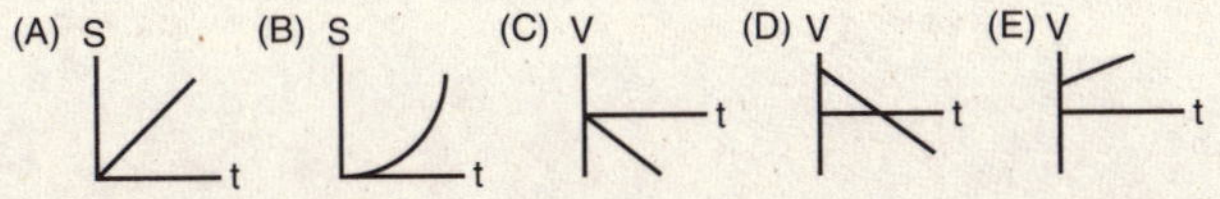

1. In which of the graphs is the velocity of the moving object constant?

2. In which of the graphs does the moving object reverse its direction?

3. Which of the graphs is equivalent to the distance vs. time graph below?

Questions 4–7 relate to the following nuclear equations:

(A) ${}^{2}_{1}H + {}^{2}_{1}H \rightarrow {}^{3}_{1}H + {}^{1}_{1}H + energy$

(B) ${}^{226}_{88}Ra \rightarrow {}^{222}_{86}Rn + {}^{4}_{2}He$

(C) ${}^{209}_{83}Bi \rightarrow {}^{209}_{84}Po + {}^{0}_{-1}e$

(D) ${}^{235}_{92}U + {}^{1}_{0}n \rightarrow {}^{140}_{54}Xe + {}^{94}_{38}Sr + {}^{1}_{0}n$

(E) ${}^{4}_{2}He + {}^{27}_{13}Al \rightarrow {}^{30}_{15}P + {}^{1}_{0}n$

4. Which of the above equations represents nuclear fission?

5. Which of the above equations represents alpha decay?

6. Which of the above equations represents beta decay?

7. Which of the above equations represents nuclear fusion?

fission：一种元素分裂成2种
fussion：2种小的合成大的
alpha decay：${}^{4}_{2}He$ 生成
beta decay：${}^{0}_{-1}e$ 生成

CONTINUE ON THE NEXT PAGE

Questions 8–11 relate to the following equations or physical principles that might be used to solve certain problems.

(A) first law of thermodynamics (conservation of energy)
(B) second law of thermodynamics (law of entropy)
(C) ideal gas law
(D) heat of fusion and heat of vaporization equation
(E) heat engine efficiency

Select the choice that should be used to provide the best and most direct solution to each of the following problems.

8. If the pressure of a gas remains constant and the temperature is doubled, what happens to the volume of the gas?

9. A new soft drink bottle is opened, allowing gas to escape into the atmosphere. As the gas escapes, how does its degree of disorder change?

10. An unknown liquid at a high temperature is safely mixed with water until an equilibrium temperature is reached. How much heat was gained by the water?

11. Heat is added to a block of ice of mass m until the entire block melts into liquid water. How much heat was required to melt the ice?

Questions 12–14 relate to the resistance-capacitance circuit and the choices below.

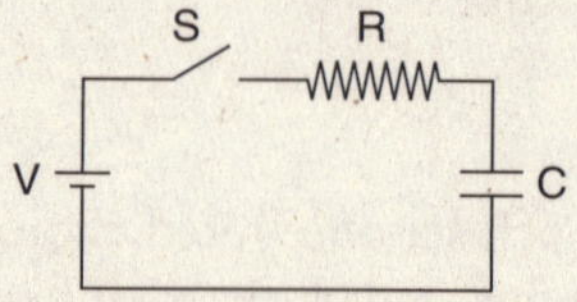

(A) zero
(B) V
(C) R
(D) C
(E) $\frac{V}{R}$

12. Immediately after the switch S is closed, what is the voltage across the resistor R?

13. Immediately after the switch S is closed, what is the current in the circuit?

14. A very long time after the switch S has been closed, what is the current in the circuit?

CONTINUE ON THE NEXT PAGE

Part B

Directions: Each of the questions or incomplete statements below is followed by five answer choices. Select the one that is best in each case.

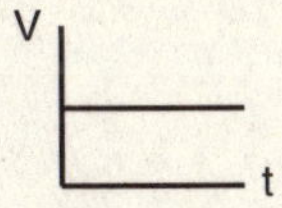

15. The graph above shows velocity v as a function of time t for a particle moving in a straight line. Graphs of displacement s versus time t that are consistent with the v vs. t graph above include which of the following?

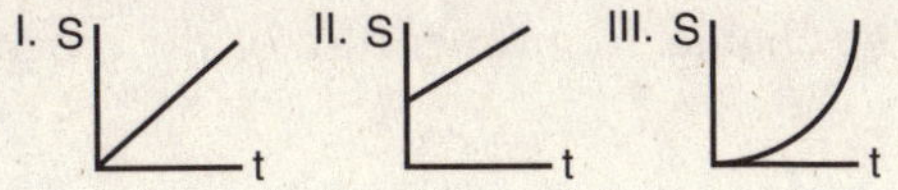

(A) I only
(B) III only
(C) I and II only
(D) II and III only
(E) I, II, and III

16. You are sitting on a seat facing forward on an airplane with its wings parallel to the ground. The window shades of the airplane are closed, and the vibration of the plane is negligible. When you place your class ring on the end of a necklace chain and hold the other end in front of you, you notice that the chain and ring hang vertically and point directly to the floor of the airplane. Which of the following could be true of the airplane?

I. The airplane is at rest.
II. The airplane is moving with a constant velocity.
III. The airplane is increasing its speed.
IV. The airplane is decreasing its speed.

(A) I only
(B) III only
(C) I or II, but not III or IV
(D) III or IV, but not I or II
(E) IV only

17. If 400 g of water at 40°C is mixed with 100 g of water at 30°C, the resulting temperature of the water is

(A) 13°C
(B) 26°C
(C) 36°C
(D) 38°C
(E) 44°C

400 × 40 × C + 100 × 30 × C = 500 × t × C
16000 + 3000 = 500t
t = 19000/500
= 38

CONTINUE ON THE NEXT PAGE

18. A 40 Ω resistor in a closed circuit has 20 volts across it. The current flowing through the resistor is

    (A) 0.5 A
    (B) 2 A
    (C) 20 A
    (D) 80 A
    (E) 800 A

19. Two blocks each weighing 50 N are connected to the ends of a light string which is passed over a pulley. The tension in the string is

    (A) 25 N
    (B) 50 N
    (C) 100 N
    (D) 200 N
    (E) 500 N

Questions 20–22: A horizontal force F acts on a block of mass m which is initially at rest on a floor of negligible friction. The force acts for a time t and moves the block a displacement d.

20. The acceleration of the block is

    (A) Ft
    (B) Fd
    (C) $\frac{F}{m}$
    (D) $\frac{m}{F}$
    (E) $\frac{d}{t}$

21. The change in momentum of the block is

    (A) $\frac{F}{t}$
    (B) $\frac{m}{t}$
    (C) Fd
    (D) Ft
    (E) mt

22. The change in the kinetic energy of the block is

    (A) Fd
    (B) $\frac{F}{d}$
    (C) Ft
    (D) $\frac{F}{t}$
    (E) $\frac{d}{t}$

CONTINUE ON THE NEXT PAGE

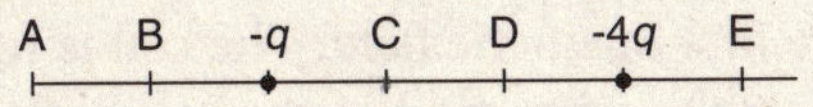

23. Two charges $-q$ and $-4q$ are located on a line as shown above. At which point could a positive charge be placed if it is to experience no force?

    (A) A
    (B) B
    (C) C
    (D) D
    (E) E

24. A charge is placed near a magnetic field. In which of the following cases would the charge experience a force?

    I. The charge is placed inside the magnetic field and is not moving.
    II. The charge is moving and its velocity is perpendicular to the magnetic field lines.
    III. The charge is moving and its velocity is parallel to the magnetic field lines.

    (A) I only
    (B) I and II only
    (C) II and III only
    (D) II only
    (E) III only

25. A bar magnet is moving through a stationary coil of wire. If the magnet suddenly stops halfway through the coil, which of the following will occur?

    (A) The magnet will lose its magnetic field.
    (B) The current induced in the coil will continue flowing in the same direction.
    (C) The current induced in the coil will stop.
    (D) The current induced in the coil will reverse direction.
    (E) The magnet will be pushed out of the coil.

26. An ideal gas is enclosed in a container which has a fixed volume. If the temperature of the gas is increased, which of the following will also increase?

    I. The pressure against the walls of the container.
    II. The average kinetic energy of the gas molecules.
    III. The number of moles of gas in the container.

    (A) I only
    (B) I and II only
    (C) II and III only
    (D) II only
    (E) III only

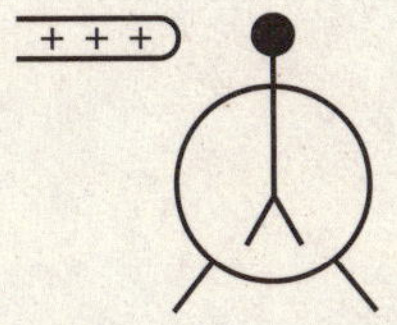

27. A neutral electroscope is shown above. If a positively charged rod is brought near the knob of the electroscope, which of the following statements is true?

    (A) The electroscope can be charged negatively without the positively charged rod touching the knob and using a only grounding wire.
    (B) The electroscope can be charged positively without the positively charged rod touching the knob and using only a grounding wire.
    (C) The leaves of the electroscope are negatively charged.
    (D) The knob of the electroscope is positively charged.
    (E) The electroscope has a net positive charge.

CONTINUE ON THE NEXT PAGE

Questions 28–30 refer to the figure below which represents a wave propagating along a string with a speed of 320 cm/s.

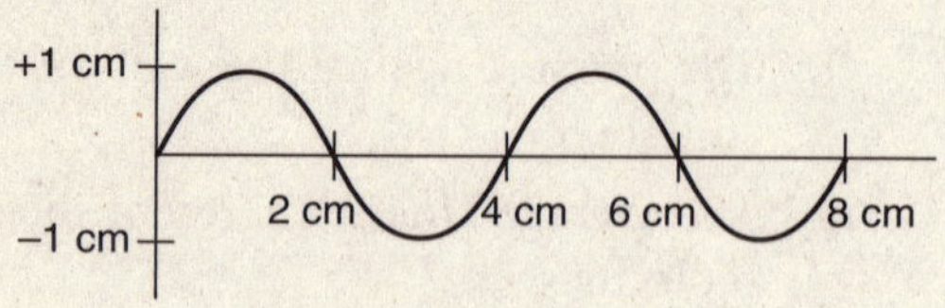

28. The wave is an example of a

(A) sound wave.
(B) transverse wave.
(C) electromagnetic wave.
(D) longitudinal wave.
(E) interference wave.

29. The frequency of the wave is

(A) 1,280 Hz
(B) 640 Hz
(C) 320 Hz
(D) 80 Hz
(E) 40 Hz

30. The amplitude of the wave is

(A) 1 cm
(B) 2 cm
(C) 4 cm
(D) 8 cm
(E) 16 cm

31. A generator is constructed by placing a coil of wire in a magnetic field. The coil is rotated in a clockwise direction in the magnetic field at a constant rate to induce a current in the wire. If the coil of wire is rotated in a counterclockwise direction at the same constant rate, which of the following will occur?

(A) The current in the coil will reverse its direction.
(B) The current in the coil will stop flowing.
(C) The current in the coil will continue to flow in the same direction as before.
(D) The current in the coil will decrease steadily.
(E) The current in the coil will increase steadily.

CONTINUE ON THE NEXT PAGE

Questions 32–33 relate to the positive charge following a circular path in a region of magnetic field in the figure below.

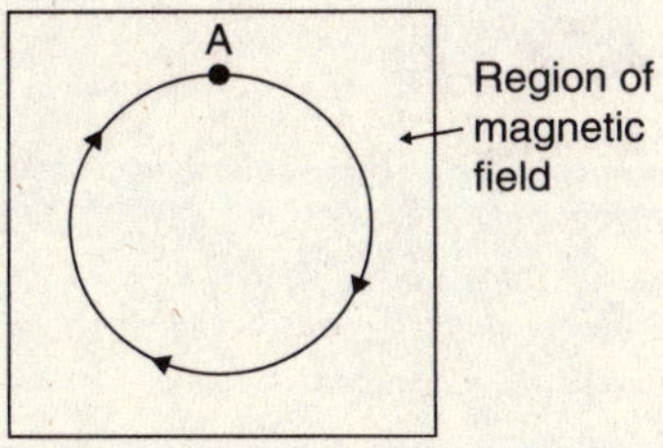

32. At point A, what is the direction of the net force acting on the positive charge?

    (A) →
    (B) ←
    (C) ↑
    (D) ↓
    (E) There is no net force acting on the charge at point A.

33. The direction of the magnetic field in the region shown is

    (A) out of the page and perpendicular to it.
    (B) into the page and perpendicular to it.
    (C) toward the top of the page.
    (D) toward the bottom of the page.
    (E) to the right.

Questions 34–35: A beam of light is incident onto a glass plate. Possible angles for the reflected and refracted rays are drawn in the figure below.

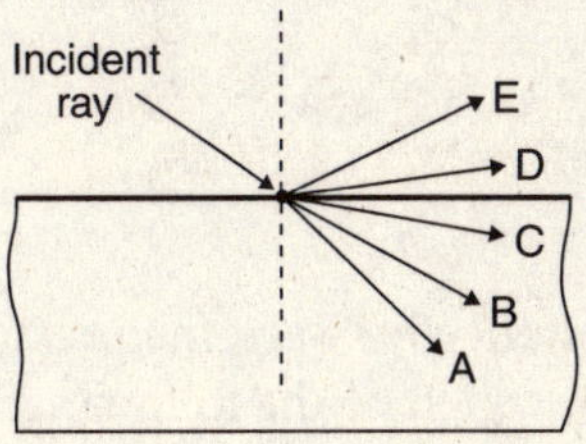

34. Which of the rays best represents the reflected ray?

    (A) A
    (B) B
    (C) C
    (D) D
    (E) E

35. Which of the rays best represents the refracted ray?

    (A) A
    (B) B
    (C) C
    (D) D
    (E) E

CONTINUE ON THE NEXT PAGE

Questions 36–38 relate to the two masses M_1 and M_2, which have a charge Q_1 and Q_2, respectively. The masses are initially separated by a distance *r*.

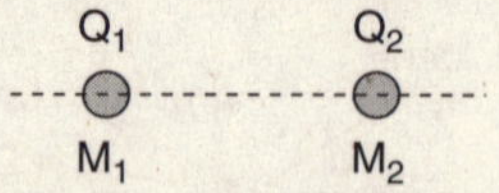

36. If the distance between the masses is doubled, which of the following is true?
    (A) The gravitational force will increase.
    (B) The electric force will increase.
    (C) The gravitational force will decrease, but the electric force will remain the same.
    (D) The electric force will decrease, but the gravitational force will remain the same.
    (E) Both the gravitational and electric forces will decrease.

37. If both masses are doubled, but the charge on each remains the same, which of the following is true?
    (A) The gravitational force will decrease.
    (B) The electric force will increase.
    (C) The gravitational force will increase, but the electric force will remain the same.
    (D) The electric force will increase, but the gravitational force will remain the same.
    (E) Both the gravitational and electric forces will decrease.

38. If the two charged masses are placed in space so that no other forces affect them, and they remain at a distance *r* apart indefinitely, which of the following is true?
    (A) Both charges are positive.
    (B) Q_1 is positive and Q_2 is negative.
    (C) Q_1 is negative and Q_1 is positive.
    (D) $Q_1 = Q_2$
    (E) $M_1 = M_2$

39. An ohm is defined as one volt per ampere and an ampere is a coulomb per second. A farad, the unit for capacitance, is a coulomb per volt. An ohm times a farad is a unit of
    (A) voltage.
    (B) resistance.
    (C) energy.
    (D) time.
    (E) charge.

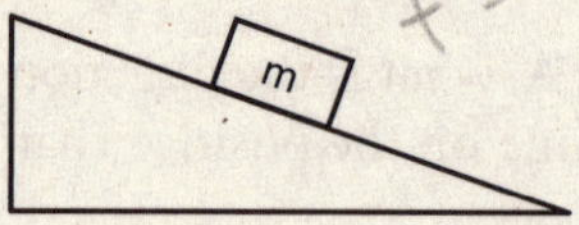

40. A block of mass *m* is at rest on a rough inclined plane. Which of the following diagrams best represents the correct directions for the forces acting on the block?

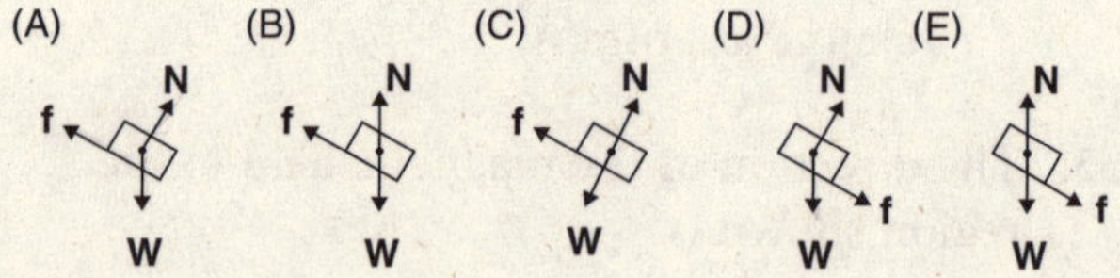

Questions 41–42:

A sealed bottle contains 100 grams of radioactive iodine. After 24 days, the bottle contains only 12.5 grams of radioactive iodine.

41. The half-life of this isotope of iodine is most nearly
    (A) 24 days
    (B) 16 days
    (C) 12 days
    (D) 8 days
    (E) 3 days

42. Which of the following graphs best represents the amount of iodine remaining versus time?

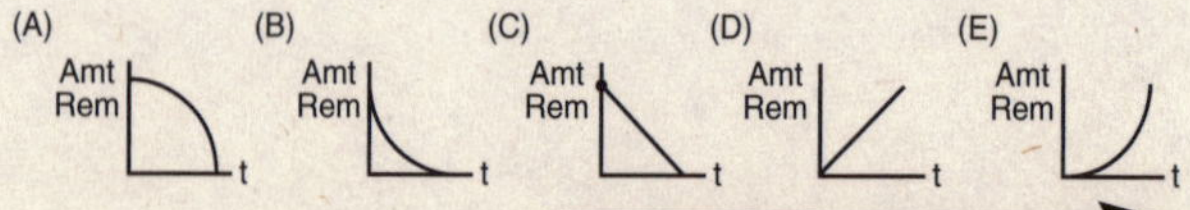

CONTINUE ON THE NEXT PAGE

Questions 43–44 relate to two waves of equal wavelength and amplitude approaching each other in the same rope as shown.

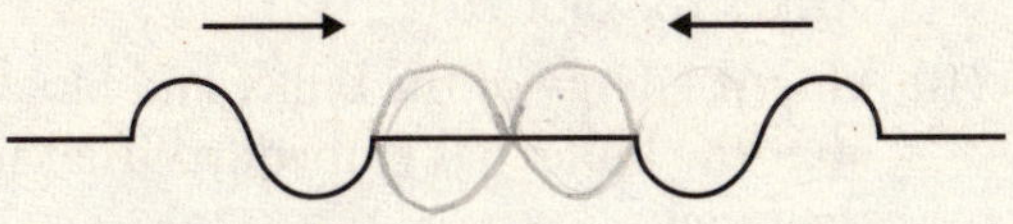

43. When the waves occupy exactly the same space at the same time, the shape of the rope will be:

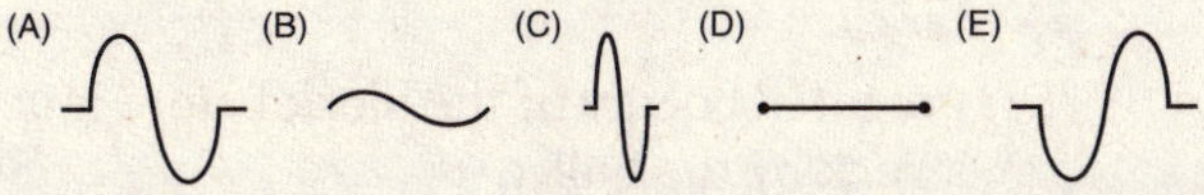

44. After the waves have completely passed each other, the shape of the rope will be

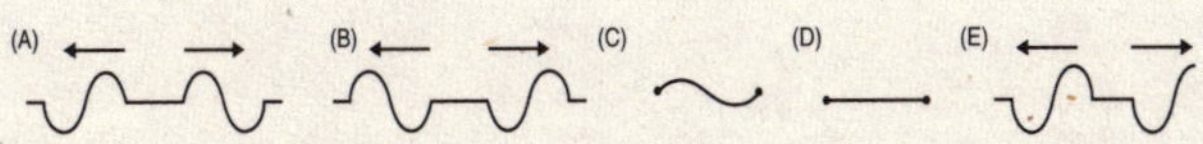

45. A converging lens forms an image primarily due to the phenomenon of

    (A) refraction.
    (B) reflection.
    (C) diffraction.
    (D) constructive interference.
    (E) destructive interference.

46. Sound is an example of a(n)

    (A) electromagnetic wave.
    (B) longitudinal wave.
    (C) transverse wave.
    (D) torsional wave.
    (E) circular wave.

47. A car has a speed of 20 m/s and a momentum of 12,000 kg m/s. The mass of the car is

    (A) 240,000 kg
    (B) 6,000 kg
    (C) 1,200 kg
    (D) 600 kg
    (E) 120 kg

48. Two metal spheres of equal size are charged and mounted on insulated stands. One sphere has a charge of +8 μC and the second sphere has a charge of –14 μC. The two spheres are touched together and then separated. The charge on the second sphere will now be

    (A) zero
    (B) +11 μC
    (C) −6 μC
    (D) −3 μC
    (E) +2 μC

CONTINUE ON THE NEXT PAGE

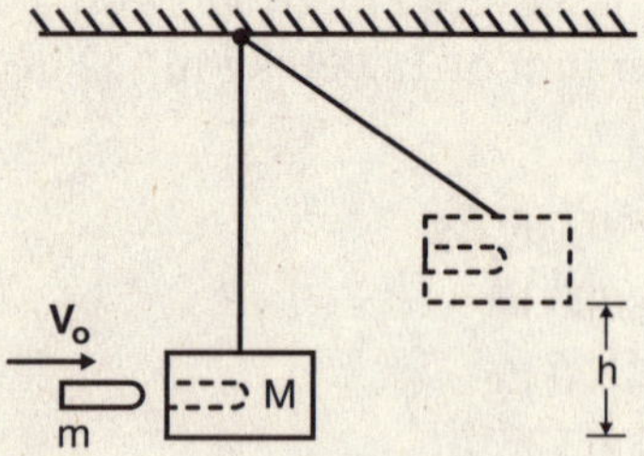

Questions 49–50 refer to the diagram above.

A bullet of mass *m* is fired with a speed v_0 into a hanging block of wood of mass *M* and embeds itself. Consequently, the bullet and block rise to a maximum height *h*.

49. As the bullet strikes and becomes embedded in the block, which of the following statements is true?

    (A) The speed of the bullet immediately before the collision is equal to the speed of the bullet and block immediately after the collision.

    (B) The momentum of the bullet immediately before the collision is equal to the momentum of the bullet and block immediately after the collision.

    (C) The kinetic energy of the bullet immediately before the collision is equal to the kinetic energy of the bullet and block immediately after the collision.

    (D) No energy is lost as the bullet enters the wood block.

    (E) The block does not move until the bullet comes to rest inside it.

50. The potential energy of the bullet and block at the maximum height *h* is equal to the

    (A) kinetic energy of the bullet before it strikes the block.

    (B) kinetic energy of the bullet and block after the bullet has embedded itself in the block.

    (C) momentum of the bullet before it strikes the block.

    (D) momentum of the bullet and block after the bullet has embedded itself in the block.

    (E) potential energy of the block before it is struck by the bullet.

CONTINUE ON THE NEXT PAGE

Questions 51–52:

A steel sphere is launched horizontally with a speed v from the edge of a table of height h above a level floor. At the same instant, another steel sphere is dropped from the edge of the same table. Air resistance may be neglected.

51. Which of the following statements is true?

    (A) The two spheres will strike the floor at the same time.
    (B) The sphere that is dropped will strike the floor first.
    (C) The sphere that is launched horizontally will strike the floor first.
    (D) The acceleration of the sphere that is dropped is greater than the acceleration of the other sphere after it is launched.
    (E) The acceleration of the sphere after it is launched is greater than the acceleration of the sphere that is dropped.

52. If the mass of the sphere which is dropped is doubled, and the mass of the launched sphere remains the same, which of the following is true?

    (A) The two spheres will strike the floor at the same time.
    (B) The sphere that is dropped will strike the floor first.
    (C) The sphere that is launched horizontally will strike the floor first.
    (D) The acceleration of the sphere that is dropped is greater than the acceleration of the other sphere after it is launched.
    (E) The acceleration of the sphere after it is launched is greater than the acceleration of the sphere that is dropped.

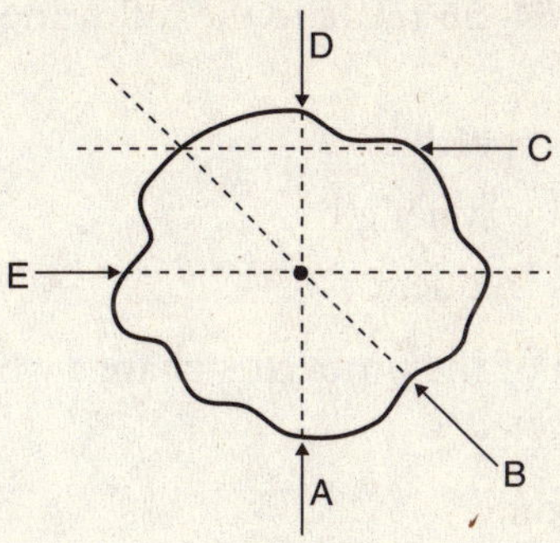

53. The figure above relates to a flat object lying on a table of negligible friction. Five forces are separately applied to the object as shown. The line of action of each force is shown by a dashed line. Which of the five forces will cause the object to rotate about point O?

    (A) A
    (B) B
    (C) C
    (D) D
    (E) E

Questions 54–56 relate to the following properties of waves.

I. speed
II. wavelength
III. frequency

54. Which of the properties change when a wave is refracted?

(A) I only
(B) II only
(C) I and II only
(D) II, and III only
(E) I, II, and III

55. Which of the properties is equal to the reciprocal of the period?

(A) I only
(B) II only
(C) III only
(D) II and III only
(E) I, II, and III

56. Which one of the three is the product of the other two?

(A) I
(B) II
(C) III
(D) Any one is the product of the other two.
(E) None of the three is the product of the other two.

57. A woman standing on a train platform listens to a train as it approaches her at a constant speed. The train engineer blows the whistle of the train which he knows to produce a frequency of 450 Hz. As the train approaches the woman on the platform, she will hear the frequency (pitch) of the whistle as

(A) greater than 450 Hz but constant.
(B) less than 450 Hz but constant.
(C) greater than 450 Hz and steadily increasing.
(D) less than 450 Hz and steadily increasing.
(E) greater than 450 Hz and steadily decreasing.

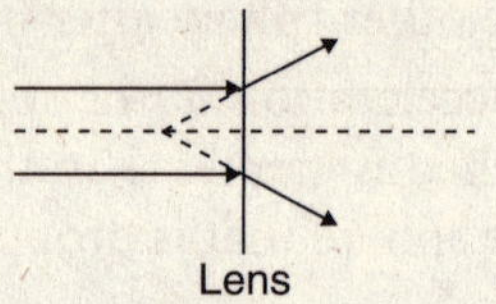

58. A light ray passes through a thin lens having a focal point *f* as shown above. Which of the following statements best describes the lens?

(A) The lens is a converging lens.
(B) The lens is thicker in the center than on the edges.
(C) The lens is thinner in the center than on the edges.
(D) The lens will always produce a real image.
(E) The lens will always produce a virtual image.

CONTINUE ON THE NEXT PAGE

Questions 59–61: An object's velocity is measured each second and the data is summarized in the table below.

velocity	time
0	0
2 m/s	1 s
4 m/s	2 s
6 m/s	3 s
8 m/s	4 s

59. The acceleration of the object is

(A) 1 m/s^2
(B) 2 m/s^2
(C) 4 m/s^2
(D) 6 m/s^2
(E) 8 m/s^2

60. The distance the object travels during the first four seconds is

(A) 8 m
(B) 16 m
(C) 18 m
(D) 24 m
(E) 32 m

61. If the object stops accelerating at 4 seconds when its speed is 8 m/s, the distance it will travel between 4 s and 7 s is

(A) 8 m
(B) 16 m
(C) 18 m
(D) 24 m
(E) 32 m

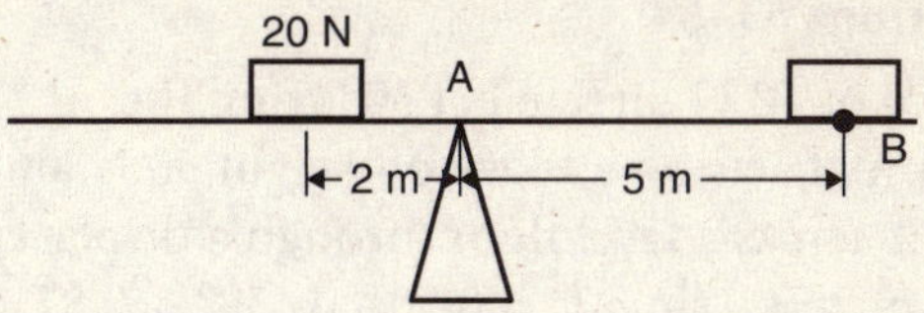

62. A board of negligible mass supports a mass weighing 20 N at a distance of 2 m from the support at point A. How much weight must be placed at point B which is 5 m from point A in order for the rod to remain horizontal?

(A) 4 N
(B) 8 N
(C) 10 N
(D) 40 N
(E) 60 N

Questions 63–64:

A force of 40 N directed at 30° from the horizontal acts on a block of weight 50 N and pulls it across a level floor through a displacement of 10 meters. (sin 30° = 0.50, cos 30° = 0.87)

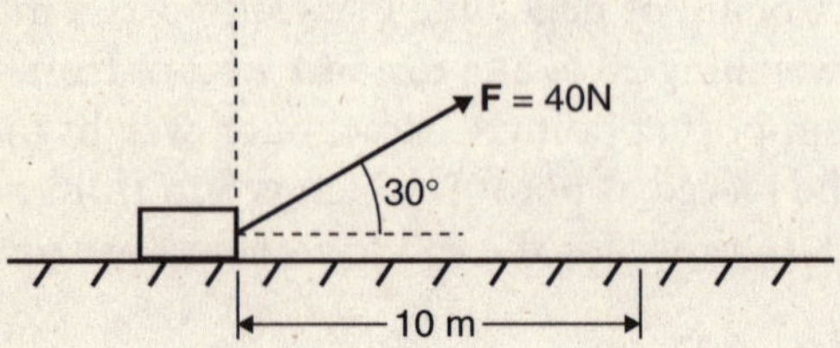

63. The work done on the block is most nearly

    (A) 160 J
    (B) 200 J
    (C) 350 J
    (D) 400 J
    (E) 1,600 J

64. The normal force between the floor and the block is most nearly

    (A) 50 N
    (B) 44 N
    (C) 30 N
    (D) 25 N
    (E) 5 N

65. Which of the following diagrams best represents the electric field around a positive charge?

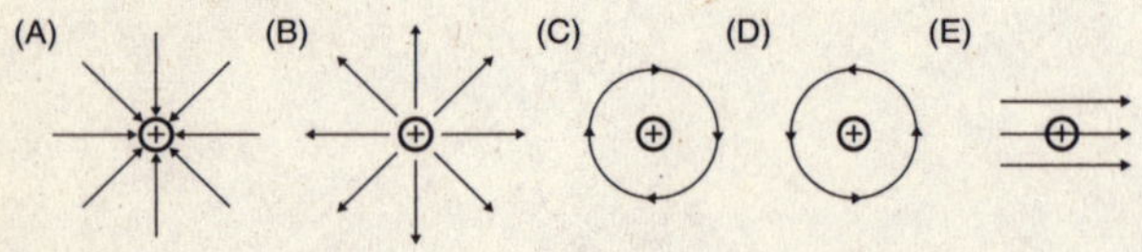

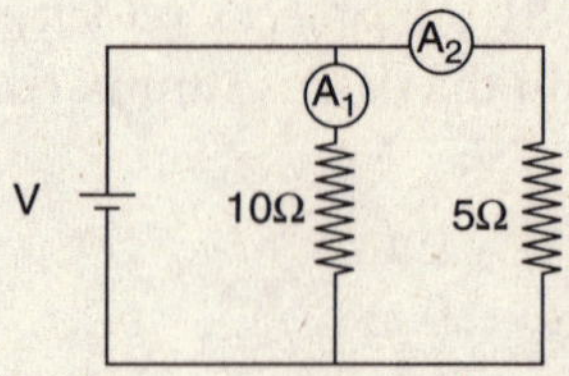

66. Two ammeters are placed in the circuit above with the two resistors shown. If A_1 is the reading on ammeter 1 and A_2 is the reading on ammeter 2, which of the following statements is true?

    (A) $A_1 = A_2$
    (B) $A_1 > A_2$
    (C) $A_1 < A_2$
    (D) $A_1 = 0$
    (E) $A_2 = 0$

67. If electrons are added or removed from an atom, the atom is said to be

    (A) polarized.
    (B) ionized.
    (C) neutralized.
    (D) an isotope.
    (E) a molecule.

68. Which of the following are emitted from a metal surface in the photoelectric effect?

    (A) electrons
    (B) protons
    (C) neutrons
    (D) photons
    (E) atoms

CONTINUE ON THE NEXT PAGE

69. Which of the following particles has the largest mass?

(A) electron
(B) proton
(C) neutron
(D) alpha particle
(E) beta particle

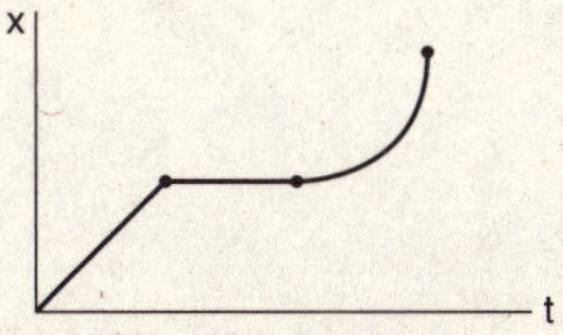

70. The graph above shows position *x* as a function of time *t* for a particle moving along the *x*–axis. Which of the following graphs of velocity *v* as a function of time *t* describes the motion of this particle?

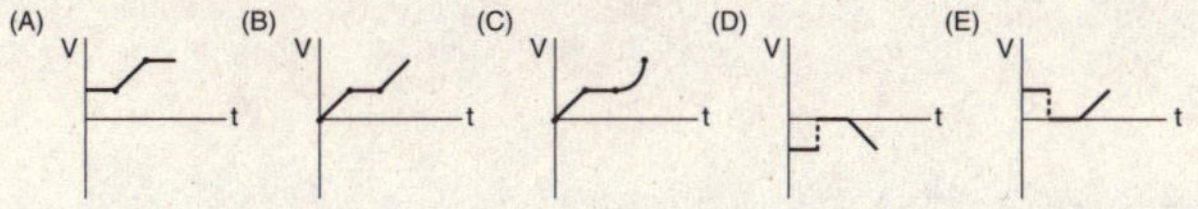

71. A stone of mass 1 kg and weight 10 N falls through the air. The air resistance acting on the stone is 2 N. What is the acceleration of the stone?

(A) 12 m/s^2
(B) 9.8 m/s^2
(C) 8 m/s^2
(D) 5 m/s^2
(E) 2 m/s^2

72. If the pressure of a gas is increased while its temperature is held constant, which of the following will occur?

(A) The average kinetic energy of the molecules of the gas will increase.
(B) The average kinetic energy of the molecules of the gas will decrease.
(C) The mass of the gas will decrease.
(D) The volume of the gas will increase.
(E) The volume of the gas will decrease.

73. A satellite is orbiting the earth in an elliptical orbit. Which of the following must be true if the satellite's speed is increasing?

(A) The satellite's distance from the Earth is increasing.
(B) The satellite's distance from the Earth is decreasing.
(C) The kinetic energy of the satellite is decreasing.
(D) The momentum of the satellite is decreasing.
(E) The gravitational force acting on the satellite is decreasing.

74. The angle above which total internal reflection will occur is called the

(A) critical angle.
(B) refracted angle.
(C) diffracted angle.
(D) incident angle.
(E) normal angle.

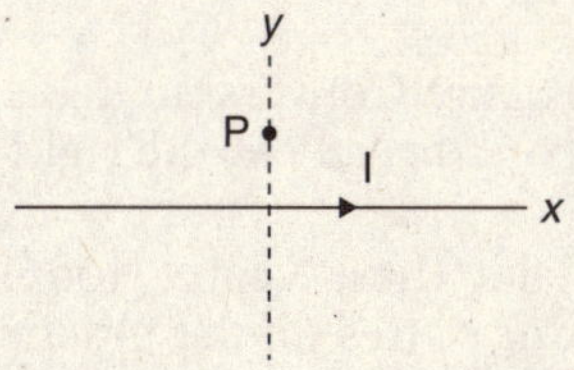

75. A wire on the *x*-axis of a coordinate system has a current I in the +*x* direction as shown above. What is the direction of the magnetic field due to the wire at point *P*?

(A) to the left
(B) to the right
(C) down into the page and perpendicular to the page
(D) up out of the page and perpendicular to the page
(E) toward the bottom of the page

END OF TEST

PRACTICE TEST 2: ANSWERS AND EXPLANATIONS

1. A
2. D
3. C
4. D
5. B
6. C
7. A
8. C
9. B
10. A
11. D
12. B
13. E
14. A
15. C
16. C
17. D
18. A
19. B
20. C
21. D
22. A
23. C
24. D
25. C
26. B
27. A
28. B
29. D
30. A
31. A
32. D
33. A
34. E
35. A
36. E
37. C
38. A
39. D
40. A
41. D
42. B
43. D
44. A
45. A
46. B
47. D
48. D
49. B
50. B
51. A
52. A
53. C
54. C
55. C
56. A
57. A
58. C
59. B
60. B
61. D
62. B
63. C
64. C
65. B
66. C
67. B
68. A
69. D
70. E
71. C
72. E
73. B
74. A
75. D

Computing Your Practice Test Scores

Step 1: Calculate your raw score.

Raw score = # of correct answers – (# of wrong answers × .25)

Step 2: Find your scaled score.

In the Score Conversion Table below, find your raw score (rounded to the nearest whole number). The score directly to the right of it will be your scaled score.

Note: Don't take these scores too literally. Practice test conditions cannot precisely mirror real test conditions. Your scores on the practice test will give you a rough idea of your range on the actual exam.

Raw	Scaled	Raw	Scaled	Raw	Scaled	Raw	Scaled	Raw	Scaled	Raw	Scaled
75	800	59	780	43	690	27	590	11	480	–5	380
74	800	58	770	42	680	26	580	10	480	–6	370
73	800	57	770	41	670	25	580	9	470	–7	370
72	800	56	760	40	670	24	570	8	470	–8	360
71	800	55	760	39	660	23	570	7	460	–9	350
70	800	54	750	38	650	22	560	6	450	–10	350
69	800	53	750	37	650	21	550	5	450	–11	340
68	800	52	740	36	640	20	540	4	440	–12	330
67	800	51	730	35	640	19	540	3	430	–13	330
66	800	50	730	34	630	18	530	2	430	–14	320
65	800	49	720	33	630	17	530	1	420	–15	310
64	800	48	720	32	620	16	520	0	410	–16	310
63	800	47	710	31	610	15	510	–1	410	–17	300
62	790	46	700	30	610	14	510	–2	400	–18	290
61	790	45	700	29	600	13	500	–3	390	–19	290
60	780	44	690	28	600	12	490	–4	390		

1. A

The slope of a displacement vs. time graph is velocity, and the slope is constant, thus the velocity is constant.

2. D

The object begins with a high velocity, then slows down until its velocity is zero, then begins speeding up again in the negative direction.

3. C

Both graphs imply that the velocity starts out as zero, then the slope of the *s* vs. *t* graph begins getting steeper and more negative, indicating that the velocity is increasing and negative.

4. D

Fission occurs when a large atom such as uranium splits into two smaller ones, such as xenon and strontium.

5. B

Radium is a radioactive element that emits an alpha particle (^4_2He), leaving radon gas.

6. C

A beta particle is an electron ($_{-1}^{0}e$) produced in this radioactive decay process where bismuth decays into polonium.

7. A

Fusion is the process in which two light elements such as deuterium (^2_1H) fuses into a heavier element, such as tritium (^3_1H).

8. C

The ideal gas law relates the pressure, volume, and temperature of a gas.

9. B

The second law of thermodynamics, or the law of entropy, states that all natural systems tend toward a state of higher disorder. As the gas escapes from the bottle, it is becoming more disordered.

10. A

The first law of thermodynamics states that the heat lost by one liquid must be gained by the other liquid, as long as no work is done.

11. D

The heat of fusion or vaporization equation allows us to calculate the heat required to change the state of a substance from a solid to a liquid or a liquid to a gas.

12. B

The voltage across the resistor is equal to the voltage in the battery, since when the switch is first closed, the capacitor has no charge on it and therefore no voltage across it.

13. E

Immediately after the switch is closed, the capacitor has no charge on it and therefore provides no resistance to the current, and so the current is simply V/R by Ohm's law.

14. A

A very long time after the switch is closed, the capacitor is full of charge and the voltage across the capacitor is equal and opposite to the voltage provided by the battery. Thus, the capacitor will not allow any current to flow through the circuit.

15. C

The velocity vs. time graph implies that the velocity is not changing. Graphs I and II both have a constant slope which also implies a constant velocity, even though one graph begins at a different position than the other graph.

16. C

The plane must be at rest or, equivalently, moving at a constant velocity for the chain to hang vertically. A reference frame which has a constant velocity (including rest) is called an inertial reference frame. If the airplane were accelerating, the chain would angle forward or backward, depending on the direction of the acceleration.

17. D

The heat lost by the warmer water must be gained by the cooler water until all of the water reaches a final equilibrium temperature T_f.

$$Q_{lost} = Q_{gained}$$
$$m_1 c\Delta T = -m_2 c\Delta T$$
$$m_1 c(T_f - T_i) = -m_2 c(T_f - T_i)$$

Since the specific heat is the same for water on both sides of the equation, we have

$$m_1(T_f - T_i) = -m_2(T_f - T_i)$$
$$(400\text{ g})(T_f - 40°) = (-100\text{ g})(T_f - 30°)$$
$$T_f = 38°$$

18. A

$$I = \frac{V}{R} = \frac{20V}{40\Omega} = 0.5A$$

19. B

If we draw the free-body force diagram for each block, we see that each block has a weight force of 50 N pulling downward which is balanced by the tension force pulling upward. Thus, the tension in the string must be 50 N.

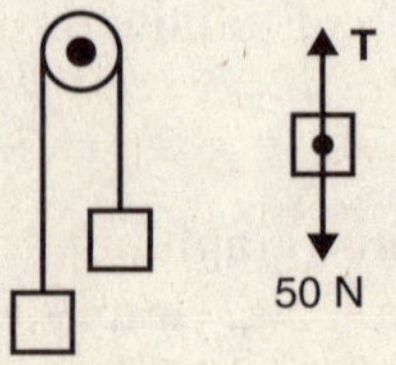

20. C

Newton's second law states that $F = ma$, so $a = F/m$.

21. D

The change in momentum of the block is equal to the impulse imparted to the block. The impulse is the product of the force and the time during which it acts, Ft.

22. A

The work-energy theorem states that the work (Fd) done on the block is equal to the change in kinetic energy of the block, since work done on an object always changes the energy of the object.

23. C

If the charge is to feel no force, it must be closer to the smaller charge, $-q$, and farther away from the larger charge, $-4q$. In other words, the charge must be at the "center of charge," similar to the center of mass of two objects. If the charge were placed at points A, B, or E, it would be repelled and accelerate away from the two negative charges. The forces acting on the charge at point C balance each other out, since electric force is proportional to the product of the charges and inversely proportional to the square of the distance between the charges.

24. D

For a charge to experience a force in a magnetic field, its velocity must have a component that is perpendicular to the magnetic field lines, which implies that it is crossing the magnetic field lines.

25. C

For a current to be induced in a coil of wire by electromagnetic induction, there must be relative motion between the magnetic field of the coil. If the magnet stops, so does the current in the coil.

26. B

If the temperature is increased but the volume remains constant, the pressure inside the container will become greater since there will be more collisions between the molecules of the gas and the walls of the container. Since temperature is proportional to the average kinetic energy of the gas molecules, a higher temperature would imply a higher average kinetic energy of the molecules of the gas.

27. A

In the process of charging by induction, a positively charged rod is brought near the knob of the electroscope, and the electroscope is grounded so that more electrons that are attracted to the positively charged rod can come up from the ground and be deposited on the electroscope, giving it a net negative charge.

28. B

A transverse wave vibrates in a direction perpendicular to the direction of motion of the wave, as this wave does.

29. D

The wavelength of the wave is 4 cm, being the length of one full vibration of the wave. The frequency of the wave is:

$$f = \frac{v}{\lambda} = \frac{320 \text{ cm/s}}{4 \text{ cm}} = 80 \text{ Hz}$$

30. A

The amplitude is the maximum displacement from the base line to the crest of the wave (or, equivalently, to the trough of the wave), which is 1 cm in this case.

31. A

The amount of current will be the same, since the coil is turned at the same constant rate, but since it is rotated in the opposite direction, the current is induced in the opposite direction.

32. D

The magnetic force acting on a charge in the magnetic field is a centripetal force causing the charge to turn its path downward in the magnetic field and point to the center of the circle, which is toward the bottom of the page from point A.

33. A

Since the charge is positive, we can use the right-hand rule to find the direction of the magnetic field. At point A, if you point your thumb in the direction of the velocity of the charge (to the right), and your palm faces the direction of the force on the charge at A (toward the bottom of the page), then your fingers point out of the page in the direction of the magnetic field.

34. E

The angle of incidence is equal to the angle of reflection, and ray E is reflected at an angle most similar to the incident angle.

35. A

Since the ray is passing from a less dense medium (air) to a more dense medium (glass), the refracted ray must bend toward the normal, causing the refracted angle to be smaller than the incident angle.

36. E

Since both the electric and gravitational forces are inverse-square laws, the farther apart the charged masses are, the smaller the forces between them.

37. C

The gravitational force is proportional to the product of the masses, so the gravitational force will be doubled. But the electric force is not related to the masses and so it will not change.

38. A

If they remain the same distance apart, then the gravitational force (attractive) is equal and opposite to the electric force, which must be repulsive in this case. Thus, the only answer choice which will produce a repulsive electric force is A, both charges are positive.

39. D

$\Omega F = \left(\frac{V}{A}\right)\left(\frac{C}{V}\right) = \frac{C}{A} = \frac{C}{C/s}$ = second, which is a unit of time.

40. A

The weight vector always points straight down, the normal force is perpendicular to the plane, and the frictional force is acting up the plane to keep the block from sliding down the plane.

41. D

The 100 gram sample of iodine is cut in half three times (50 g, 25 g, 12.5 g), which took 24 days. Thus, the time for the iodine sample to be cut in half one time is 24/3 = 8 days.

42. B

The amount of iodine starts off at 100 grams, and is reduced exponentially over time.

43. D

When the two waves meet, the crest of one wave is superposed onto the trough of the other, creating destructive interference, and the waves cancel each other out for the instant they occupy the same space at the same time.

44. A

After the waves completely pass each other, they behave exactly as they did before they met. They've just switched places.

45. A

As a ray of light passes through a lens, it refracts, changing its direction.

46. B

Sound waves vibrate in a direction which is parallel to the direction of motion of the wave, and are thus longitudinal.

47. D

Momentum is mass times velocity, so

$$m = \frac{p}{v} = \frac{12{,}000 \text{ kgm/s}}{20 \text{ m/s}} = 600 \text{ kg}$$

48. D

When the two spheres are touched, they exchange charge until they both have the same charge. Since the total charge is -6 μC, they will each retain -3 μC when they are again separated.

49. B

The bullet collides with and sticks to the block in an inelastic collision. Momentum is conserved in an inelastic collision, but kinetic energy is not conserved, since some of the energy is lost to heat and deformation of the wood and bullet.

50. B

After the bullet embeds itself in the block, the bullet and block have a new kinetic energy (since the collision is not elastic) which is converted into potential energy at the height *h*.

51. A

The two spheres will strike the floor at the same time since they have the same initial vertical velocity (zero) and the same vertical acceleration (10 m/s^2).

52. A

Once again, the spheres will strike the floor at the same time, since mass has no effect on the acceleration of a falling object.

53. C

The force applied at C is the only force whose line of action does not pass through the point of rotation A, and thus it is the only force which can cause a torque and a rotation about point A.

54. C

When a wave passes from one medium to another, it refracts, changing its speed and wavelength. The frequency does not depend on the medium, and therefore it is not changed when the wave is refracted.

55. C

Frequency (cycles/second) is the reciprocal of period (seconds/cycle).

56. A

The speed is the product of frequency and wavelength.

57. A

As the train approaches the stationary woman on the platform, she hears a higher pitch (frequency) than the pitch associated with 450 Hz since the waves appear to be reaching her more often due to the Doppler effect. The pitch she hears will be constant, since the train is not accelerating.

58. C

Since the parallel rays diverge as they exit the lens, this lens is a diverging (concave) lens, which is thinner in the center and thicker on the edges.

59. B

The speed of the object is increasing by 2 m/s for each second it moves.

60. B

$$s = v_i t + \tfrac{1}{2}at^2 = (0) + \tfrac{1}{2}(2\text{ m/s})^2\,(4\text{ s})^2 = 16\text{ m}$$

61. D

The distance traveled at a constant speed between 4 s and 7 s is $d = vt = (8\text{ m/s})(3\text{ s}) = 24\text{ m}$.

62. B

The torque on the left must equal the torque on the right for the board to be balanced.

$$(F_{left})(r_{left}) = (F_{right})(r_{right})$$
$$(20N)(2m) = (F_{right})(5m)$$
$$F_{right} = 8N$$

63. C

$$W = Fs\cos\theta = (40N)(10m)\cos 30 = 350J$$

64. C

The normal force is equal to the difference between the weight of the block and the vertical component of the applied force.

$$N = W - F\sin\theta = 50N - (40N)\sin 30 = 30N$$

65. B

Electric field lines are always drawn in the direction a small positive test charge would experience a force, which is radially outward anywhere around a positive source charge.

66. C

Ammeter 2 will read twice as much current as ammeter 1, since ammeter 2 is in series with a resistor which is only half as much as the resistor on the branch with ammeter 1.

67. B

An atom with extra electrons is a negative ion, and an atom with a deficiency of electrons is a positive ion.

68. A

Photons above the threshold frequency are absorbed by electrons, which in turn are emitted by the metal.

69. D

An alpha particle is equivalent to a helium nucleus, and is composed of two protons and two neutrons.

70. E

The x vs. t graph and the v vs. t graph both imply that the object is initially moving at a constant positive velocity, then remains at rest for a time, then accelerates positively.

71. C

The net force acting on the falling stone is 10 N – 2 N = 8 N. The stone's acceleration, then, is:

$$a = \frac{F_{net}}{m} = \frac{8\ \text{N}}{1\ \text{kg}} = 8\ \text{m/s}^2$$

72. E

The ideal gas law states that if temperature is held constant, pressure and volume are inversely proportional to each other. Thus, if pressure increases, volume must decrease. This relationship is sometimes referred to as Boyle's law.

73. B

Conservation of angular momentum states that speed and orbital distance are inversely proportional to each other. Thus, the satellite moves faster as its orbital distance gets smaller (closer to the Earth).

74. A

A ray directed at an angle greater than the critical angle as measured from the normal line will not exit the medium, but will be reflected internally.

75. D

By the first right-hand rule, place your thumb in the direction of the current, and your fingers will curl around the wire in the direction of the magnetic field the current produces, which is out of the page at point A.

Answer Sheet for Practice Test 3

1. Ⓐ Ⓑ Ⓒ Ⓓ Ⓔ
2. Ⓐ Ⓑ Ⓒ Ⓓ Ⓔ
3. Ⓐ Ⓑ Ⓒ Ⓓ Ⓔ
4. Ⓐ Ⓑ Ⓒ Ⓓ Ⓔ
5. Ⓐ Ⓑ Ⓒ Ⓓ Ⓔ
6. Ⓐ Ⓑ Ⓒ Ⓓ Ⓔ
7. Ⓐ Ⓑ Ⓒ Ⓓ Ⓔ
8. Ⓐ Ⓑ Ⓒ Ⓓ Ⓔ
9. Ⓐ Ⓑ Ⓒ Ⓓ Ⓔ
10. Ⓐ Ⓑ Ⓒ Ⓓ Ⓔ
11. Ⓐ Ⓑ Ⓒ Ⓓ Ⓔ
12. Ⓐ Ⓑ Ⓒ Ⓓ Ⓔ
13. Ⓐ Ⓑ Ⓒ Ⓓ Ⓔ
14. Ⓐ Ⓑ Ⓒ Ⓓ Ⓔ
15. Ⓐ Ⓑ Ⓒ Ⓓ Ⓔ
16. Ⓐ Ⓑ Ⓒ Ⓓ Ⓔ
17. Ⓐ Ⓑ Ⓒ Ⓓ Ⓔ
18. Ⓐ Ⓑ Ⓒ Ⓓ Ⓔ
19. Ⓐ Ⓑ Ⓒ Ⓓ Ⓔ
20. Ⓐ Ⓑ Ⓒ Ⓓ Ⓔ
21. Ⓐ Ⓑ Ⓒ Ⓓ Ⓔ
22. Ⓐ Ⓑ Ⓒ Ⓓ Ⓔ
23. Ⓐ Ⓑ Ⓒ Ⓓ Ⓔ
24. Ⓐ Ⓑ Ⓒ Ⓓ Ⓔ
25. Ⓐ Ⓑ Ⓒ Ⓓ Ⓔ
26. Ⓐ Ⓑ Ⓒ Ⓓ Ⓔ
27. Ⓐ Ⓑ Ⓒ Ⓓ Ⓔ
28. Ⓐ Ⓑ Ⓒ Ⓓ Ⓔ
29. Ⓐ Ⓑ Ⓒ Ⓓ Ⓔ
30. Ⓐ Ⓑ Ⓒ Ⓓ Ⓔ
31. Ⓐ Ⓑ Ⓒ Ⓓ Ⓔ
32. Ⓐ Ⓑ Ⓒ Ⓓ Ⓔ
33. Ⓐ Ⓑ Ⓒ Ⓓ Ⓔ
34. Ⓐ Ⓑ Ⓒ Ⓓ Ⓔ
35. Ⓐ Ⓑ Ⓒ Ⓓ Ⓔ
36. Ⓐ Ⓑ Ⓒ Ⓓ Ⓔ
37. Ⓐ Ⓑ Ⓒ Ⓓ Ⓔ
38. Ⓐ Ⓑ Ⓒ Ⓓ Ⓔ
39. Ⓐ Ⓑ Ⓒ Ⓓ Ⓔ
40. Ⓐ Ⓑ Ⓒ Ⓓ Ⓔ
41. Ⓐ Ⓑ Ⓒ Ⓓ Ⓔ
42. Ⓐ Ⓑ Ⓒ Ⓓ Ⓔ
43. Ⓐ Ⓑ Ⓒ Ⓓ Ⓔ
44. Ⓐ Ⓑ Ⓒ Ⓓ Ⓔ
45. Ⓐ Ⓑ Ⓒ Ⓓ Ⓔ
46. Ⓐ Ⓑ Ⓒ Ⓓ Ⓔ
47. Ⓐ Ⓑ Ⓒ Ⓓ Ⓔ
48. Ⓐ Ⓑ Ⓒ Ⓓ Ⓔ
49. Ⓐ Ⓑ Ⓒ Ⓓ Ⓔ
50. Ⓐ Ⓑ Ⓒ Ⓓ Ⓔ
51. Ⓐ Ⓑ Ⓒ Ⓓ Ⓔ
52. Ⓐ Ⓑ Ⓒ Ⓓ Ⓔ
53. Ⓐ Ⓑ Ⓒ Ⓓ Ⓔ
54. Ⓐ Ⓑ Ⓒ Ⓓ Ⓔ
55. Ⓐ Ⓑ Ⓒ Ⓓ Ⓔ
56. Ⓐ Ⓑ Ⓒ Ⓓ Ⓔ
57. Ⓐ Ⓑ Ⓒ Ⓓ Ⓔ
58. Ⓐ Ⓑ Ⓒ Ⓓ Ⓔ
59. Ⓐ Ⓑ Ⓒ Ⓓ Ⓔ
60. Ⓐ Ⓑ Ⓒ Ⓓ Ⓔ
61. Ⓐ Ⓑ Ⓒ Ⓓ Ⓔ
62. Ⓐ Ⓑ Ⓒ Ⓓ Ⓔ
63. Ⓐ Ⓑ Ⓒ Ⓓ Ⓔ
64. Ⓐ Ⓑ Ⓒ Ⓓ Ⓔ
65. Ⓐ Ⓑ Ⓒ Ⓓ Ⓔ
66. Ⓐ Ⓑ Ⓒ Ⓓ Ⓔ
67. Ⓐ Ⓑ Ⓒ Ⓓ Ⓔ
68. Ⓐ Ⓑ Ⓒ Ⓓ Ⓔ
69. Ⓐ Ⓑ Ⓒ Ⓓ Ⓔ
70. Ⓐ Ⓑ Ⓒ Ⓓ Ⓔ
71. Ⓐ Ⓑ Ⓒ Ⓓ Ⓔ
72. Ⓐ Ⓑ Ⓒ Ⓓ Ⓔ
73. Ⓐ Ⓑ Ⓒ Ⓓ Ⓔ
74. Ⓐ Ⓑ Ⓒ Ⓓ Ⓔ
75. Ⓐ Ⓑ Ⓒ Ⓓ Ⓔ

Practice Test 3

Part A

Directions: Each set of lettered choices below relates to the numbered questions immediately following it. Select the one lettered choice that best answers each question. A choice may be used once, more than once, or not at all in each set.

Questions 1–3 relate to the following:

A ball of mass m on the end of a string is swung clockwise in a horizontal circle at a constant radius r. When the ball is a point P as shown below, some of the choices that follow represent the directions of the vectors associated with the motion of the ball.

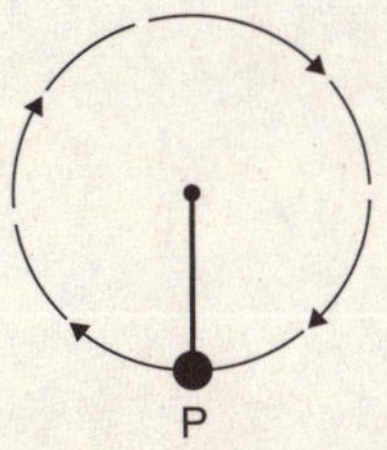

(A) → (B) ← (C) ↓ (D) ↑ (E) ↔

1. What is the direction of the velocity of the ball at point P?

2. What is the direction of the acceleration of the ball at point P?

3. What is the direction of the net force acting on the ball at point P?

Questions 4–7 relate to the following series of images of a moving ball. Each image represents a time interval of one second, and the motion of the ball is not necessarily horizontal.

4. In which of the choices above is the ball increasing its speed?

5. In which of the choices above is the ball moving at a constant velocity?

6. Which of the choices above could represent the ball being thrown upward before it begins falling back down?

7. Which of the choices above could represent the ball being rolled up an incline and then allowed to roll back down again?

CONTINUE ON THE NEXT PAGE

Questions 8–10 relate to the following diagram of two charges, +Q and −4Q.

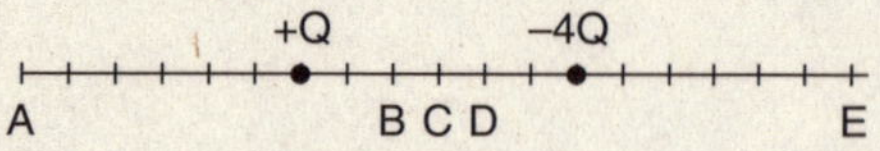

8. The net electric field is zero nearest which point?

9. At which point does the net electric field vector point to the left?

10. At which point would a small positive charge q feel the greatest force?

Questions 11–13 relate to the following graphs.

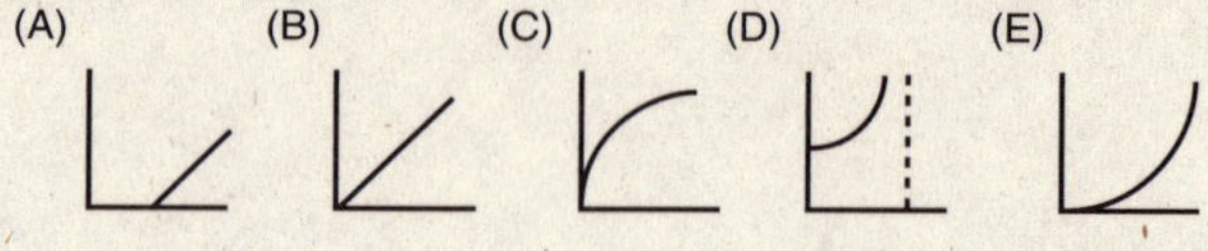

11. Which of the graphs above represent the energy of a photon vs. its frequency?

12. Which of the graphs above represent the maximum kinetic energy of electrons emitted in the photoelectric effect vs. frequency of incoming light?

13. Which of the graphs above represent the mass of a relativistic particle vs. its speed?

注：一个光子的能量和频率成正比

注：电场力被定义为正电荷所在电场中受的力.

CONTINUE ON THE NEXT PAGE

Part B

Directions: Each of the questions or incomplete statements below is followed by five answer choices. Select the one that is best in each case.

Questions 14–15:

A red car and a blue car have the same mass and are moving on the highway. The red car is traveling at 60 miles per hour and the blue car is traveling at 30 miles per hour.

14. The ratio of the red car's momentum to the blue car's momentum is

(A) 4
(B) 2
(C) 1
(D) ½
(E) ¼

15. The ratio of the red car's kinetic energy to the blue car's kinetic energy is

(A) 4
(B) 2
(C) 1
(D) ½
(E) ¼

16. The waves on a lake cause a buoy to oscillate up and down 90 times per minute. The frequency of the waves in hertz is

(A) 90 Hz
(B) 60 Hz
(C) 1.5 Hz
(D) 0.6 Hz
(E) 0.67 Hz

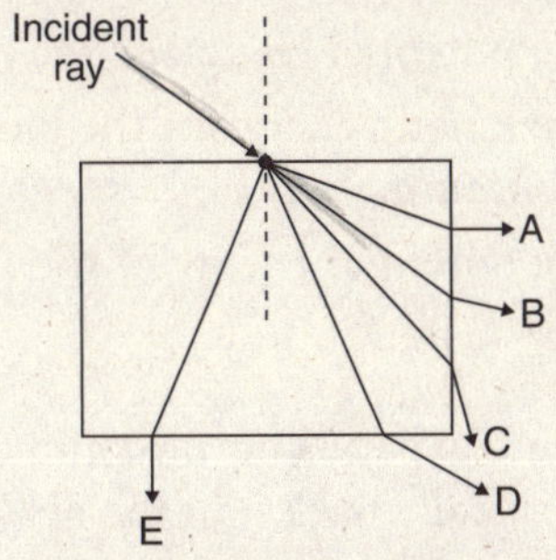

17. A ray of light passes through a piece of glass at the angle of incidence shown above. As the light passes through the glass and exits the glass into the air, the path of the ray is represented by which of the rays above?

(A) A
(B) B
(C) C
(D) D
(E) E

CONTINUE ON THE NEXT PAGE

Questions 18–19:

A block falls onto a vertical spring from a height h and compresses the spring a distance y. Air resistance and friction may be neglected.

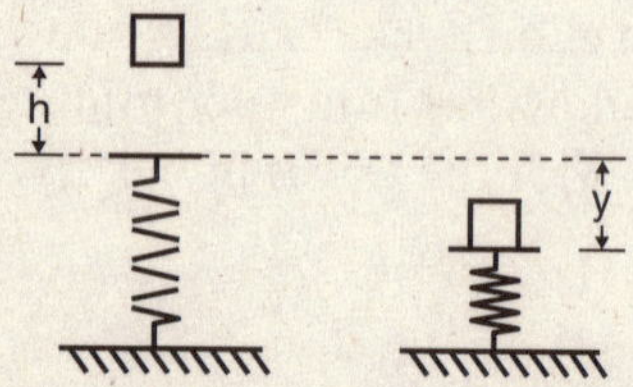

18. At the point of maximum compression, which of the following must be true?

    (A) The speed of the block is maximum.
    (B) The acceleration of the block is not zero.
    (C) The potential energy of the block is zero.
    (D) The kinetic energy of the block is maximum.
    (E) The potential energy of the spring is zero.

19. The spring expands upward and throws the block vertically back into the air. Which of the following is true after the block leaves the spring?

    (A) The block has no kinetic energy.
    (B) The block has no potential energy.
    (C) The block will rise to exactly half the height from which it was dropped.
    (D) The block reaches its maximum acceleration just after leaving the spring.
    (E) The block will rise to the height from which it was dropped.

Questions 20–21: The velocity and acceleration vectors associated with the motion of three particles are shown below.

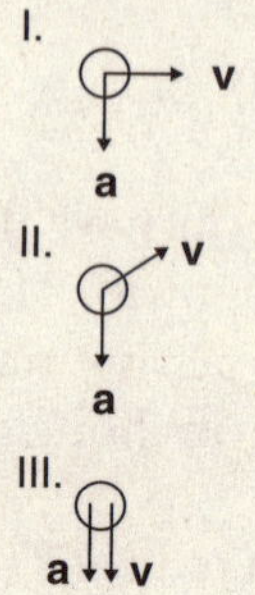

20. Which of the above could represent the velocity and acceleration vectors for an object in uniform circular motion?

    (A) I only
    (B) II only
    (C) II only
    (D) I and II only
    (E) II and III only

21. Which of the above could represent the velocity and acceleration vectors for a projectile following a parabolic path?

    (A) I only
    (B) II only
    (C) II only
    (D) I and II only
    (E) II and III only

22. Two stars are separated by a distance r and are moving away from each other. Which of the graphs below best represents the gravitational force between the stars as a function of r, the distance between them?

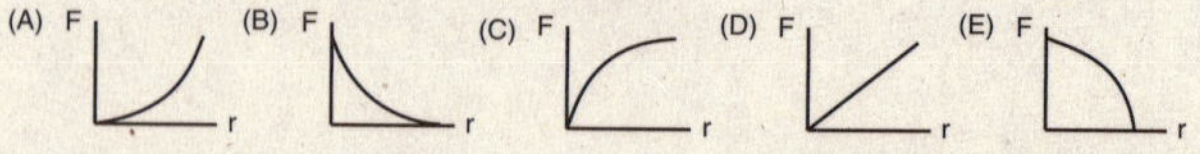

CONTINUE ON THE NEXT PAGE

Questions 23–24:

A 15 kg block rests on a surface of negligible friction and is pulled by a string which is passed over a pulley of negligible mass and connected to a hanging 5 kg block.

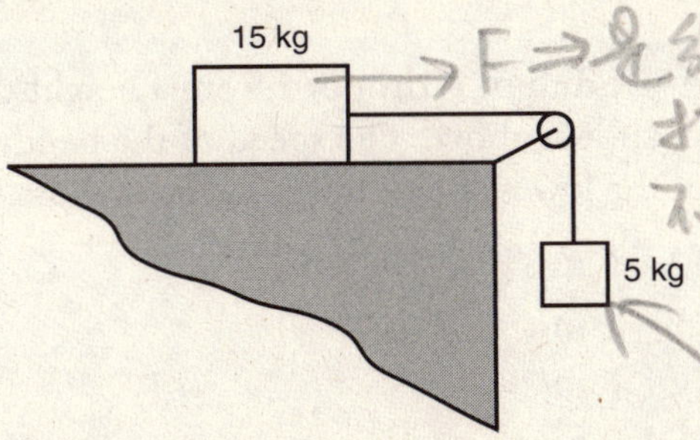

23. The net force acting on the 15 kg block is equal to

    (A) the weight of the 5 kg block.
    (B) the tension in the string.
    (C) the difference between the weight of the 15 kg block and the 5 kg block.
    (D) the sum of the weight of the 15 kg block and the 5 kg block.
    (E) the weight of the 15 kg block.

24. In terms of the acceleration due to gravity g, the acceleration of the system is

    (A) $\frac{g}{5}$
    (B) $\frac{g}{4}$
    (C) $\frac{g}{3}$
    (D) g
    (E) $3g$

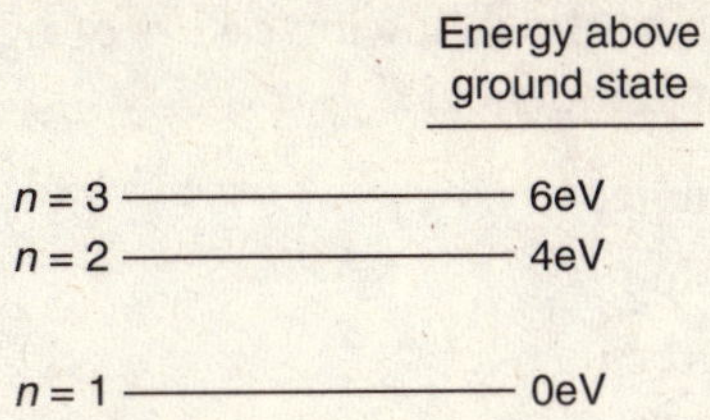

25. The energy levels above ground state for a hypothetical atom are shown above. What is the energy of a photon emitted as a result of an electron in this atom making a transition from the third energy level to the second energy level?

    (A) 10 eV
    (B) 6 eV
    (C) 4 eV
    (D) 2 eV
    (E) 1 eV

26. Consider the three statements below:

    I. The atom is mostly empty space.
    II. Electrons orbit the nucleus of the atom.
    III. The atom has a dense, positively charged nucleus.

Which of the above statements were conclusions drawn by Rutherford about the structure of the atom after studying the results of his alpha-scattering experiment?

    (A) I only
    (B) II only
    (C) I and II only
    (D) II and III only
    (E) I, II, and III

CONTINUE ON THE NEXT PAGE

27. Which of the following colors of light has the lowest frequency?

(A) violet
(B) blue
(C) green
(D) yellow
(E) orange

28. A heat pump warms a house by absorbing 90 joules of heat and doing 60 J of work on the pump. The heat then delivered to the house is

(A) 5,400 J
(B) 150 J
(C) 30 J
(D) 1.5 J
(E) 0.67 J

Questions 29–30: A resistor in a closed circuit has 2 A of current flowing through it and 12 volts across it.

29. The value of the resistor is

(A) 6 Ω
(B) 10 Ω
(C) 14 Ω
(D) 24 Ω
(E) 144 Ω

30. The power dissipated in the resistor is

(A) 6 watts
(B) 10 watts
(C) 14 watts
(D) 24 watts
(E) 144 watts

31. Which of the following is NOT a unit of energy?

(A) joule
(B) kilowatt-hour
(C) calorie
(D) watt
(E) eV

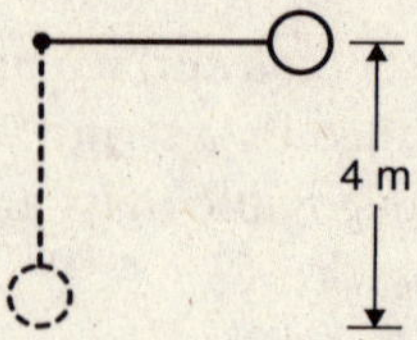

32. A pendulum is dropped from a height of 4 m as shown above. The speed of the pendulum at the lowest point in the swing is most nearly

(A) 6 m/s
(B) 7 m/s
(C) 8 m/s
(D) 9 m/s
(E) 10 m/s

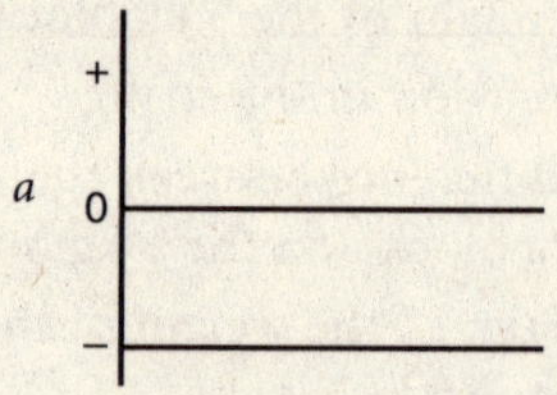

33. Consider the acceleration graph shown above, and the following choices below.

I. A stone is falling toward the earth.
II. A car is decreasing its speed.
III. A bicycle is rolling up a hill.

Which of the choices above could be represented by the acceleration vs. time graph?

(A) I only
(B) II only
(C) II and III only
(D) I and III only
(E) I, II, and III

CONTINUE ON THE NEXT PAGE

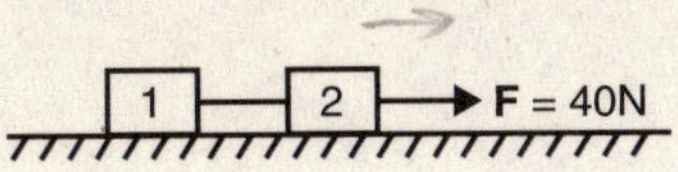

34. Two blocks of equal mass are connected by a string. Block 1 is pulled by a force of 40 N which accelerates both blocks at 3 m/s^2 along a surface of negligible friction. The tension in the string connecting the blocks is

(A) 120 N
(B) 80 N
(C) 40 N
(D) 20 N
(E) 13 N

35. A ball dropped from a tower will strike the ground below in 3 s. If the ball is launched horizontally from the tower at a speed of 10 m/s, how far horizontally from the base of the tower will the ball land on the level ground?

(A) 100 m
(B) 45 m
(C) 30 m
(D) 10 m
(E) 3.3 m

36. A cluster of magnetically aligned atoms in a magnetic material is called a

(A) domain.
(B) molecule.
(C) pole.
(D) charge.
(E) electron cloud.

37. On the microscopic scale, friction is caused by which of the following fundamental forces?

(A) gravitational
(B) electrostatic
(C) strong nuclear
(D) weak nuclear
(E) proton and neutron

38. The time for one complete swing of a pendulum is called its period. The period of a pendulum for small swings near the Earth depends on its

(A) length and amplitude of swing only.
(B) amplitude of swing and mass only.
(C) length only.
(D) amplitude only.
(E) length, mass, and amplitude of swing.

CONTINUE ON THE NEXT PAGE

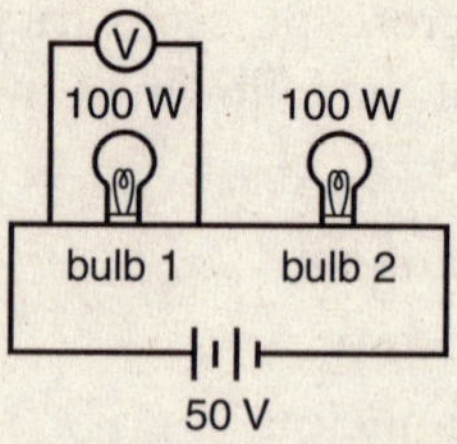

Questions 39–41 refer to the circuit shown above. A 50-volt battery supplies 100 watts of power to each of the two identical light bulbs. The current passing through bulb 1 is 4 A.

39. The voltmeter across bulb 1 will read

(A) 100 V
(B) 50 V
(C) 25 V
(D) 4 V
(E) zero

40. The current through bulb 2 is

(A) 2 A
(B) 4 A
(C) 8 A
(D) 100 A
(E) zero

41. The resistance of each bulb is most nearly

(A) 25 Ω
(B) 20 Ω
(C) 6 Ω
(D) 4 Ω
(E) 2 Ω

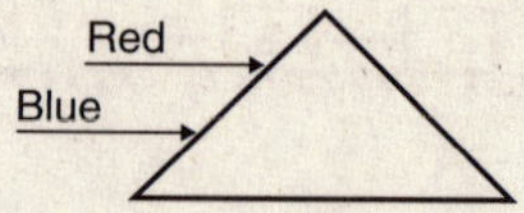

42. Two beams of light, red and blue, enter a prism as shown above. Which of the following statements is true concerning the light as it passes through the prism?

(A) The blue light will bend more than the red light, since the blue light has a longer wavelength.
(B) The red light will bend more than the blue light, since the red light has a longer wavelength.
(C) The blue light will bend more than the red light, since the blue light has a shorter wavelength.
(D) The red light will bend more than the blue light, since the red light has a shorter wavelength.
(E) The red and blue light will bend by the same amount, since all colors of light refract equally.

43. Green light is passed through two narrow slits and forms a pattern of bright and dark lines on a screen. The phenomena primarily responsible for this pattern is

(A) refraction.
(B) reflection.
(C) polarization.
(D) interference.
(E) intensity.

CONTINUE ON THE NEXT PAGE

44. An ultraviolet photon has twice the frequency of a red photon. The ratio of the energy of the ultraviolet photon to the energy of the red photon is

光子的能量和频率成正比!!!

(A) 4
(B) 2
(C) 1
(D) ½
(E) ¼

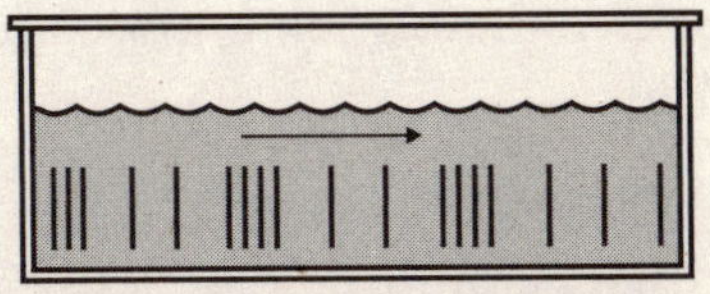

Questions 45–47 refer to the figure above which represents a longitudinal wave moving through water in a glass tank of length 9 meters. The frequency of the wave is 500 Hz.

45. This wave could be which of the following types of waves?

(A) visible light
(B) radio wave
(C) microwave
(D) sound wave
(E) x-ray

46. What is the wavelength of the longitudinal wave?

(A) 1.5 m
(B) 3 m
(C) 4.5 m
(D) 9 m
(E) 18 m

47. The speed of the wave is most nearly

(A) 500 m/s
(B) 750 m/s
(C) 1,500 m/s
(D) 3,000 m/s
(E) 4,500 m/s

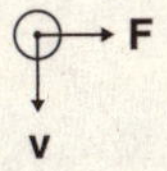

48. A ball is moving downward with a velocity **v** as shown in the figure above, with a force **F** directed to the right acting on the ball. Which of the statements is true?

(A) The ball is moving with a constant velocity.
(B) The ball is moving with a constant speed.
(C) The force **F** is doing work on the ball.
(D) The force **F** is changing the kinetic energy of the ball.
(E) The velocity and the acceleration of the ball are in the same direction.

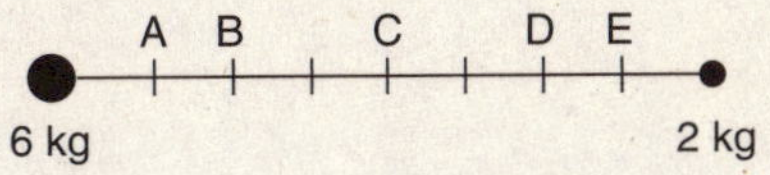

49. The bar of negligible mass shown above is marked in eight equal parts. The bar has a 2 kg mass attached to one end and a 6 kg mass attached to the other end. The center of mass of the two masses is most nearly at which point?

(A) A
(B) B
(C) C
(D) D
(E) E

CONTINUE ON THE NEXT PAGE

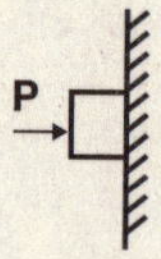

Questions 50–51 refer to the diagram above. A block of mass m is pushed against a rough, fixed wall by a force **P**. The frictional force between the block and the surface is **f**, and the normal force is **N**.

50. Which of the following diagrams correctly shows all of the forces acting on the block?

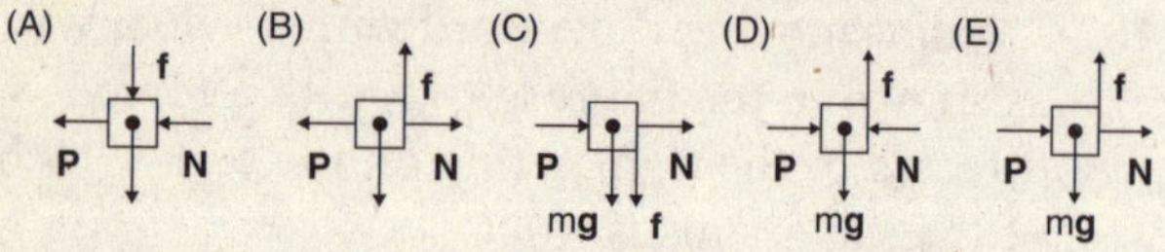

51. The magnitude of the force of friction acting on the block is equal to the magnitude of the force

    (A) **P**
    (B) **mg**
    (C) **N**
    (D) **N + P**
    (E) **N − P**

52. Which of the following is NOT a vector quantity?

    (A) kinetic energy
    (B) force
    (C) displacement
    (D) velocity
    (E) momentum

53. The molecules of an ideal gas are increasing their collisions per unit time with the walls of a sealed container of constant volume. Which of the three choices below must also be increasing?

    I. pressure
    II. temperature
    III. number of moles of gas

    (A) I only
    (B) II only
    (C) I and II only
    (D) II and III only
    (E) I and III only

CONTINUE ON THE NEXT PAGE

注："+"极的那块板就带"+"电!!!

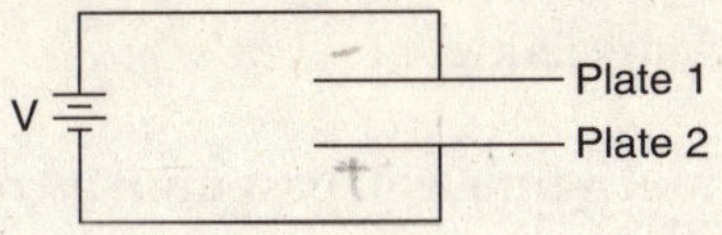

Questions 54–55 relate to the diagram above. A battery of voltage V is connected to two parallel conducting plates, 1 and 2.

54. The device connected to the battery is generally called a

(A) generator.
(B) capacitor.
(C) motor.
(D) charger.
(E) resistor.

55. An electron is placed in the center of the space between the plates. The subsequent motion of the electron is described by which of the following?

(A) The electron accelerates toward plate 1.
(B) The electron accelerates toward plate 2.
(C) The electron moves with constant velocity toward plate 1.
(D) The electron moves with constant velocity toward plate 2.
(E) The electron remains at rest halfway between the plates.

56. The current in a wire is directed out of the page and perpendicular to the page. Which of the following best represents the magnetic field produced by the current in the wire?

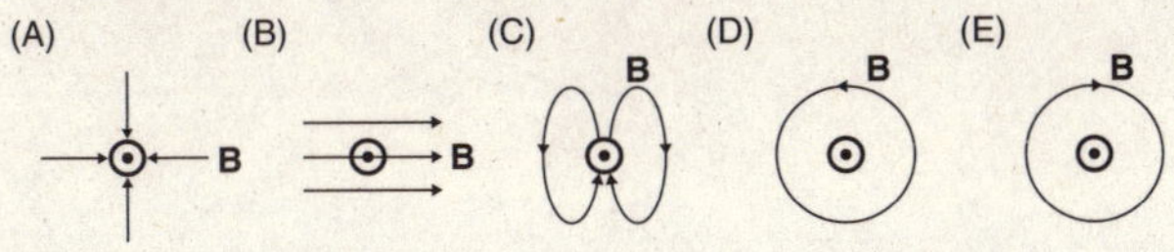

Questions 57–58 refer to a candle burning inside the metal container shown above. Three types of heat transfer are listed below:

I. radiation
II. conduction
III. convection

57. Heat can be transferred to the inside surface of the walls of the container by which of the above?

(A) I only
(B) I and II only
(C) I and III only
(D) II only
(E) I, II, and III only

58. If you touch the outside surface of the metal container, your hand will become warmer directly by which of the above choices?

(A) I only
(B) I and II only
(C) I and III only
(D) II only
(E) III only

CONTINUE ON THE NEXT PAGE

59. Let g be the acceleration due to gravity at the surface of the Earth. What would be the acceleration due to gravity at a distance of four Earth radii from the center of the Earth?

    (A) $16g$
    (B) $4g$
    (C) $2g$
    (D) $\frac{1}{4}g$
    (E) $\frac{1}{16}g$

60. The graph above shows the number of kilograms of sand which falls into a truck per hour as a function of time in hours. The shaded area under the curve has units of

    (A) volume.
    (B) mass.
    (C) mass per hour.
    (D) volume per hour.
    (E) inverse hours.

Questions 61–62: An object starts with a speed of 10 m/s and accelerates at –2 m/s^2.

61. How much time will pass until it comes to rest?

    (A) 2 s
    (B) 4 s
    (C) 5 s
    (D) 12 s
    (E) 20 s

62. How far does the object travel before coming to rest?

    (A) 10 m
    (B) 25 m
    (C) 50 m
    (D) 75 m
    (E) 100 m

CONTINUE ON THE NEXT PAGE

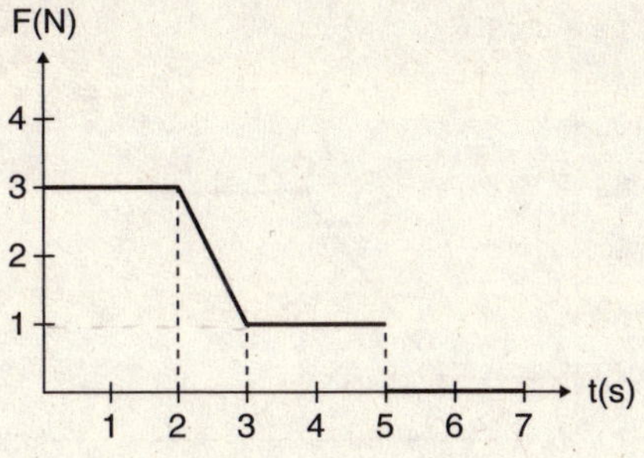

Questions 63–65 relate to the graph of the net force **F** exerted on a 2 kg block as a function of time in seconds, shown above.

63. The acceleration of the block between 0 and 2 seconds is

(A) 1.5 m/s^2
(B) 3 m/s^2
(C) 4 m/s^2
(D) 6 m/s^2
(E) 9 m/s^2

64. The speed of the block is greatest at

(A) 2 s
(B) 3 s
(C) 4 s
(D) 5 s
(E) The speed is constant at all times.

65. If the block starts from rest, its momentum at 5 seconds is

(A) 1 kg m/s
(B) 3 kg m/s
(C) 6 kg m/s
(D) 8 kg m/s
(E) 10 kg m/s

Questions 66–67 refer to a positive charge moving in a straight line through space.

66. A person who is at rest as the charge moves by him will measure

(A) a magnetic field and an electric field due to the moving charge.
(B) a magnetic field, but not an electric field due to the moving charge.
(C) an electric field, but not a magnetic field due to the moving charge.
(D) neither an electric field nor a magnetic field due to the moving charge.
(E) a decrease in the amount of charge.

67. A person who is moving parallel to the charge and at the same velocity as the charge will measure

(A) a magnetic field and an electric field due to the moving charge.
(B) a magnetic field, but not an electric field due to the moving charge.
(C) an electric field, but not a magnetic field due to the moving charge.
(D) neither an electric field nor a magnetic field due to the moving charge.
(E) a decrease in the amount of charge.

68. In 1887, Heinrich Hertz discovered the photoelectric effect. When ultraviolet light is shined on a zinc metal surface, he found that the metal

(A) emitted protons.
(B) emitted neutrons.
(C) became negatively charged.
(D) became positively charged.
(E) emitted photons.

CONTINUE ON THE NEXT PAGE

69. Which of the following is NOT a postulate or result of Bohr's model of the hydrogen atom?

(A) Excited gases emit a bright-line emission spectrum.
(B) The electron in the hydrogen atom can be found at any radius around the nucleus of the atom.
(C) When the electron makes a transition to a higher energy level, it has absorbed energy.
(D) When the electron makes a transition to a lower energy level, it has emitted energy.
(E) The energy levels of the electron are quantized.

70. The Heisenberg uncertainty principle states that

(A) we can measure the position of an electron only if we know its momentum.
(B) we can measure the momentum of an electron only if we know its position.
(C) only one electron can exist in the lowest energy level of an atom.
(D) light can behave like a wave or a photon.
(E) there is a limit to the accuracy of the measurement of subatomic particles.

71. If electricity costs \$0.12 per kilowatt-hour, how much does it cost for electricity to operate a 150-watt light bulb for 5 hours?

(A) \$0.009
(B) \$0.09
(C) \$0.90
(D) \$9.00
(E) \$90.00

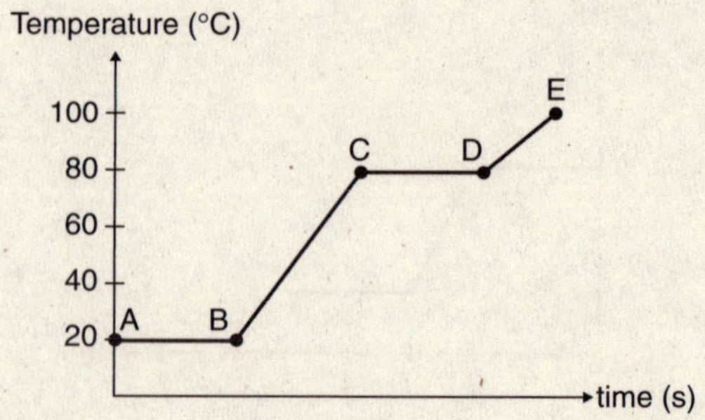

Questions 72–74 refer to the temperature vs. time graph above for 2 kg of a particular substance. The substance is in the solid state at a temperature of 20°C.

72. The interval during which the substance is changing from the liquid phase to the gas phase is

(A) AB
(B) AC
(C) BC
(D) CD
(E) DE

73. If 100 joules of heat is needed to completely change the substance from a solid to a liquid state, what is the heat of fusion for the substance?

(A) 1,000 J/kg
(B) 100 J/kg
(C) 50 J/kg
(D) 20 J/kg
(E) 5 J/kg

74. If 200 joules of heat is added during the time interval from B to C, the specific heat of the substance is

(A) 200 J/kg°C
(B) 100 J/kg°C
(C) 60 J/kg°C
(D) 17 J/kg°C
(E) 1.67 J/kg°C

CONTINUE ON THE NEXT PAGE

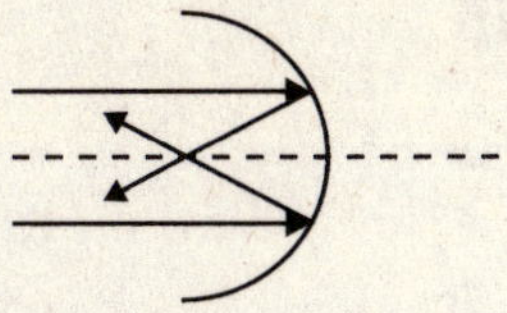

75. The diagram above best represents the ray diagram for a

    (A) converging lens.
    (B) diverging lens.
    (C) converging mirror.
    (D) diverging mirror.
    (E) plane mirror.

END OF TEST

YOU MAY GO BACK AND CHECK YOUR WORK ON THIS TEST IF TIME HAS NOT BEEN CALLED.

PRACTICE TEST 3: ANSWERS AND EXPLANATIONS

1. B	16. C	31. D	46. B	61. C
2. D	17. D	32. D	47. C	62. B
3. D	18. B	33. E	48. B	63. A
4. B	19. E	34. D	49. B	64. D
5. A	20. A	35. C	50. D	65. E
6. C	21. D	36. A	51. B	66. A
7. E	22. B	37. B	52. A	67. C
8. A	23. B	38. C	53. C	68. D
9. E	24. B	39. C	54. B	69. B
10. D	25. D	40. B	55. A	70. E
11. B	26. E	41. C	56. D	71. B
12. A	27. E	42. C	57. C	72. D
13. D	28. C	43. D	58. D	73. C
14. B	29. A	44. B	59. E	74. E
15 A	30. D	45. D	60. B	75. C

Computing Your Practice Test Scores

Step 1: Calculate your raw score.

Raw score = # of correct answers – (# of wrong answers × .25)

Step 2: Find your scaled score.

In the Score Conversion Table below, find your raw score (rounded to the nearest whole number). The score directly to the right of it will be your scaled score.

Note: Don't take these scores too literally. Practice test conditions cannot precisely mirror real test conditions. Your scores on the practice test will give you a rough idea of your range on the actual exam.

Raw	***Scaled***	***Raw***	***Scaled***	***Raw***	***Scaled***	***Raw***	***Scaled***	***Raw***	***Scaled***	***Raw***	***Scaled***
75	800	59	780	43	690	27	590	11	480	–5	380
74	800	58	770	42	680	26	580	10	480	–6	370
73	800	57	770	41	670	25	580	9	470	–7	370
72	800	56	760	40	670	24	570	8	470	–8	360
71	800	55	760	39	660	23	570	7	460	–9	350
70	800	54	750	38	650	22	560	6	450	–10	350
69	800	53	750	37	650	21	550	5	450	–11	340
68	800	52	740	36	640	20	540	4	440	–12	330
67	800	51	730	35	640	19	540	3	430	–13	330
66	800	50	730	34	630	18	530	2	430	–14	320
65	800	49	720	33	630	17	530	1	420	–15	310
64	800	48	720	32	620	16	520	0	410	–16	310
63	800	47	710	31	610	15	510	–1	410	–17	300
62	790	46	700	30	610	14	510	–2	400	–18	290
61	790	45	700	29	600	13	500	–3	390	–19	290
60	780	44	690	28	600	12	490	–4	390		

1. B

The velocity vector is tangent to the path of the circle and in the direction the ball is moving at the instant the ball is at point P.

2. D

The ball is constantly accelerating toward the center of the circle as it accelerates away from a straight-line path. We say that the acceleration is centripetal.

3. D

The net force and the acceleration must be in the same direction according to Newton's second law. The string is always pulling the ball toward the center of the circle.

4. B

Since the distance between the constant time intervals is increasing, the ball is covering more distance per second each second, which is an increase in speed (acceleration).

5. A

The distance covered per one second time interval is constant, therefore the velocity is constant.

6. C

When a ball is thrown upward, the acceleration is in the opposite direction of the velocity, and so the speed of the ball decreases. This means that the distance covered per time interval decreases.

7. E

The dots represent a ball which starts out at a high velocity (large distance per unit time), slowing down to a small velocity (small distance per unit time), and then the speed increases again as the ball rolls down the incline.

8. A

The electric field is defined as the force per unit charge acting on a positive charge. If we place a small positive charge at point A to test the electric field, it would be repelled from +Q and attracted by −4Q. The electric force is proportional to the product of the charges and inversely proportional to the square of the distance between the charges (Coulomb's law), and since the negative charge is four times greater than +Q, the positive test charge must be farther from −4Q than from +Q for the forces from each to cancel each other out and produce a net electric field of zero at point A.

9. E

Even though the electric field at point E due to +Q points to the right, +Q is very far away, and −4Q applies a greater force to a charge placed at E, and the electric field due to −4Q points to the left at E.

10. D

A small charge placed at point D would experience a strong attractive force (to the right) from −4Q and a repulsive force (to the right) from +Q. The force on the small charge is greatest at the closest distance from the large charge.

11. B

The energy of a photon is proportional to its frequency.

12. A

The emissions of electrons in the photoelectric effect does not occur until a minimum frequency is met, indicated by the graph starting at a particular frequency called the threshold frequency.

13. D

When the object is at rest, the mass has a value that increases as the object begins to move with a velocity. As the object moves closer to the speed of light, the mass gets larger and larger, but can't quite achieve the speed of light *c*.

14. B

Momentum is mass times velocity. Since the masses are the same but the red car has twice the velocity of the blue car, the red car has twice the momentum.

15 A

Kinetic energy is proportional to the square of the speed ($KE = \frac{1}{2}mv^2$), so the car with twice the speed will have four times the kinetic energy.

16. C

A hertz is a cycle per second. The waves oscillate at 90 cycles per minute, or 90 cycles/60 s = 1.5 cycles/sec or 1.5 Hz.

17. D

As the ray passes from a less dense medium (air) to a more dense medium (glass), the ray bends toward the normal line. As the ray exits the glass back into the air, it must bend back to the original angle in the air.

18. B

The block is momentarily at rest, but it will not stay at rest because the spring is accelerating the block upward with a spring force. This is similar to a ball thrown up into the air when it comes to rest at the top of the flight; it has no velocity but has an acceleration at that point.

19. E

As long as there is no energy lost to friction, the gravitational potential energy of the block at height *h* will be converted into spring potential energy, and then back to gravitational potential energy at the maximum height again.

20. A

For an object in uniform circular motion, the force (toward the center of the circle) is perpendicular to the velocity (tangent to the circular path).

21. D

Figure I could represent the velocity and acceleration vectors of the projectile when it is at the top of the trajectory, and figure II could represent the velocity and acceleration vectors when the projectile is rising during the first half of its flight.

22. B

The force between the stars is inversely proportional to the square of the distance *r* between them, so a small *r* would produce a large force, and a large *r* would produce a small force.

23. B

The weight of the 15 kg block and the normal force the table exerts on the block are equal and opposite, so the only force that can accelerate the 15 kg block is the tension in the string.

24. B

If we look at the entire system as a whole, then the weight of the 5 kg block, which is 50 N, is the force that accelerates both the 5 kg block and the 15 kg block. In other words, the weight of one block is accelerating the equivalent of four 5 kg blocks, giving only one-fourth as much acceleration as the block would have if it were falling freely.

25. D

The energy of the emitted photon is equal to the difference in the two energy levels, so 6 eV – 4 eV = 2 eV.

26. E

All three of the statements describe the conclusions reached by Rutherford as a result of his alpha-scattering experiment.

27. E

The color with the lowest frequency corresponds to the longest wavelength, which is orange in this case.

28. C

Since 60 of the 90 joules of heat are used to do work, there are only 30 joules of heat left to deliver to the house, 90 J – 60 J = 30 J.

29. A

$$R = \frac{V}{I} = \frac{12V}{2A} = 6\Omega$$

30. D

$$P = IV = (2A)(12V) = 24W$$

31. D

A watt is a unit of power, which is energy per unit time.

32. D

By conservation of energy, the pendulum's initial potential energy is equal to the kinetic energy at the bottom of the swing:

$$mgh = \tfrac{1}{2}mv^2$$

$$v = \sqrt{2gh} = \sqrt{2(10\text{ m/s}^2)(4\text{m})}$$

$$= 9\text{ m/s (approximately)}$$

33. E

All three choices could be represented by the graph, which depicts constant negative acceleration.

34. D

The string between the blocks has to accelerate only half the mass (block 1) as the 40 N force at an acceleration of 3 m/s^2, so the tension in the string between the blocks is half of 40 N, or 20 N.

35. C

The horizontal distance $x = vt = (10\text{ m/s})(3\text{ s}) = 30\text{ m}$.

36. A

Each atom has its own magnetic field, and if a group of atoms can align their magnetic fields, this group or cluster of atoms is called a magnetic domain.

37. B

Friction is caused by the electrons in two surfaces repelling each other.

38. C

The period of a pendulum does not depend of the mass of the pendulum or the amplitude of swing (for small swings). The period is proportional to the square root of the length of the pendulum.

39. C

The identical bulbs have equal resistance, so they will divide the total voltage equally, each bulb having 25 V across it. Another way to calculate the voltage across bulb 1 is power divided by current, also yielding 25 V.

40. B

Since the two bulbs are in series with each other, the same current passes through both of them.

41. C

$$R = \frac{V}{I} = \frac{25V}{4A} = 6\Omega \text{ (approximately)}$$

42. C

Blue light has a shorter wavelength than red light, and shorter wavelengths refract more than longer wavelengths in the glass.

43. D

The waves pass through the double-slit and interfere with each other, creating nodes and antinodes on the screen, which we see as alternating bright and dark lines.

44. B

Since the energy of a photon is proportional to its frequency, twice the frequency will give twice the energy.

45. D

Sound is the only longitudinal wave listed. All of the other waves listed are transverse.

46. B

The wavelength is the length of one compression and one expansion. Since there are three wavelengths shown in the 9 m tank, each wavelength must be 3 m.

47. C

The speed of the wave is the frequency times the wavelength, so (500 Hz)(3 m) = 1,500 m/s.

48. B

The ball is moving with constant speed because the force is perpendicular to the velocity and cannot change the magnitude of the velocity (speed) but only its direction, like we would see in uniform circular motion. The force does no work on the ball and therefore cannot change its kinetic energy.

49. B

The product of mass and distance from the center of mass on the left must equal the product of mass and distance from the center of mass on the right: (6 kg)(2) = (2 kg)(6)

50. D

The normal force is equal and opposite to the push P, and the weight is equal and opposite to the frictional force.

51. B

The frictional force must be equal and opposite to the weight (mg) of the block, since the block is not sliding up or down the wall.

52. A

Kinetic energy is not a vector, since it has no direction associated with it. Work and energy are scalars.

53. C

If there are more collisions between the molecules and the walls of the container, there would be more pressure against the walls. If there are more collisions, the molecules must have a higher average kinetic energy, and since kinetic energy is proportional to temperature, the temperature is also increasing.

54. B

Two conductors which are oppositely but equally charged are the capacitor.

55. A

The negatively charged electron will be attracted to plate I since it is the positive plate (connected to the positive terminal of the battery). The electron will accelerate, since the electric force acts on it until it reaches the top plate.

56. D

By the first right-hand rule, your thumb points out of the page in the direction of the current, and your fingers curl around in the direction of the magnetic field produced by the wire.

57. C

As the candle is burning, it radiates heat in the form of infrared radiation, and also heats the air around it, creating convection currents that reach the inner walls of the container.

58. D

Heat is conducted through the metal to your hand.

59. E

Like the gravitational fore, the gravitational acceleration depends on the inverse square of the distance from the center of the Earth, so four times the distance gives 1/16 of the force and acceleration.

60. B

The area under the curve is $\left(\frac{\text{kg}}{\text{hr}}\right)(\text{hr}) = \text{kg}$, the unit for mass.

61. C

$v_f = 0 = v_i + at = 10\text{ m/s} + (-2\text{ m/s}^2)(t)$
gives $t = 5$ s

62. B

$s = v_i t + \frac{1}{2}at^2 =$
$(10\text{ m/s})(5\text{ s}) + \frac{1}{2}(-2\text{ m/s}^2)(5\text{s}^2) = 25\text{ m}$

63. A

$$a = \frac{F}{m} = \frac{3\text{N}}{2\text{kg}} = 1.5\text{ m/s}^2$$

64. D

Since there is a force acting on the block at all times up to 5 s, then the block accelerates up to 5 s, where its speed is the greatest.

65. E

The change in momentum is equal to the impulse ($F\Delta t$), which is the area under the curve. The change in momentum is equal to the final momentum, since the initial momentum is zero, and the area under the curve is 10 kg m/s.

66. A

All charges produce an electric field whether they are moving or not, and magnetic fields are created by moving charges, so the person would measure both.

67. C

The person will measure an electric field, but since the charge is at rest relative to the person making the measurement, no magnetic field will be detected.

68. D

Hertz found that the metal became positively charged when ultraviolet light was shined on it, and it was later discovered that the metal was emitting electrons.

69. B

Bohr showed that the electron in the hydrogen atom could only be found in certain selected (quantized) orbits, and no others.

70. E

The more we know about the momentum of a subatomic particle, the less we can be certain of its position, and vice versa.

71. B

$$\text{cost} = (0.150kW)(5hr)\left(12\,\frac{\text{cents}}{kWh}\right) = 9\text{ cents}$$
$$= \$0.09$$

72. D

The temperature does not change during a phase change, since the heat is used to change the state of the substance and not its temperature. Since the substance is in the solid state at 20°C, it is changing from a solid to a liquid during interval AB and a liquid to a gas during the interval CD.

73. C

$Q = mL_f$ implies that:

$$L_f = \frac{Q}{m} = \frac{100\text{J}}{2\text{kg}} = 50\frac{\text{J}}{\text{kg}}$$

74. E

$Q = mc\Delta T$ implies that:

$$c = \frac{Q}{m\Delta T} = \frac{200\text{J}}{(2\text{kg})(80°\text{C} - 20°\text{C})} = 1.67\frac{\text{J}}{\text{kg}°\text{C}}$$

75. C

The rays strike the concave side of the mirror and converge at the focal point.

Answer Sheet for Practice Test 4

1. A B C D E
2. A B C D E
3. A B C D E
4. A B C D E
5. A B C D E
6. A B C D E
7. A B C D E
8. A B C D E
9. A B C D E
10. A B C D E
11. A B C D E
12. A B C D E
13. A B C D E
14. A B C D E
15. A B C D E
16. A B C D E
17. A B C D E
18. A B C D E
19. A B C D E
20. A B C D E
21. A B C D E
22. A B C D E
23. A B C D E
24. A B C D E
25. A B C D E
26. A B C D E
27. A B C D E
28. A B C D E
29. A B C D E
30. A B C D E
31. A B C D E
32. A B C D E
33. A B C D E
34. A B C D E
35. A B C D E
36. A B C D E
37. A B C D E
38. A B C D E
39. A B C D E
40. A B C D E
41. A B C D E
42. A B C D E
43. A B C D E
44. A B C D E
45. A B C D E
46. A B C D E
47. A B C D E
48. A B C D E
49. A B C D E
50. A B C D E
51. A B C D E
52. A B C D E
53. A B C D E
54. A B C D E
55. A B C D E
56. A B C D E
57. A B C D E
58. A B C D E
59. A B C D E
60. A B C D E
61. A B C D E
62. A B C D E
63. A B C D E
64. A B C D E
65. A B C D E
66. A B C D E
67. A B C D E
68. A B C D E
69. A B C D E
70. A B C D E
71. A B C D E
72. A B C D E
73. A B C D E
74. A B C D E
75. A B C D E

Practice Test 4

Part A

Directions: Each set of lettered choices below relates to the numbered questions immediately following it. Select the one lettered choice that best answers each question. A choice may be used once, more than once, or not at all in each set.

Questions 1–3 relate to the following.

Two masses m_1 and m_2 are separated by a distance R so that there is a gravitational force F between them. The following choices refer to the gravitational force on m_1 due to m_2.

(A) It is quadrupled.
(B) It is doubled.
(C) It remains the same.
(D) It is halved.
(E) It is quartered.

1. What happens to the magnitude of the force on m_1 if the mass of m_2 is doubled?

2. What happens to the magnitude of the force on m_1 if the distance between the centers of the masses is doubled?

3. What happens to the magnitude of the force on m_1 if the distance between the centers of the masses is halved?

Questions 4–6 relate to the following circuit and the choices that follow. The resistors R_1, R_2, and R_3 each have a different value.

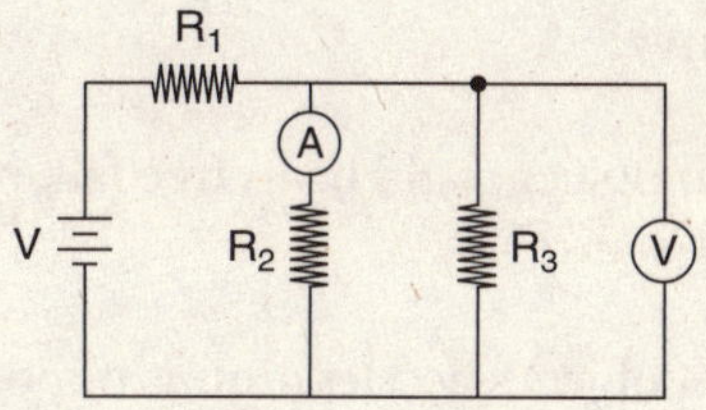

(A) R_1, R_2, and R_3
(B) R_1 only
(C) R_2 only
(D) R_3 only
(E) R_2 and R_3 only

4. Through which resistor(s) will the total current in the circuit pass?

5. Through which resistor(s) will the ammeter read the current?

6. Across which resistor(s) will the voltmeter correctly read the voltage?

Questions 7–9 relate to five objects which are moving in straight-line paths. The objects all cross a starting line at the instant a clock is started. The distances from the starting line in meters after 1, 2, 3, 4, and 5 equal time units are as follows:

	Time (units)				
Object	1	2	3	4	5
(A)	1 m	4 m	9 m	16 m	25 m
(B)	2 m	2 m	4 m	4 m	6 m
(C)	2 m	4 m	6 m	8 m	10 m
(D)	6 m	11 m	15 m	18 m	20 m
(E)	4 m	9 m	15 m	22 m	29 m

7. Which object is moving with a constant velocity?

8. Which object could be in free fall, neglecting air resistance?

9. Which object's acceleration is opposite to its velocity?

Questions 10–12 relate to the converging lens and principal axis shown and the choices that follow. The focal length *f* and twice the focal length *2f* are marked on either side of the lens.

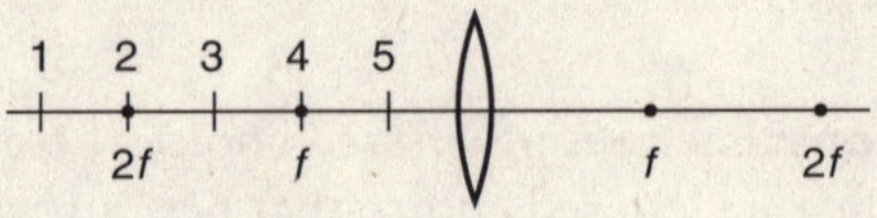

(A) Position 1

(B) Position 2

(C) Position 3

(D) Position 4

(E) Position 5

10. At which position could a candle be placed so that a virtual image would be formed?

11. At which position could a candle be placed so that an image smaller than the candle would be formed?

12. At which position could a candle be placed so that neither a real nor virtual image would be formed?

CONTINUE ON THE NEXT PAGE

Part B

Directions: Each of the questions or incomplete statements below is followed by five answer choices. Select the one that is best in each case.

13. Which of the following radioactive processes, when occurring separately, must alter the atomic number of the radioactive nucleus?

    I. alpha decay
    II. beta decay
    III. gamma decay

    (A) I only
    (B) II only
    (C) III only
    (D) I and II only
    (E) I, II, and III

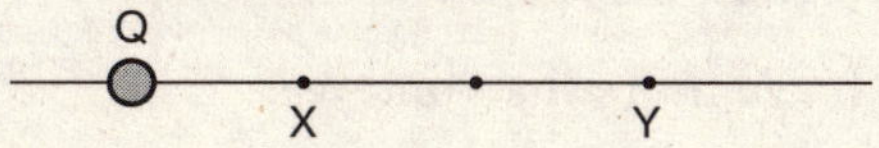

14. The diagram above shows an isolated positive charge Q. Point Y is three times as far away from Q as point X. The ratio of the electric force that would act on a small charge placed at point Y compared to the charge placed at point X is

    (A) 1 to 9
    (B) 1 to 3
    (C) 1 to 1
    (D) 3 to 1
    (E) 9 to 1

15. Which of the following particles would experience the greatest electric force in the same uniform electric field?

    (A) proton
    (B) electron
    (C) alpha particle
    (D) neutron
    (E) photon

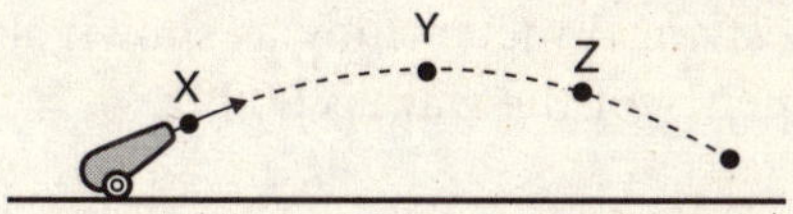

Questions 16–18 refer to the figure above which shows a ball that is projected and follows a parabolic path. Point Y is the highest point on the path, and air resistance is negligible.

16. Which of the following is true of how the speeds of the ball at the three points compare?

    (A) $v_X > v_Y > v_Z$
    (B) $v_X > v_Z > v_Y$
    (C) $v_Y > v_X > v_Z$
    (D) $v_Z > v_X = v_Y$
    (E) $v_Y > v_X = v_Z$

17. Which of the following best shows the direction of the acceleration of the ball at point Z?

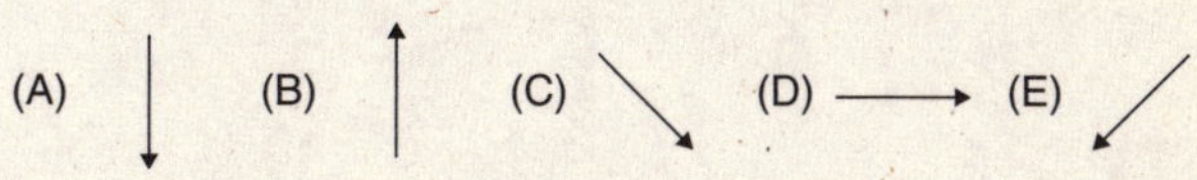

18. What is the direction of the net force acting on the ball at point Y?

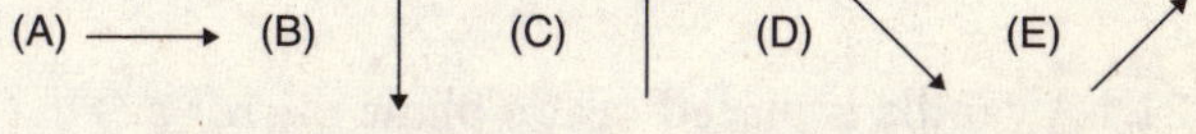

19. Neutral isotopes of the same element may have different numbers of

    (A) protons.
    (B) neutrons.
    (C) electrons.
    (D) alpha particles.
    (E) positrons.

CONTINUE ON THE NEXT PAGE

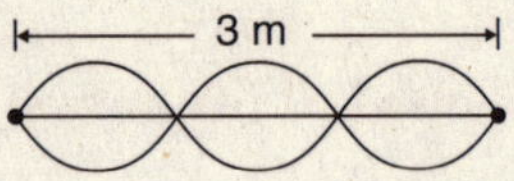

Questions 20–21 refer to the standing wave in a vibrating string 3 meters long as shown above. The frequency of the wave is 60 Hz.

20. The wavelength of the wave is

(A) 1 m
(B) 2 m
(C) 3 m
(D) 6 m
(E) 9 m

21. The speed of the wave is

(A) 180 m/s
(B) 120 m/s
(C) 60 m/s
(D) 30 m/s
(E) 6 m/s

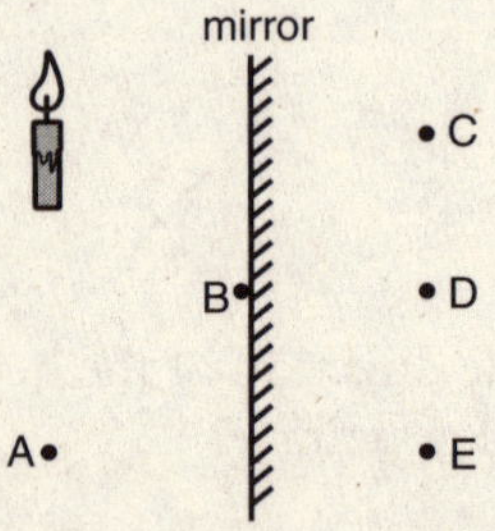

22. A candle is placed near a plane mirror as shown above. At which point above is the image of the candle formed?

(A) A
(B) B
(C) C
(D) D
(E) E

23. Which of the following was one of the two basic postulates of special relativity formulated by Albert Einstein?

(A) Accelerating reference frames are equivalent to reference frames which are moving with a constant velocity.
(B) The speed of light has the same constant value in all inertial reference frames.
(C) Photons move slightly faster than light waves.
(D) When moving the speed of light, mass will be converted to pure energy.
(E) All physical laws are relative.

24. A space traveler is moving relative to the Earth at 0.6 times the speed of light, as measured in the Earth's frame of reference. In one year as measured in the Earth's frame of reference, the space traveler will

(A) age more than one year.
(B) age less than one year.
(C) age exactly one year.
(D) not age at all.
(E) become younger.

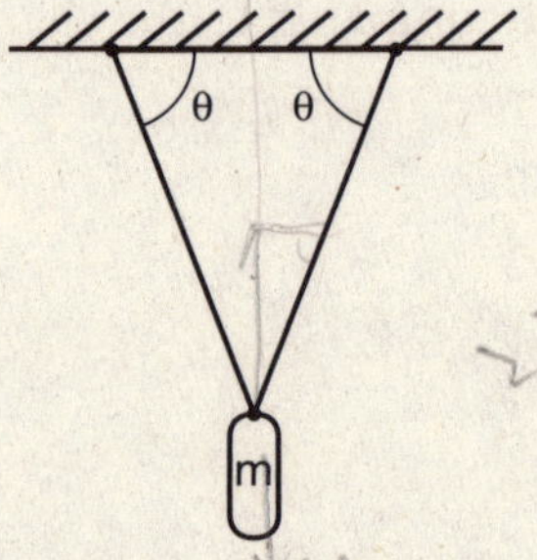

25. A mass m is hung from two light strings as shown above. The tension in each string is

(A) $mg \cos\theta$
(B) $2mg \cos\theta$
(C) $mg \sin\theta$
(D) $\frac{1}{2} mg \sin\theta$
(E) $\frac{mg}{2 \sin\theta}$

CONTINUE ON THE NEXT PAGE

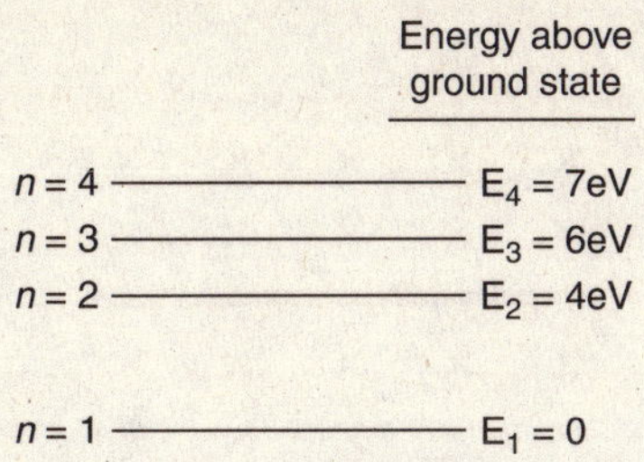

Questions 26–27 refer to the energy level diagram for a hypothetical atom shown above. The energy for each level is given above ground state.

26. Which of the following photon energies could NOT be emitted from this atom after it has been excited to the 4th energy level?

(A) 1 eV
(B) 2 eV
(C) 3 eV
(D) 4 eV
(E) 5 eV

27. Which of the following transitions will produce the photon with the highest frequency?

(A) $n = 2$ to $n = 1$
(B) $n = 3$ to $n = 1$
(C) $n = 3$ to $n = 2$
(D) $n = 4$ to $n = 1$
(E) $n = 4$ to $n = 3$

28. If the mass of an object is doubled, and the net force acting on the object is quadrupled, the acceleration of the object is

(A) quartered.
(B) halved.
(C) unchanged.
(D) doubled.
(E) quadrupled.

29. An object starts from rest and accelerates at 5 m/s^2. How far will it travel during the first 4 seconds?

(A) 20 m
(B) 40 m
(C) 80 m
(D) 120 m
(E) 160 m

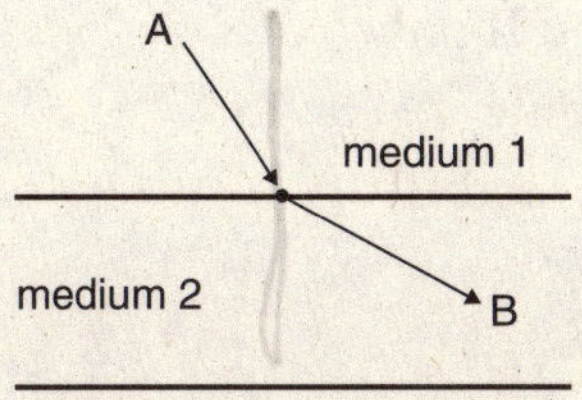

30. Light passes from medium 1 to medium 2 and bends along the path between points A and B as shown above. Which of the following statements is correct?

(A) Medium 1 is more dense than medium 2.
(B) Medium 1 is less dense than medium 2.
(C) The speed of the light in medium 1 is greater than the speed of the light in medium 2.
(D) The frequency of the light in medium 1 is greater than the frequency of the light in medium 2.
(E) The wavelength is the same in medium 1 and medium 2.

CONTINUE ON THE NEXT PAGE

31. Consider the three lenses below:

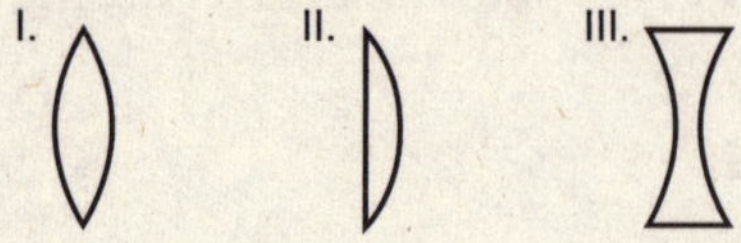

Which of the above will cause light rays to diverge?

(A) I only
(B) II only
(C) I and II only
(D) III only
(F) I, II, and III

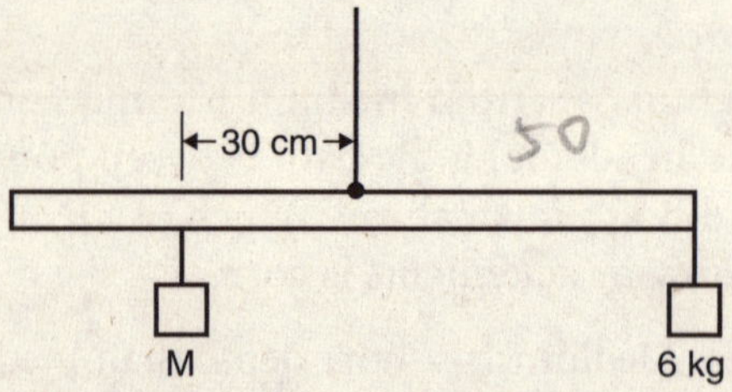

32. A meter stick of negligible mass is hung from a rope attached to its center, as shown above. A 6-kg mass is hung from one end of the meter stick, and another mass *M* is hung at a distance of 30 cm from the center of the meter stick, but on the opposite side of the 6-kg mass. What is the value of *M*?

(A) 3 kg
(B) 5 kg
(C) 6 kg
(D) 10 kg
(E) 30 kg

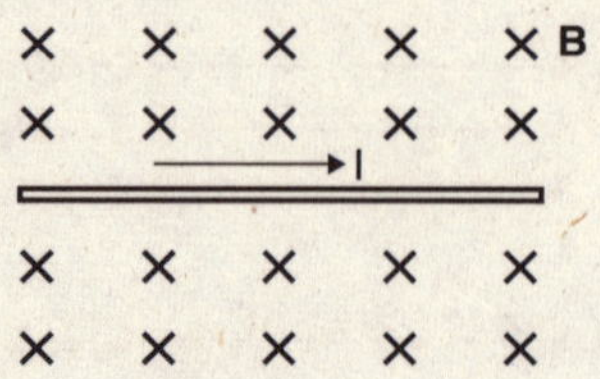

33. A wire in the plane of the page carries a current *I* directed toward the right, as shown. If the wire is located in a uniform magnetic field which is directed into the page, the force on the wire resulting from the magnetic field is

(A) zero.
(B) directed out of the page.
(C) directed to the bottom of the page.
(D) directed to the left.
(E) directed to the top of the page.

34. An ideal gas in a closed container initially has a volume *V*, a pressure *P*, and Kelvin temperature *T*. If the temperature is changed to $4T$, which of the following pairs of pressure and volume is possible?

(A) P and V
(B) P and $\frac{1}{2}V$
(C) $4P$ and $4V$
(D) $4P$ and V
(E) $\frac{1}{4}P$ and V

35. A gas in a container absorbs 300 joules of heat, then has 100 joules of work done on it, then does 50 joules of work, the increase in the internal energy of the gas is

(A) 450 J
(B) 400 J
(C) 350 J
(D) 200 J
(E) 100 J

CONTINUE ON THE NEXT PAGE

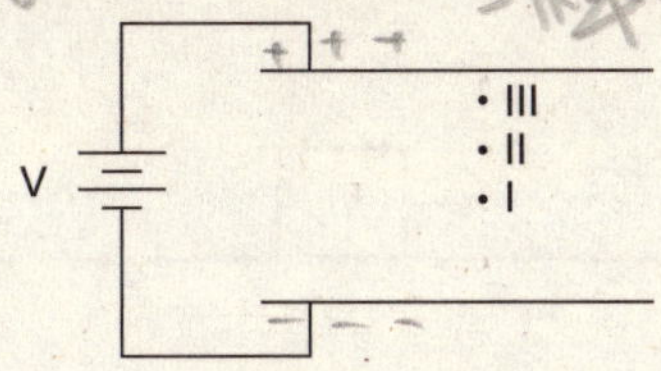

Questions 36–37 refer to the two large parallel conducting plates which are connected to a battery of voltage *V*, as shown above. Point I is halfway between the plates. Edge effects may be neglected.

36. If a proton is placed at one of the three points above, which of the following statements is true?

    (A) The proton will experience a greater force at point I than at point II.
    (B) The proton will experience a greater force at point II than at point I.
    (C) The proton will experience a greater force at point III than at point II.
    (D) The proton will experience the same force at points I, II, and III.
    (E) The proton will experience no force at point I.

37. If an electron is placed at one of the three points above, which of the following statements is true?

    (A) The electron will experience a greater upward acceleration at point I than at point II.
    (B) The electron will experience a greater downward acceleration at point II than at point I.
    (C) The electron will experience the same upward acceleration at points I, II, and III.
    (D) The electron will experience the same downward acceleration at points I, II, and III.
    (E) The electron will experience no acceleration at points I, II, or III.

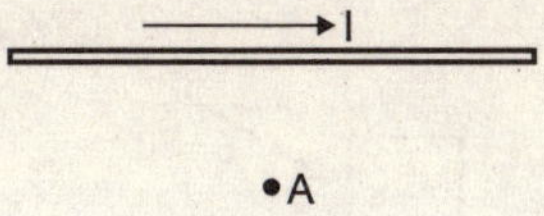

38. The direction of the magnetic field at point A caused by the current I in the wire shown above is

    (A) to the right.
    (B) to the left.
    (C) toward the wire.
    (D) into the page.
    (E) out of the page.

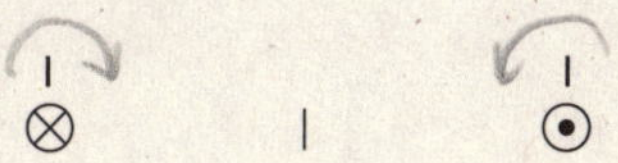

39. Two long parallel wires are separated by a distance *d* as shown above. One wire carries a steady current *I* into the plane of the page, and the other wire carries an equal steady current *I* which is out of the page. The net magnetic field at a point halfway between the wires points in which of the following directions?

    (A) → (B) ← (C) ↑ (D) ↓ (E) zero

CONTINUE ON THE NEXT PAGE

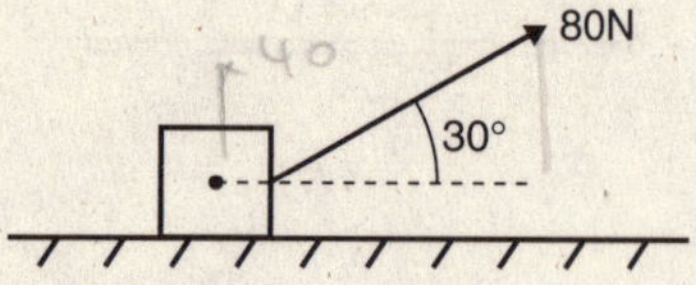

Questions 40–42 refer to a block of weight 180 N which is pulled along a horizontal surface at a constant speed by a force of 80 N. The force acts at an angle of 30° with the horizontal, as shown above. (sin 30° = 0.50 , cos 30° = 0.87)

40. The normal force by the surface on the block is most nearly

(A) 20 N
(B) 40 N
(C) 60 N
(D) 80 N
(E) 140 N

41. The frictional force acting between the block and the surface is most nearly

(A) 20 N
(B) 40 N
(C) 70 N
(D) 80 N
(E) 92 N

42. The coefficient of friction between the block and the surface is most nearly

(A) 5
(B) 2
(C) 0.87
(D) 0.50
(E) 0.13

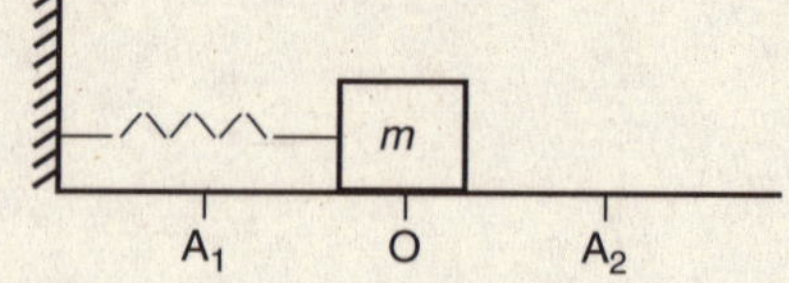

Questions 43–45 refer to the diagram of a mass on a spring above. The mass is oscillating about point O on a surface of negligible friction, and the maximum distances on either side of point O are labeled A_1 and A_2. The graphs below can represent quantities associated with the oscillation as a function of length x of the spring.

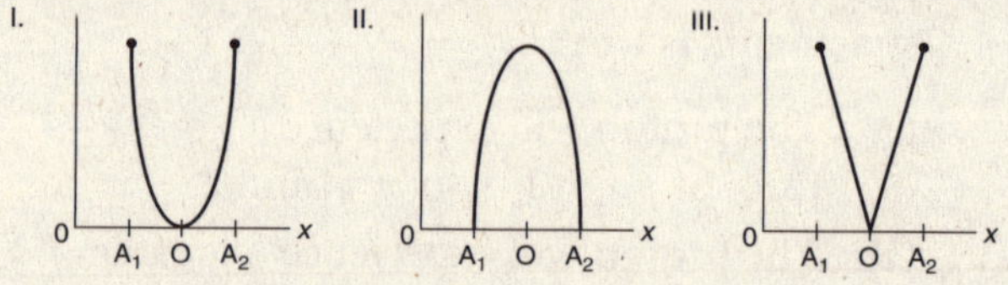

43. Which graph(s) can represent the potential energy of the spring and mass as a function of distance x?

(A) I only
(B) I and II only
(C) II only
(D) III only
(E) II and III only

44. Which graph(s) can represent the kinetic energy of the spring and mass as a function of distance x?

(A) I only
(B) I and II only
(C) II only
(D) III only
(E) II and III only

CONTINUE ON THE NEXT PAGE

45. Which graph(s) can represent the net force acting on the mass as a function of distance *x*?
    - (A) I only
    - (B) I and II only
    - (C) II only
    - (D) III only
    - (E) II and III only

Questions 46–47:

Two moons orbit a planet in circular orbits. The first moon orbits at a radius R from the center of the planet, and the second moon orbits at a radius 2R.

46. The ratio of the gravitational force between the planet and the first moon to the gravitational force to the second moon is equal to
    - (A) $\frac{1}{4}$
    - (B) $\frac{1}{2}$
    - (C) 1
    - (D) 2
    - (E) 4

47. Which of the following statements is true?
    - (A) The closer moon has a shorter period than the farther moon.
    - (B) The farther moon has a shorter period than the closer moon.
    - (C) The two moons must have the same period.
    - (D) The two moons must have the same linear speed.
    - (E) The two moons must have the same period and the same linear speed.

Questions 48–50 relate to a ball thrown straight up, reaching its maximum height in 3 s.

Air resistance may be neglected.

48. The initial velocity of the ball is most nearly
    - (A) 10 m/s
    - (B) 15 m/s
    - (C) 30 m/s
    - (D) 45 m/s
    - (E) 60 m/s

49. The magnitude of the acceleration of the ball at the instant it reaches its maximum height is most nearly
    - (A) zero
    - (B) 5 m/s^2
    - (C) 10 m/s^2
    - (D) 12 m/s^2
    - (E) 20 m/s^2

50. The maximum height of the ball is most nearly
    - (A) 10 m
    - (B) 15 m
    - (C) 30 m
    - (D) 45 m
    - (E) 60 m

CONTINUE ON THE NEXT PAGE

51. The half-life of thorium is 24 days. If you start with a 60 microgram sample of thorium, approximately how many micrograms of thorium remains after 120 days?

(A) zero
(B) 2
(C) 4
(D) 8
(E) 15

52. The three statements below relate to ammeters and voltmeters used in an electrical circuit.

I. An ammeter must be placed in series with a resistor to correctly measure the current.
II. A voltmeter must be placed in series with a resistor to correctly measure the voltage across the resistor.
III. A voltmeter must have a high resistance.

Which of the above statements are true?

(A) I only
(B) II only
(C) I and II only
(D) II and III only
(E) I and III only

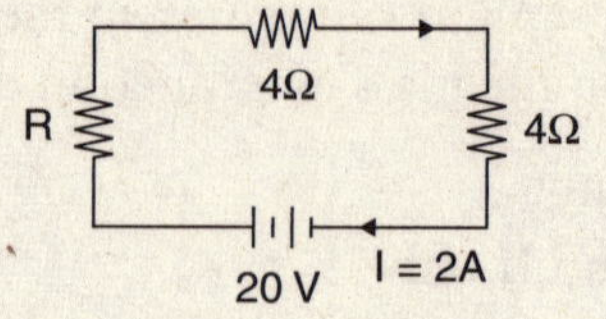

Questions 53–54 relate to the circuit shown above.

53. What is the value of the resistor *R*?

(A) 1 Ω
(B) 2 Ω
(C) 4 Ω
(D) 6 Ω
(E) 12 Ω

54. What is the power dissipated through one of the 4 Ω resistors?

(A) 4 W
(B) 8 W
(C) 16 W
(D) 32 W
(E) 36 W

CONTINUE ON THE NEXT PAGE

Questions 55–56:

A piece of metal with a mass of 2 kg and specific heat of 200 J/kg°C is initially at a temperature of 120°C. The metal is placed into an insulated container that contains a liquid of mass 4 kg, a specific heat of 600 J/kg°C, and an initial temperature of 20°C.

55. After a long time, the final equilibrium temperature of the metal and liquid is

(A) 24°C
(B) 34°C
(C) 48°C
(D) 68°C
(E) 100°C

56. In actuality, some of the heat is lost to the surroundings during the heat transfer from the metal to the liquid. This heat loss would result in

(A) a lower equilibrium temperature.
(B) a higher equilibrium temperature.
(C) a lower specific heat for the liquid.
(D) a lower specific heat for the metal.
(E) more heat gained by the liquid.

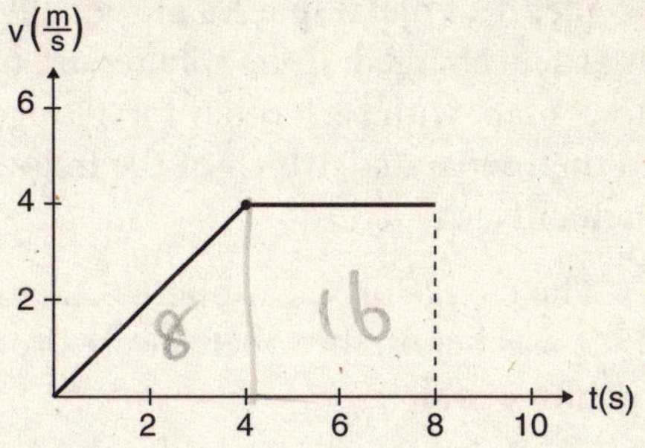

Questions 57–58 refer to the velocity vs. time graph above.

57. The acceleration of the object between t = 0 and t = 4 s is

(A) zero
(B) 1 m/s^2
(C) 2 m/s^2
(D) 4 m/s^2
(E) 8 m/s^2

58. The displacement of the object between t = 0 and t = 8 s is

(A) 64 m
(B) 32 m
(C) 24 m
(D) 16 m
(E) 8 m

59. Two identical metal spheres are equally but oppositely charged. If the spheres are brought into contact with each other for a long time, and then separated, which of the following statements is true?

    (A) The charge on each sphere is the same as it was before the spheres were brought into contact.
    (B) The spheres will be equally charged and positive.
    (C) The spheres will be equally charged and negative.
    (D) The spheres will both be neutral.
    (E) The positive sphere must have more charge than the negative sphere.

60. A pencil is placed in a beaker of water and appears to be bent or broken. The property of light responsible for the pencil appearing to be bent is

    (A) refraction.
    (B) reflection.
    (C) diffraction.
    (D) interference.
    (E) polarization.

61. The second law of thermodynamics (law of entropy) explains which of the following?

    (A) The heat lost by one object must be gained by another object.
    (B) Heat flows naturally from a hotter body to a cooler body.
    (C) Celsius degrees and Kelvin degrees are equivalent.
    (D) Heat can be transformed into work.
    (E) The average kinetic energy of molecules is proportional to temperature.

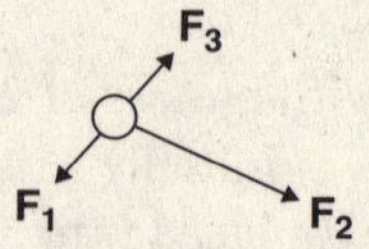

62. Three forces which are in the same plane act on the same point above. Which of the following diagrams best represents the resultant of the forces?

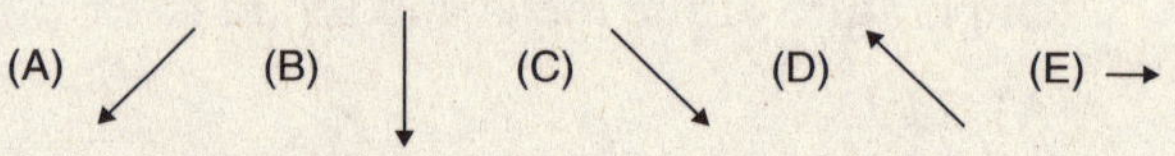

63. Two electrons, each with a charge of magnitude 10^{-19} C are separated by a distance of magnitude 10^{-10} m. If the Coulomb constant is 10^{10} N m^2/C^2, the magnitude of the force between the electrons is most nearly

    (A) 10^{-8} N
    (B) 10^{-28} N
    (C) 10^{-38} N
    (D) 10^{-48} N
    (E) 10^{-58} N

64. A toy railroad car of mass 3 kg moving east collides with a 6 kg railroad car at rest, and the two cars lock together on impact and move away together toward the east at 2 m/s. The speed of the first car before the collision is

    (A) 2 m/s
    (B) 3 m/s
    (C) 4 m/s
    (D) 6 m/s
    (E) 9 m/s

CONTINUE ON THE NEXT PAGE

65. A machine can lift a 500 N weight up to a height of 2 m in 20 seconds. The power developed by this machine is

(A) 10,000 W
(B) 1,000 W
(C) 50 W
(D) 40 W
(E) 25 W

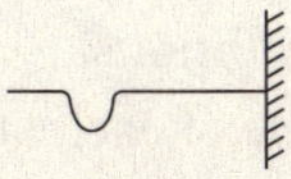

66. One end of a horizontal string is fixed to a wall, as shown above. A transverse wave pulse is generated at the other end, moves toward the wall, and is reflected at the wall. Which of the diagrams below best represents the pulse after it is reflected from the wall?

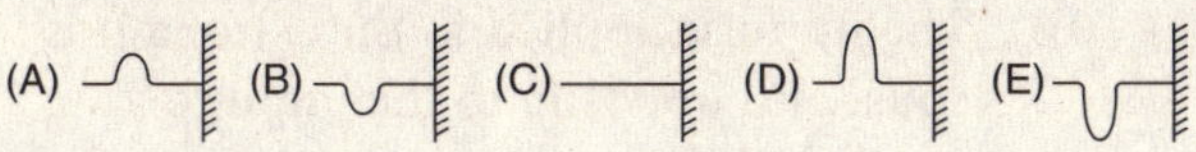

67. Electromagnetic induction is the basic principle for

(A) a resistor.
(B) an electric generator.
(C) a capacitor.
(D) a battery.
(E) an electroscope.

68. In a radioactive decay process, the parent atom undergoes the following radioactive decays: alpha, alpha, beta. The atomic number of the final daughter element is how many atomic mass units less than the parent atom?

(A) 2
(B) 3
(C) 4
(D) 5
(E) 7

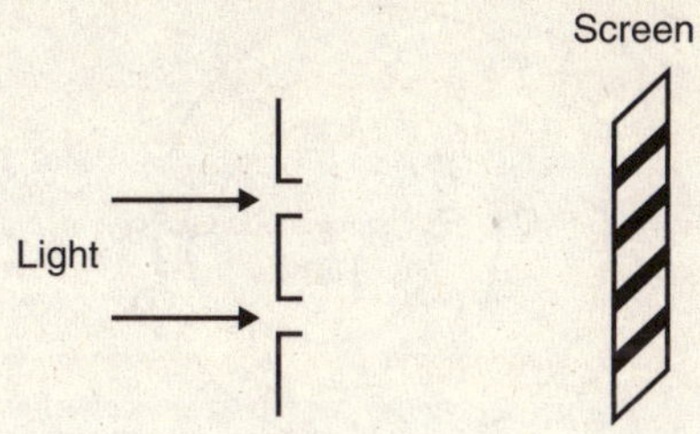

Questions 69–70 refer to the diagram above. A pattern of alternating bright and dark lines is produced on the screen when light is passed through the two narrow slits.

69. The property of light which is responsible for the bright and dark lines on the screen is

(A) refraction.
(B) reflection.
(C) polarization.
(D) frequency.
(E) interference.

70. If the screen is moved a greater distance from the slits, the center bright band of light will

(A) become wider.
(B) become narrower.
(C) remain unchanged.
(D) disappear completely.
(E) rotate 90°.

CONTINUE ON THE NEXT PAGE

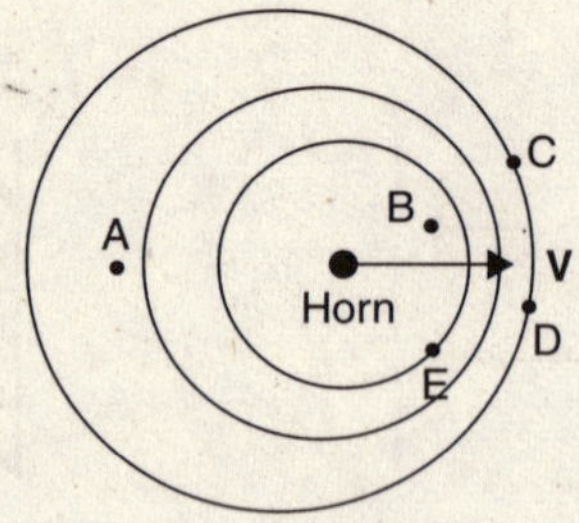

71. Sound waves of a constant frequency are being emitted by a horn as it moves to the right, as shown above. At which point would a listener hear a pitch which is lower than she would hear if the horn were at rest?

(A) A
(B) B
(C) C
(D) D
(E) E

72. In a photoelectric effect experiment, light of a particular frequency is shined on a metal surface, and as a result electrons are emitted from the metal surface. Which of the following actions would produce more electrons per second?

(A) Increase the frequency of the light.
(B) Decrease the frequency of the light.
(C) Increase the intensity (brightness) of the light.
(D) Decrease the intensity (brightness) of the light.
(E) Increase the wavelength of the light.

A more intense / brighter light has more photons
⇓
emit more electrons

73. Which of the following statements is NOT a correct assumption of the model of an ideal monatomic gas?

(A) The atoms are constantly moving.
(B) The atoms' collisions with one another create pressure directly on the container of the gas.
(C) The collisions between atoms are elastic.
(D) The only significant forces acting on the atoms are those which are applied as a result of collisions.
(E) The volume of each atom can be neglected.

74. An astronaut floating in space throws a small hand tool away from his space shuttle. Which of the following statements is true?

(A) The velocity of the astronaut is equal and opposite to the velocity of the tool.
(B) The impulse applied to the astronaut is equal and opposite to the impulse applied to the tool.
(C) The mass of the astronaut is equal to the mass of the tool.
(D) The momentum of the astronaut is greater than the momentum of the tool.
(E) The force applied to the tool is greater than the force applied to the astronaut.

CONTINUE ON THE NEXT PAGE

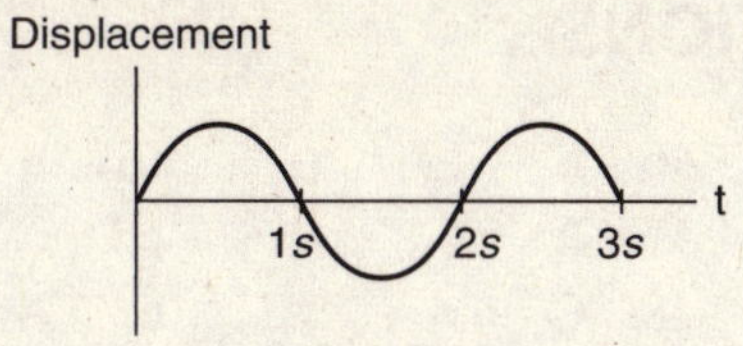

75. A pendulum oscillates in simple harmonic motion as represented by the figure above. The frequency of the pendulum is

(A) 3 Hz
(B) 2 Hz
(C) 1 Hz
(D) $^1/_2$ Hz
(E) $^1/_3$ Hz

END OF TEST

YOU MAY GO BACK AND CHECK YOUR WORK ON THIS TEST IF TIME HAS NOT BEEN CALLED.

PRACTICE TEST 4: ANSWERS AND EXPLANATIONS

1. B
2. E
3. A
4. B
5. C
6. E
7. C
8. A
9. D
10. E
11. A
12. D
13. D
14. A
15. C
16. B
17. A
18. B
19. B
20. B
21. B
22. C
23. B
24. B
25. E
26. E
27. D
28. D
29. B
30. A
31. D
32. D
33. E
34. D
35. C
36. D
37. C
38. D
39. D
40. E
41. C
42. D
43. A
44. C
45. D
46. E
47. A
48. C
49. C
50. D
51. B
52. E
53. B
54. C
55. B
56. A
57. B
58. C
59. D
60. A
61. B
62. C
63. A
64. D
65. C
66. A
67. B
68. B
69. E
70. A
71. A
72. C
73. B
74. B
75. D

Computing Your Practice Test Scores

Step 1: Calculate your raw score.

Raw score = # of correct answers – (# of wrong answers × .25)

Step 2: Find your scaled score.

In the Score Conversion Table below, find your raw score (rounded to the nearest whole number). The score directly to the right of it will be your scaled score.

Note: Don't take these scores too literally. Practice test conditions cannot precisely mirror real test conditions. Your scores on the practice test will give you a rough idea of your range on the actual exam.

Raw	***Scaled***	***Raw***	***Scaled***	***Raw***	***Scaled***	***Raw***	***Scaled***	***Raw***	***Scaled***	***Raw***	***Scaled***
75	800	59	780	43	690	27	590	11	480	–5	380
74	800	58	770	42	680	26	580	10	480	–6	370
73	800	57	770	41	670	25	580	9	470	–7	370
72	800	56	760	40	670	24	570	8	470	–8	360
71	800	55	760	39	660	23	570	7	460	–9	350
70	800	54	750	38	650	22	560	6	450	–10	350
69	800	53	750	37	650	21	550	5	450	–11	340
68	800	52	740	36	640	20	540	4	440	–12	330
67	800	51	730	35	640	19	540	3	430	–13	330
66	800	50	730	34	630	18	530	2	430	–14	320
65	800	49	720	33	630	17	530	1	420	–15	310
64	800	48	720	32	620	16	520	0	410	–16	310
63	800	47	710	31	610	15	510	–1	410	–17	300
62	790	46	700	30	610	14	510	–2	400	–18	290
61	790	45	700	29	600	13	500	–3	390	–19	290
60	780	44	690	28	600	12	490	–4	390		

1. B

The gravitational force is proportional to the product of the masses, so doubling one of the masses would double the force:

$$\frac{G(m)(2m)}{r^2} = 2F$$

2. E

The force is inversely proportional to the square of the distance between the two masses, so twice the distance would result in one-quarter the force:

$$\frac{Gm_1m_2}{(2r)^2} = {}^1\!/\!_4F$$

3. A

If the masses are brought closer together the force between them will be greater. Since the force is inversely proportional to the square of the distance between the masses, cutting the distance between them in half would increase the force by four times.

4. B

The total current will pass through R_1 since the current from the battery has not split up into any branches.

5. C

The ammeter is in series only with R_2, so the same current will pass through the ammeter and R_2.

6. E

The voltmeter, R_2, and R_3 are all in parallel with each other, and thus will have the same voltage across them.

7. C

The distance covered during each equal time unit is the same, 2 m. Thus, the object must be moving at a constant velocity.

8. A

For an object in free fall, the distance fallen is proportional to the square of the time.

9. D

If the change in distance between time intervals is decreasing, the object is slowing down, and thus the acceleration is opposite to the velocity.

10. E

An enlarged virtual image is formed when an object is placed inside the focal length of a converging lens. This is the principle behind a magnifying glass.

11. A

When the candle is placed at a distance greater than twice the focal length, the image is smaller than the object and is real and inverted.

12. D

When an object is placed at the focal length of a converging lens, the rays pass through the lens parallel to each other, and no image is formed.

13. D

Alpha decay reduces the atomic number of the parent atom by 2 atomic mass units, and beta decay increases the atomic number by one atomic mass unit. The gamma ray is an electromagnetic wave which has no mass.

14. A

Coulomb's law states that the force between two charges is inversely proportional to the square of the distance between the charges, and thus a charge placed at three times the distance will experience one-ninth as much force.

15. C

The alpha particle is made up of two protons and two neutrons, and therefore has twice the charge of either a proton or an electron and will experience twice the force in the same electric field.

16. B

The velocity is greatest at point X, and although the horizontal component of the velocity remains constant, the vertical component of the velocity decreases on the way up and increases on the way down. Thus, the velocity is greater at Z than at Y.

17. A

The only acceleration the ball has after it is launched is the acceleration due to gravity, which points straight downward.

18. B

The net force acting on the ball at point Y is the gravitational force, which points straight downward.

19. B

Isotopes of a particular element have the same number of protons, but may have a different number of neutrons, which do not change the charge on the nucleus.

20. B

The wavelength is the length of one full cycle of the wave, which includes two loops of the standing wave, and is two-thirds of the length of the string in this case, or 2 m.

21. B

$v = f\lambda = (60 \text{ Hz})(2 \text{ m}) = 120 \text{ m/s}$

22. C

Someone looking into the mirror would see an image of the candle at point C, the same distance behind the mirror as the candle is in front of the mirror. The reflected rays from the light of the candle appear to originate at point C.

23. B

The two postulates are: (1) All inertial reference frames are equivalent; and (2) All observers will measure the same value for the speed of light regardless of any relative motion between the observer and the light source.

24. B

The astronaut's moving clock will run more slowly according to the Earth's frame of reference, and the astronaut will age more slowly as well.

25. E

If we break the tensions in the strings into their x and y-components, we see that the x-components cancel each other out, and the sum of the two y-components of the tension must equal the weight of the block.

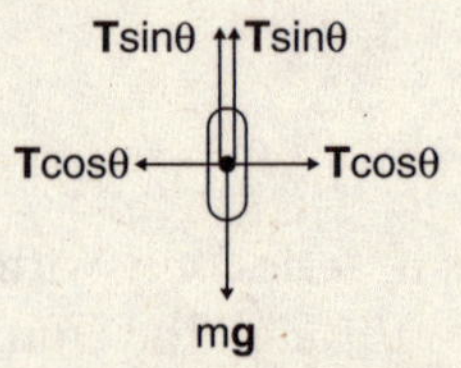

Thus, $2T \sin\theta = mg$, and $T = \dfrac{mg}{2 \sin\theta}$.

26. E

The energy of the emitted photon must equal the difference between two levels, whether the electron jumps down from the 4th level to a lower level and emits a photon, or jumps down from the 4th level to a lower level and emits a photon and then jumps to an even lower level, emitting another photon. None of the energy differences in the diagram corresponds to 5 eV.

27. D

Since the energy of the emitted photon is proportional to its frequency, the highest energy difference corresponds to the highest frequency photon.

28. D

By Newton's second law, the acceleration is proportional to the force, and inversely proportional to the mass:

$$\frac{4F}{2m} = 2a$$

29. B

$$s = v_i t + \tfrac{1}{2}at^2 = (0) + \tfrac{1}{2}(5\text{ m/s}^2)(4\text{s}^2) = 40\text{ m}$$

30. A

Light bends away from the normal when it passes into a less dense medium, or, equivalently, the angle of refraction is greater than the angle of incidence when light enters a less dense medium.

31. D

Lens III is a diverging (concave) lens, which will diverge parallel rays passing through it. The other two lenses will converge parallel rays passing through them.

32. D

For the meter stick to remain horizontal, the torque on the left of the string must equal the torque on the right of the string.

$$Mgr_1 = mgr_2$$
$$(M)(30\text{ cm}) = (6\text{ kg})(50\text{ cm})$$

Thus, $M = 10$ kg.

33. E

By the second right-hand rule, if you put your fingers in the direction of the magnetic field (into the page), and your thumb in the direction of the current, the force on the wire comes out of your palm, toward the top of the page.

34. D

By the ideal gas law,

$$\frac{PV}{T} = \frac{4PV}{4T}$$

35. C

The internal energy of the gas rises to 400 J when 100 J of work is done on it, then loses internal energy when the gas itself does 50 J of work, leaving an increase in internal energy of 350 J.

36. D

Since the electric field is uniform everywhere between the plates, the proton will experience the same force (downward) everywhere between the plates.

37. C

Since the electric field is uniform everywhere between the plates, the electron will experience the same force (upward) everywhere between the plates.

38. D

Using the first right-hand rule, we place our thumb in the direction of the current and our fingers curl around the wire in the direction of the magnetic field, which is out of the page above the wire and into the page at point A below the wire.

39. D

Once again, applying the first right-hand rule, the wire on the left creates a magnetic field halfway between the wires which points down toward the bottom of the page, as does the current on the right. Thus, the net magnetic field at that point is downward.

40. E

The free-body force diagram for the block looks like this:

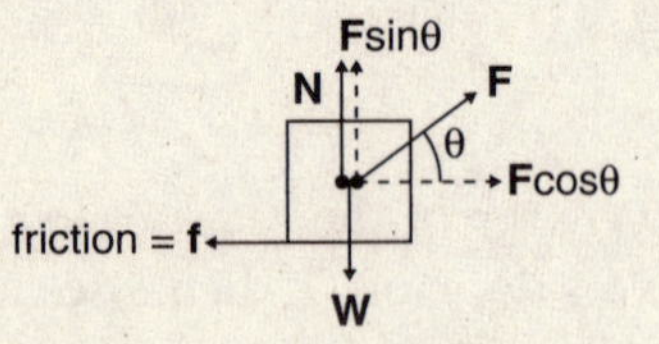

The forces in the y-direction must balance each other, and so N + Fsin θ = W, or N = 180 N – 40 N = 140 N

41. C

Since the block is moving with a constant velocity in the horizontal direction, the forces in the horizontal direction must balance each other.

friction = Fcos θ = (80 N)(0.87) = 70 N

42. D

The coefficient of friction is the ratio of the frictional force to the normal force:

$$\mu = \frac{f}{N} = \frac{70\text{N}}{140\text{N}} = 0.5$$

43. A

The potential energy is maximum at the amplitudes A_1 and A_2, and decreases parabolically to zero at the equilibrium position O.

44. C

The kinetic energy is always doing the opposite of the potential energy. When the potential energy is maximum, the kinetic energy is minimum, and vice-versa, with the maximum kinetic energy occurring at the equilibrium position O.

45. D

For an ideal spring, the spring force acting on the mass is proportional to the stretch of the spring x (Hooke's law), and therefore a force vs. distance graph for this spring would be linear. The force is maximum at the amplitudes, and zero at the equilibrium position.

46. E

The gravitational force is inversely proportional to the square of the distance between the moons and the planet, so the closer moon, being twice as close to the planet, will experience four times the force as the farther moon, which is twice as far away.

47. A

The shorter the orbital radius, the more quickly a satellite will complete an orbit.

48. C

At the ball's maximum height, its speed is zero.

$v = 0 = v_i - gt$

$v_i = gt = (10\ \text{m/s}^2)(3\text{s}) = 30\ \text{m/s}$

49. C

Since gravity causes the acceleration of the ball, and gravity is still acting on the ball at the top of its flight, the acceleration is still 10 m/s^2. If there were no acceleration of the ball at the top, the ball would never come back down.

50. D

The maximum height of the ball is the same distance as if we dropped the ball from its maximum height and let it fall for 3 seconds:

$s = \frac{1}{2}gt^2 = \frac{1}{2}(10\ \text{m/s}^2)(3\ \text{s})^2 = 45\ \text{m}$

51. B

The 60 μg sample is cut in half five times in 120 days: 60 μg, 30 μg, 15 μg, 7.5 μg, 3.75 μg, and 1.88 μg, rounded to 2 micrograms of thorium remaining after 120 days.

52. E

An ammeter must be placed in series with the resistor through which it is measuring the current, so that the same current passes through the ammeter and the resistor. A voltmeter is placed in parallel with a resistor, so that the voltmeter and resistor will have the same voltage across them, but the resistance of the voltmeter should be high, so that very little current will flow through it.

53. B

Since the three resistors are in series, the total resistance is the sum of the resistances.
By Ohm's law:

$$R + 4\Omega + 4\Omega = \frac{20V}{2A} \text{ implies that } R = 2\Omega.$$

54. C

Power dissipated is equal to the product of the current squared and the resistance:

$$P = I^2R = (2A)^2(4\Omega) = 16 \text{ W}$$

55. B

The heat lost by the metal equals the heat gained by the liquid:

$$m_{metal}\, c_{metal}(T_f - T_i) = m_{liquid}\, c_{liquid}\, (T_f - T_i)$$

$$(2\text{kg})(200\text{J/kg°C})(T_f - 120°\text{C}) = (4\text{kg})(600\text{J/kg°C})(T_f - 20°\text{C})$$

$$T_f = 34°\text{C}$$

56. A

Since heat is lost to the surroundings, the water is not gaining as much heat as it would in a totally insulated container, so the final temperature would be lower than our previous calculation.

57. B

The acceleration is the slope of the graph from a time of zero to 4 s, which is 1 m/s^2.

58. C

The displacement of the object is the area under the graph from a time of zero to 8 s, which is 24 m.

59. D

If the spheres begin by being equally and oppositely charged, the total charge on both together is zero. When they come into contact with each other, they will exchange charge in such a way as to keep a total charge of zero, but each must take on the same amount of charge, which would have to be zero. Thus, each sphere will be neutral.

60. A

Light enters the water, bends, reflects off of the pencil, and then comes back out of the water to your eye. Because the light changes media, it refracts, and you see a distorted image of the pencil.

61. B

The law of entropy states that all natural systems tend toward a state of higher disorder. Heat leaving a hotter body implies a move to higher disorder, because the vibrating molecules in the hot body impart momentum and energy to the molecules in the cold body, causing them to be in a state of higher disorder.

62. C

The forces $\mathbf{F_1}$ and $\mathbf{F_3}$ are equal and opposite and therefore cancel each other out, leaving only $\mathbf{F_2}$ as the resultant.

63. A

Coulomb's law gives $F = \frac{Kq_1q_2}{r^2} =$

$$\frac{(10^{10}\, Nm^2/C^2)(10^{-19}\, C)(10^{-19}C)}{(10^{-10}\, m)^2} = 10^{-8}\, N$$

64. D

Conservation of momentum:

$m_1v_1 = (m_1 + m_2)v'$

$(3\text{ kg})v_1 = (3\text{kg} + 6\text{ kg})(2\text{ m/s})$

$v_1 = 6\text{ m/s}.$

65. C

$$\text{Power} = \frac{\text{Work}}{\text{time}} = \frac{Fs}{t} = \frac{(500\text{N})(2\text{m})}{20\text{s}} = 50\text{ W}$$

66. A

A wave pulse, when reflected off a fixed end, will invert and return on the opposite side of the string with the same amplitude.

67. B

A coil of wire is rotated in a magnetic field, inducing a current in the wire by electromagnetic induction.

68. B

The atomic number of an element is the number of protons it has in its nucleus. When an element emits an alpha particle, it loses two protons (and two neutrons), and therefore loses 2 atomic mass units. When an element emits a beta particle, a neutron in the nucleus has ejected an electron, leaving one more proton in the nucleus and increasing the atomic number by one. Thus, this element loses 4 protons and then gains one, leaving the nucleus 3 atomic mass units less than before.

69. E

The bright and dark lines are called an interference pattern, and are created by the interference of the light waves after they have passed through the double-slit.

70. A

If the screen is moved back away from the slits, the bright band will become wider, because the light waves will have a little more space to spread out before striking the screen.

71. A

The waves trailing the sound source appear to the listener to have a lower frequency, and thus a lower pitch, since the waves reach the listener's ears less often.

72. C

A more intense (brighter) light would contain more photons, which can eject more electrons.

73. B

The gas atoms create pressure on the container by striking the container walls, not each other.

74. B

By Newton's third law, the law of action and reaction, both the astronaut and the tool experience the same force during the same time interval, and thus the same impulse acts on each of them. According to the law of conservation of momentum, they each have different accelerations and resulting velocities, since their masses are different.

75. D

The period for one oscillation of the pendulum is 2 s, and the frequency is the inverse of its period, $\frac{1}{2}$ cycle per second, or $\frac{1}{2}$ Hz.

Glossary

absolute zero
the lowest possible temperature, at which all molecular motion would cease and a gas would have no volume

acceleration
the rate of change in velocity

action-reaction forces
the pair of forces which are equal in magnitude and opposite in direction; these forces are involved in the same interaction, but act on different objects

adiabatic
the expansion or compression of a gas without a gain or loss of heat

air resistance
the opposing force the air exerts on an object moving through it

alpha decay
the process by which a radioactive element emits an alpha particle

alpha particle
positively charged particle consisting of two protons and two neutrons

alternating current
electric current that rapidly reverses its direction

ammeter
device used to measure electrical current

ampere
unit of electrical current equal to one coulomb per second

amplitude
maximum displacement from equilibrium position; the distance from the midpoint of a wave to its crest or trough

angle of incidence
the angle between the normal line to a surface and the incident ray or wave

angle of reflection
the angle between the normal line to a surface and the reflected ray or wave

angle of refraction

the angle between the normal line to a surface and the refracted ray or wave at the boundary between two media

angular momentum

the conserved rotational quantity which is equal to the product of the mass, velocity, and radius of motion

antinode

point of maximum displacement of two or more waves constructively interfering

atom

the smallest particle of an element that can be identified with that element; the atom consists of protons and neutrons in the nucleus, and electrons orbiting around the nucleus.

atomic mass unit

the unit of mass equal to $1/12$ the mass of a carbon-12 nucleus

atomic number

the number of protons in the nucleus of an atom

average acceleration

the acceleration of an object measured over a time interval

average velocity

the velocity of an object measured over a time interval; the total change in displacement divided by the total time taken to achieve that displacement

battery

device that converts chemical energy into electrical energy, creating a potential difference (voltage)

beta decay

the emission of an electron from the nucleus of a radioactive element as a result of a neutron decaying into a proton

beta particle

high speed electron emitted from a radioactive element when a neutron decays into a proton

binding energy

the nuclear energy that binds protons and neutrons in the nucleus of the atom

boiling point

temperature at which a substance boils at normal atmospheric pressure, changing from a liquid to a vapor state

calorie

the amount of heat required to raise the temperature of one gram of water by one degree Celsius

calorimeter

device that isolates objects to measure temperature changes due to heat flow

capacitance

ratio of the charge stored on a conductor per unit voltage

capacitor

electrical device used to store charge and energy in an electric field

Carnot efficiency

the ideal efficiency of a heat engine or refrigerator working between two constant temperatures

center of mass

the point at which all of the mass of an object can be considered concentrated

centripetal acceleration

the acceleration of an object moving in circular motion which is directed toward the center of the circular path

centripetal force

the central force causing an object to move in a circular path

chain reaction

nuclear process producing more neutrons which in turn can create more nuclear processes, usually applied to fission

charge

the fundamental quantity which underlies all electrical phenomena

charging by conduction

transfer of charge by actual contact between two objects

charging by induction

transfer of charge by bringing a charged object near a conductor, then grounding the conductor

circular motion

motion of an object moving at a constant radius in a circular path

closed system

a system that can exchange energy but cannot exchange matter with the surroundings

coefficient of friction

ratio of the frictional force between two surfaces and the normal force between the surfaces

component

the shadow or projection of a vector on a particular coordinate axis

compound

chemical substance consisting of two or more different elements combined in a fixed proportion

concave lens

a lens which is thinner in the center than at the edges; also known as a diverging lens

concave mirror

a mirror that converges light rays to a focal point; also known as a converging mirror

conductor

a material through which heat or electric current can easily flow

conservation of charge

law stating that the total charge in a system must remain constant during any process

conserved properties

any properties which remain constant during a process

constant acceleration

acceleration which does not change during a time interval

constant velocity

velocity which does not change during a time interval

constructive interference

addition of two or more waves which are in phase, resulting in a wave of increased amplitude

convection

heat transfer by the movement a heated substance, such as currents in a fluid

conventional current

the movement of positive charges through a conductor

converging lens

a lens which converges light rays to a focal point; also known as a convex lens

convex lens

a lens which is thicker in the center than at the edges; also known as a converging lens

convex mirror

a mirror that diverges light rays; also known as a diverging mirror

cosine

the ratio of the adjacent side of an angle in a right triangle to the hypotenuse

coulomb

the unit for electric charge

Coulomb's law

the electric force between two charges is proportional to the product of the charges and inversely proportional to the square of the distance between them

crest of a wave

the highest point on a wave

critical angle

the minimum angle for light entering a different medium at which total internal reflection will occur

critical mass

the minimum amount of mass of fissionable material necessary to sustain a nuclear chain reaction

de Broglie wavelength

the wavelength associated with a moving particle with a momentum *mv*

density

the ratio of the mass to the volume of a substance

destructive interference

addition of two or more waves which are out of phase, resulting in a wave of decreased amplitude

diffraction

the spreading of a wave beyond the edge of a barrier or through an opening

diffraction grating

material containing many parallel lines which are very closely spaced so that when light is passed through the lines, an interference pattern is produced

diffuse reflection

reflection of light in many directions by a rough surface

direct current

electric current whose flow of charges is in one direction only

dispersion of light

the separation of light into its component colors using a prism or diffraction grating

displacement

change in position in a particular direction (vector)

distance

the length moved between two points (scalar)

diverging lens

a lens that is thinner in the middle than at the edges which diverges light rays passing through it; also known as a concave lens

Doppler effect

the apparent change in frequency of a sound or light source due to relative motion between the source and the observer

dynamics

the study of the causes of motion (forces)

elastic collision

a collision in which both momentum and kinetic energy are conserved

electric circuit

a continuous closed path in which electric charges can flow

electric current

flow of charged particles; conventionally, the flow of positive charges

electric field

the space around a charge in which another charge will experience a force; electric field lines always point from positive charge to negative charge

electric generator

a device that uses electromagnetic induction to convert mechanical energy into electrical energy

electric potential

the amount of work per unit charge to move a charge from a very distant point to another point in an electric field

electric potential difference

the difference in potential between two points in an electric field; also known as voltage

electromagnet

a magnet with a field produced by an electric current

electromagnetic induction

inducing a voltage in a conductor by changing the magnetic field around the conductor

electromagnetic wave

a wave produced by vibrating charges which propagates itself through space by the mutual generation of changing electric and magnetic fields

electron

the smallest negatively charged particle

electrostatics

the study of electric charge, field, and potential at rest

element

a substance made of only one kind of atom

elementary charge

the smallest existing charge; the charge on one electron or one proton (1.6×10^{-19} C)

ellipse

an oval-shaped curve which is the path taken by a point that moves such that the sum of its distances from two fixed points (foci) is constant; the planets move in elliptical orbits around the Sun

emf

electromotive force; another name for voltage, particularly voltage induced in a conductor by electromagnetic induction

energy

the nonmaterial quantity which is the ability to do work on a system

energy level

amount of energy an electron has while in a particular orbit around the nucleus of an atom

entropy

the measure of the amount of disorder in a system

equilibrant

the vector which can balance a resultant vector; the force which can put a system in equilibrium

equilibrium

condition of a system in which the vector sum of the forces and torques are equal to zero

equivalent resistance

the single resistance that could replace the individual resistances in a circuit and produce the same result

evaporation

the process by which a liquid changes into a gas

excited state

the energy level of an electron in an atom after the electron has absorbed energy

external force

force exerted from outside the defined system

farad

the unit for capacitance equal to one coulomb per volt

Faraday's law of induction

law stating that a voltage can be induced in a conductor by changing the magnetic field around the conductor

first law of thermodynamics

the heat lost by a system is equal to the heat gained by the system minus any work done by the system; conservation of energy

fission

the splitting of a large nucleus into two smaller ones

fluid

any substance that flows, typically a liquid or a gas

focal length

the distance between the center of a lens or mirror to the point at which the rays converge at the focal point

focal point

the point at which light rays converge or appear to originate

force

any influence that tends to accelerate an object; a push or a pull

frame of reference

point of view or coordinate system used to study motion

free fall

motion under the influence of gravity

frequency

the number of vibrations or revolutions per unit of time

friction

the force that acts to resist the relative motion between two rough surfaces which are in contact with each other

fundamental particles

the particles (quarks and leptons) of which all matter is composed

fundamental overtone

lowest frequency sound produced by a musical instrument

fusion

the combining of two light nuclei into a heavier one with a release of energy

galvanometer
device used to measure small electrical currents

gamma decay
process by which a radioactive element emits a gamma ray

gamma ray
high energy electromagnetic wave emitted by a radioactive nucleus

gas
a state of matter that expands to fill a container

gravitational field
space around a mass in which another mass will experience a force

gravitational force
the force of attraction between two objects due to their masses

gravitational potential energy
the energy a mass has because of its position in a gravitational field

ground state
the lowest energy state of an electron in an atom

grounding
the process of connecting a charged object to the earth or a large conductor to remove its excess charge

half-life
the time it takes for half the atoms in a radioactive sample to decay

heat
the energy which is transferred from one body to another because of a temperature difference

heat engine
device that changes internal energy into mechanical work

heat of fusion
energy needed to change a unit mass of a substance from a solid to a liquid state at the melting point

heat of vaporization
energy needed to change a unit mass of a substance from a liquid to a gaseous state at the boiling point

Heisenberg uncertainty principle
the more accurately one determines the position of a subatomic particle, the less accurately its momentum is known, and vice-versa

hertz
the unit for frequency equal to one cycle or vibration per second

Hooke's law
the displacement (stretch) of a spring is proportional to the force applied to it

hypotenuse
the side opposite to the right angle in a right triangle

image
reproduction of an object using lenses or mirrors

impulse
the product of the force acting on an object and the time over which it acts

impulse-momentum theorem
the impulse imparted to an object is equal to the change in momentum it produces

incandescent body
an object that emits light because of its high temperature

index of refraction

the ratio of the speed of light in a vacuum to the speed of light in another medium

inelastic collision

a collision in which only momentum is conserved, and not kinetic energy

inertia

the tendency of an object not to change its state of rest or motion at a constant velocity; mass is a measure of inertia

inertial reference frame

a reference frame which is at rest or moving with a constant velocity; Newton's laws are valid within any inertial reference frame

infrared

electromagnetic waves of frequencies just below those of red visible light

initial velocity

the velocity at which an object starts at the beginning of a time interval

in phase

term applied to two or more waves whose crests and troughs arrive at a place at the same time in such a way as to produce constructive interference

instantaneous acceleration

the acceleration of an object at a particular instant of time

instantaneous position

the position of an object at a particular instant of time

instantaneous velocity

the velocity of an object at a particular instant of time

insulator

a material that is a poor conductor of heat or electric current due to a poor supply of free electrons

interference of waves

displacements of two or more waves in the same medium at the same time producing either larger or smaller waves

internal forces

forces between objects in the same system

inverse relationship

relationship between two variables that change in opposite directions, so that if one is doubled the other is reduced to one-half

inverse square law

situation where one physical quantity varies as the inverse square of the distance from its source

isolated system

a collection of objects not acted upon by any external forces and which energy neither enters or exits

isotope

a form of an element which has a particular number of neutrons, that is, has the same atomic number but a different mass number than the other elements which occupy the same place on the periodic table

joule

the unit for energy equal to one Newton-meter

joule heating

the increase in temperature in an electrical conductor due to the conversion of electrical energy into heat energy

Joule's law of heating

the heating power of an electric current through a resistance is equal to the product of the current and the voltage across the resistor

Kelvin (absolute) temperature scale

scale in which zero Kelvins is defined as absolute zero, the temperature at which all molecular motion ceases

Kepler's laws

the three laws of motion for objects, such as the planets and the sun, attracted to each other by the gravitational force

kilogram

the fundamental unit of mass

kilowatt-hour

amount of energy equal to 3.6×10^6 joules, usually used in electrical measurement

kinematics

the study of how motion occurs, including distance, displacement, speed, velocity, acceleration, and time

kinetic energy

the energy a mass has by virtue of its motion

kinetic molecular theory

the description of matter as being made up of extremely small particles that are in constant motion

laser

light amplification by stimulated emission of radiation; laser light is coherent and monochromatic (i.e., all the waves are of the same wavelength and are in phase with each other)

law of conservation of energy

the total energy of a system remains constant during a process

law of conservation of momentum

the total momentum of a system remains constant during a process

law of inertia

every body continues in its state of rest or constant velocity unless acted on by an unbalanced force (Newton's first law)

law of reflection

the angle of incidence of an incoming wave is equal to the angle of reflection measured from a line normal (perpendicular) to the surface

law of universal gravitation

the gravitational force between two masses is proportional to the product of the masses and inversely proportional to the square of the distance between them

lens

a piece of transparent material that can bend light rays to converge or diverge

lever arm

perpendicular distance from the axis of rotation to a line of force producing or potentially producing the rotation

light

the range of electromagnetic waves which is visible to the human eye; alternately used by physicists to refer to the entire range of electromagnetic radiation

line spectrum

discrete lines emitted by a cool excited gas

linear accelerator

device used to accelerate subatomic particles to high energies so that they may be more easily studied

linear relationship

relationship between two variables that are proportional to each other

linear restoring force

a force such as a spring force in which the force is proportional to the displacement (stretch)

liquid

substance which has a fixed volume, but assumes the shape of its container

lodestone

a naturally occurring magnetic rock made principally of iron

longitudinal wave

wave in which the vibration of the medium is parallel to the direction of motion of the wave

loudness

the quality of a sound wave which is measured by its amplitude

magnetic domain

cluster of magnetically aligned atoms

magnetic field

the space around a magnet in which another magnet or moving charge will experience a force

magnification

ratio of the size of an optical image to the size of the object

mass defect

the mass equivalent of the binding energy in the nucleus of an atom by $E = mc^2$

mass number

the total number of protons and neutrons in the nucleus of a particular atom

matter wave

the wave associated with a mass having a momentum according to the de Broglie hypothesis

mechanical energy

the sum of the potential and kinetic energies in a system

mechanical resonance

condition in which natural oscillation frequency equals frequency of driving force

mechanical wave

a wave which needs a medium through which to travel

melting point

the temperature at which a sustance changes from a solid to a liquid state

meter

the SI unit of length

molecule

two or more atoms joined to form a larger particle

momentum

the product of the mass of an object and its velocity

monochromatic light

light having a single color or frequency

net force

the vector sum of the forces acting on an object

neutral

having no net charge

neutron

an electrically neutral subatomic particle found in the nucleus of an atom

newton

the SI unit for force equal to the force needed to accelerate one kilogram of mass by one meter per second squared

Newton's first law of motion

the law of inertia

Newton's second law of motion

a net force acting on a mass causes an acceleration that is inversely proportional to the mass and in the direction of the force

Newton's third law of motion

for every action force there is an equal and opposite reaction force

node

the point of no displacement in a standing wave

noninertial reference frame

a reference frame which is accelerating; Newton's laws are not valid within a noninertial reference frame

normal

a line which is perpendicular to a surface

normal force

the reaction force of a surface on an object when an object is in contact with the surface

nuclear fission

the splitting of a heavy nucleus into smaller fragments with the release of energy

nuclear fusion

the combining of two light nuclei into one larger one with the release of energy

nuclear reaction

any process in the nucleus of an atom that causes the number of protons and/or neutrons to change

nuclear reactor

device in which nuclear fission or fusion is used to generate electricity

nucleon

either a proton or a neutron in the nucleus of an atom

object (optics)

the source of diverging light rays

ohm

the SI unit for resistance equal to one volt per ampere

Ohm's law

the ratio of voltage to current in a circuit is a constant called resistance

opaque

material that will not transmit light

open system

a system that can exchange both matter and energy with the surroundings

out of phase

term applied to two or more waves for which the crest of one wave arrives at a point at the same time as the trough of a second wave arrives, producing destructive interference

parallel circuit

an electric circuit which has two or more paths for the current to follow, allowing each branch to function independently of the others

Pascal

the SI unit for pressure equal to one newton of force per square meter of area

period

the time for one complete cycle or revolution

periodic motion
motion that repeats itself at regular intervals of time

photoelectric effect
the ejection of electrons from certain metals when exposed to light of a minimum frequency

photon
the smallest particle (quantum) of light

physics
the study of matter and energy and all their relationships

pitch
the perceived sound characteristic equivalent to frequency

Planck's constant
the quantity that results when the energy of a photon is divided by its frequency

plane mirror
smooth, flat surface that reflects light regularly

polarized light
light in which the electric fields are all in the same plane

position
the distance between an object and a reference point

position-time graph
the graph of the motion of an object that shows how its position varies with time

power
the rate at which work is done or energy is dissipated

pressure
force per unit area

primary light colors
red, green, or blue light

principal axis
the line connecting the center of curvature of a curved mirror with its geometrical vertex; the line perpendicular to the plane of a lens passing through its center

principle of superposition
the displacement due to two or more interfering waves is equal to the sum of the displacement of the individual waves

projectile
any object that is projected by a force and continues to move by its own inertia in a gravitational field

proton
subatomic particle with positive charge that is the nucleus of a hydrogen atom

quantized
a quantity that cannot be divided into smaller increments forever, for which there exists a minimum, quantum increment

quantum mechanics
the study of the properties of matter using its wave properties

quantum model of the atom
atomic model in which only the probability of locating an electron is known

quark
one of the elementary particles of which all protons and neutrons are made

radiation
the transmission of energy by electromagnetic waves, or the particles given off by radioactive atoms

radioactive decay
the spontaneous change of unstable nuclei into other nuclei

range of a projectile
the horizontal distance between the launch point of a projectile and where it returns to its launch height

ray model of light
light may be represented by a straight line along the direction of motion

ray optics
study of light using the ray model

real image
an image that can be projected onto a screen

reference point
zero location in a coordinate system

refraction
the bending of a wave due to a change in medium

resistance
the ratio of the voltage across a device to the current running through it

resistor
device designed to have a specific resistance to electric current

resultant
the vector sum of two or more vectors

right-hand rules
used to find the magnetic field around a current-carrying wire or the force acting on a wire or charge in a magnetic field

Rutherford's model of the atom
the first nuclear model of the atom (1911)

scalar
a quantity in physics, such as mass, which can be completely described by its magnitude or size, without regard to direction

schematic diagram
a diagram using special symbols to represent a circuit

scientific notation
numbers represented as a mantissa times a power of ten, such as 3×10^8

second
the SI unit of time

second law of thermodynamics
heat flows naturally from a region of higher temperature to a lower temperature; all natural systems tend toward a state of higher disorder

series circuit
an electric circuit in which devices are arranged so that charge flows through each equally.

SI
Systeme Internationale; agreed-upon method of using the metric system of measurement

simple harmonic motion
periodic motion in which the restoring force is proportional to the displacement of the oscillating object, and this force is directed toward the point of equilibrium

sine

the ratio of the opposite side of a right triangle to its hypotenuse

sliding (or kinetic) friction

resistive force between two surfaces which are moving relative to each other

slope

on a graph, the ratio of the vertical separation (rise) to the horizontal separation (run)

solid

state of matter with fixed volume and shape

special relativity

theory that describes how mass, length, and time, and the equivalence of matter and energy

specific heat capacity

the quantity of heat required to raise the temperature of one gram of water by one degree Celsius

spectrum

the range of electromagnetic waves from low frequency to high frequency, or colors when white light is passed through a prism

speed

the ratio of distance to time

speed of light

3×10^8 m/s in a vacuum

standing wave

wave with stationary nodes produced by two identical waves traveling in opposite directions in the same medium at the same time

static friction

the resistive force that opposes the start of motion between two surfaces in contact

strong nuclear force

the force that binds protons and neutrons together in the nucleus of an atom

superconductor

material which has practically no resistance to the flow of current at low temperatures

symmetry

property which is not changed when the reference frame is changed

system

defined collection of objects

tangent

touching at only one point on a curve; in a right triangle, the ratio of the opposite side to an adjacent side for either of the acute angles

tangential velocity

the velocity tangent to the path of an object moving in a curved path

temperature

the property of a body which indicates how hot or cold a substance is with respect to a standard

terminal velocity

the constant velocity of a falling object when the force of air resistance equals the object's weight

test charge

the very small charge used to test the strength of an electric field

thermal energy

the sum of the internal potential and kinetic energies of the random motion of the molecules making up an object

thermal equilibrium
state between two or more objects or systems in which temperature doesn't vary

thermal expansion
increase in length or volume of a material due to an increase in temperature

thermodynamics
the study of heat transfer

torque
the tendency of a force to cause rotation about an axis; the product of the force and the lever arm

total internal reflection
the complete reflection of light that strikes the boundary between two media at an angle greater than the critical angle

trajectory
the path followed by a projectile

transformer
device that uses electromagnetic induction to transfer energy from one circuit to another

translucent
a material that passes light but distorts its path

transparent
material through which light can pass without distorting the direction of the rays

transverse wave
a wave in which the vibration is perpendicular to the velocity of the wave

trigonometry
the study of the relationships among the angles and sides of triangles

trough of a wave
the low point of wave motion

ultraviolet
electromagnetic waves of frequencies higher than those of violet light

uniform acceleration
constant acceleration

uniform velocity
constant velocity

uniform circular motion
motion in a circular path of constant radius at a constant speed

vector quantity
a quantity, such as displacement, having both magnitude and direction

velocity
ratio of the displacement of an object to a time interval

velocity-time graph
plot of the velocity of an object as a function of time, the slope of which is acceleration

virtual image
an image which cannot be projected onto a screen; point at which diverging light rays appear to originate

volt
the SI unit of potential or potential difference

voltage
amount of electric potential or potential difference

watt
the SI unit for power equal to one joule of energy per second

wavelength

the distance between successive identical parts of a wave

weight

the result of a gravitational force acting on a mass; the product of mass and the acceleration due to gravity at a location

weightlessness

the state of free fall where an object experiences only the force of gravity

white light

visible light consisting of all colors

work

the scalar product of force and displacement

work function

the minimum energy required to release an electron from a metal

work-energy theorem

the work done on a system is equal to the change in energy of the system

x-ray

high-frequency, high-energy electromagnetic waves or photons

Index

NOTES

NOTES

NOTES

NOTES

How Did We Do? Grade Us.

Thank you for choosing a Kaplan book. Your comments and suggestions are very useful to us. Please answer the following questions to assist us in our continued development of high-quality resources to meet your needs.

The title of the Kaplan book I read was: ______________________

My name is: ______________________

My address is: ______________________

____

My e-mail address is: ______________________

What overall grade would you give this book? (A) (B) (C) (D) (F)

How relevant was the information to your goals? (A) (B) (C) (D) (F)

How comprehensive was the information in this book? (A) (B) (C) (D) (F)

How accurate was the information in this book? (A) (B) (C) (D) (F)

How easy was the book to use? (A) (B) (C) (D) (F)

How appealing was the book's design? (A) (B) (C) (D) (F)

What were the book's strong points? ______________________

How could this book be improved? ______________________

Is there anything that we left out that you wanted to know more about?

Would you recommend this book to others? ☐ YES ☐ NO

Other comments: ______________________

____

____

Do we have permission to quote you? ☐ YES ☐ NO

Thank you for your help.
Please tear out this page and mail it to:

Managing Editor
Kaplan, Inc.
888 Seventh Avenue
New York, NY 10106

KAPLAN®

Thanks!